Climate Change Governance in Asia

Asian countries are among the largest contributors to climate change. China, India, Japan, and South Korea are among the top ten largest carbon emitters in the world, with South Korea, Japan, and Taiwan also some of the largest on a per capita basis. At the same time, many Asian countries, notably India, Taiwan, Japan, the Philippines, and Thailand are among those most affected by climate change, in terms of economic losses attributed to climate-related disasters. Asia is an extremely diverse region, in terms of the political regimes of its constituent countries, and of their level of development and the nature of their civil societies. As such, its countries are producing a wide range of governance approaches to climate change. Covering the diversity of climate change governance in Asia, this book presents cosmopolitan governance from the perspective of urban and rural communities, local and central governments, state-society relations, and international relations. In doing so it offers both a valuable overview of individual Asian countries' approaches to climate change governance, and a series of case studies for finding solutions to climate change challenges.

Kuei-Tien Chou is Professor at the Graduate Institute of National Development, National Taiwan University and Lead Principal Investigator of the Risk Society and Policy Research Center, National Taiwan University.

Koichi Hasegawa is Professor of Sociology at the Graduate School of Arts and Letters, Tohoku University, Japan.

Dowan Ku is Director of the Environment and Society Research Institute, Seoul, South Korea.

Shu-Fen Kao is Associate Professor at the Department of Sociology and Social Work, Fo Guang University, Taiwan.

Routledge Contemporary Asia Series

For more information about this series, please visit: https://www.routledge.
com/Routledge-Contemporary-Asia-Series/book-series/SE0794

Climate Change Governance in Asia

Edited by Kuei-Tien Chou,
Koichi Hasegawa, Dowan Ku,
and Shu-Fen Kao

Routledge
Taylor & Francis Group

LONDON AND NEW YORK

First published in English 2020
by Routledge
2 Park Square, Milton Park, Abingdon, Oxon OX14 4RN

and by Routledge
605 Third Avenue, New York, NY 10017

First issued in paperback 2022

Routledge is an imprint of the Taylor & Francis Group, an informa business

Publisher's Note
The publisher has gone to great lengths to ensure the quality of this reprint but points out that some imperfections in the original copies may be apparent.

British Library Cataloguing-in-Publication Data
A catalogue record for this book is available from the British Library

Library of Congress Cataloging-in-Publication Data
A catalog record for this book has been requested

ISBN: 978-0-367-51221-7 (pbk)
ISBN: 978-0-367-22700-5 (hbk)
ISBN: 978-0-429-27645-3 (ebk)

DOI: 10.4324/9780429276453

Typeset in Galliard
by Apex CoVantage, LLC

Contents

Figures

Tables

Contributors

Leah Abayao is Professor of History and former Director of Cordillera Studies Center at the University of the Philippines, Baguio.

Midori Aoyagi is Principle Researcher at the Center for Social & Environmental Systems Research, National Institute for Environmental Studies, Japan.

Chia-Wei Chao is Postdoctoral Fellow at the Risk Society and Policy Research Center, College of Social Science, National Taiwan University.

Kang Chen is a PhD graduate of the Department of Geography, the University of Hong Kong.

Roger S. Chen is Associate Professor at the Graduate Institute of Construction Engineering and Management, National Central University, Taiwan.

Kuei-Tien Chou is Professor at the Graduate Institute of National Development, National Taiwan University and Lead Principal Investigator of Risk Society and Policy Research Center, National Taiwan University.

Koichi Hasegawa is Professor of Sociology at the Graduate School of Arts and Letters, Tohoku University, Japan.

Shu-Fen Kao is Associate Professor at the Department of Sociology and Social Work, Fo Guang University, Taiwan.

Hajime Kimura already has a PhD in Science, Waseda University in Tokyo, and now he is studying for his PhD in Sociology, Meiji University, Japan.

Dowan Ku is Director of the Environment and Society Research Institute, Seoul, South Korea.

Ho-Ching Lee is Professor at the Center for General Education, National Central University, Taiwan.

Seejae Lee is Professor Emeritus at the Catholic University of Korea.

So-Young Lee is Senior Policy Researcher and Research Manager at the Sustainability Governance Centre, Institute for Global Environmental Strategies, Japan.

Hwa-Meei Liou is Professor at the Graduate Institute of Technology Management, National Taiwan University of Science & Technology.

Alex Y. Lo is Senior Lecturer in Climate Change at the School of Geography, Environment and Earth Sciences, Victoria University of Wellington, New Zealand.

Anshu Ogra is a Post-Doctoral Research Associate in the Landslip Project at the Department of Geography, King's College London.

Chaya Vaddhanaphuti is Lecturer in the Department of Geography, Faculty of Social Sciences, Chiang Mai University, Thailand.

Sun-Jin Yun is Professor of Environmental and Energy Policy at the Graduate School of Environmental Studies, Seoul National University.

Eric Zusman is Senior Policy Researcher/Area Leader at the Institute for Global Environmental Strategies, Japan.

1 Introduction

Kuei Tien Chou and Dowan Ku

Climate change is a cross-border risk that is global in nature. It exhibits the characteristics of being cross-scale, cross-spatial, and cross-border (Bulkeley 2005), and therefore compels people to develop new research methods to manage these highly complex and transdisciplinary issues. Many studies have pointed to how climate change has opened up new analytical orientations which have created new challenges to existing research. For example, Hannigan (1995), Reusswig (2010), and Heinrichs and Gross (2010) highlighted the challenges that acid rain and climate change bring to environmental sociology. In particular, Jasanoff (2010) pointed out that while global and large-scale representations of scientific knowledge articulated by the Intergovernmental Panel on Climate Change (IPCC) belong to the confines of the *scale of knowledge*, this should instead be changed to one of the *scale of meaning*. Such a viewpoint goes straight to the heart of the matter, that is, large-scale scientific knowledge should be translated into forms that would enable us to understand the impact on humans, life, production, consumption, and even on health, and how social science should also enter such research, in order to look at how it can be used to interpret, analyze, and construct astute governance, systems, and actions.

This book will discuss climate change governance in countries where a Western linear path to modernization has been adopted, and which have thus developed into high-carbon economies with a special focus on Asia, whether it be later-comer East Asian countries such as Japan, Taiwan, or South Korea, or developing countries like China, Southeast Asian, or South Asian countries. According to the Global Carbon Project (2018), China's population has grown to over 1.4 billion people and its carbon emissions are in excess of 9.8 billion metric tons, making it the world's largest carbon emitting country; Japan ranks fifth in carbon emissions, emitting more than 1.2 billion metric tons of emissions with a population of 127 million, and even though South Korea has a population of only about 51 million people, its carbon emissions rank seventh at 610 million metric tons, while Taiwan's population of 23 million with carbon emissions of 270 million metric tons (and per capita emissions of 10.8 metric tons) ranks eighth in carbon emissions among countries with a population of more than 10 million people. Not to be outdone by these East Asian countries, rapidly developing countries too have increasing carbon emissions. India, which has a population of 1.33 billion and carbon emissions of 2.46 billon metric tons, ranks third in emissions. Thailand with a population of 69 million and carbon emissions of 330 million metric tons ranks twentieth in emissions, while the Philippines ranks thirty-seventh with 1.27 billion metric tons of carbon emissions. Basically, the post-war development in Asia has relied on a high-carbon society and a high-carbon economy model, on the back of a developmentalism model grounded in a high-carbon regime.

There is therefore an urgent need to change the frame of discussion around climate change research from one held within the confines of traditional national boundaries to that of a cross-border framework, whether it be about international norms, systems, or actions or the alignment of national policies to international standards. As such, since the 2000s, research on cosmopolitanism has gradually received attention, with many academics arguing for the need for a new framework to understand transboundary risk issues. Western scholars, such as Beck (1996, 2002, 2008, 2009), Grande (2006), Delanty (2006), Hulme (2010), Zurn (2016), and Beck and Levy (2013), have also emphasized the need for such development, and Asian scholars, such as Chang (2010), Han and Shim (2010), Zhang (2015), Chou and Liou (2012), and Chou (2018), have also pointed to the importance of adopting a new methodology to understand these issues. Fundamentally, there is a need in the social sciences to move toward the adoption of methodological cosmopolitanism as a replacement for methodological nationalism (Beck and Sznaider 2006; Beck and Grande 2010), which is to say that, even as scholars are focused on transboundary and cross-national events, the research cannot be confined to the traditional social science concept of studying them from a domestic perspective, but should instead be grounded in universalism and the synchronicity of global events, with an eye on the specific political and economic contexts of each country, in order to understand their commonalities and differences.

Although the changes in research methods have resulted in paradigm shifts, there have also been various research developments, the first of which is international comparative research, but such research has been focused on risk shocks, the decision-making of governments, and the social resistance faced by individual countries. The second approach focused at the level of the country, but where the discussion is taken from a global framework, of global norms, systems, and governance, to the corresponding development of the country. The third approach analyzes systems, governance, and norms from a trans- and cross-national perspective, to understand how these impact on countries, and the adjustments countries subsequently make. The fourth approach uses the trans- and cross-national perspective to understand the non-governmental organization (NGO) network and actions, and to study them in terms of their identities, collectivities, and communities. These various approaches involve the exploration of normative research and action research in cosmopolitanism studies. Beck (2014) stressed that the analytical units of study in methodological cosmopolitanism should be *embedded in the national systems and processes*, and should *replace these national systems and processes* with the cosmopolitanism governance approach, which would be in line with the first and second approaches mentioned above.

We start the discussion in this book by seeking to understand how people and governments in Asia address climate change, for instance whether it is being treated as a global, national, or local agenda, or whether it is seen as a scientific agenda, or an issue that touches on everyday life. We also look at the types of governances that have been constructed in Asia in order to tackle climate change. However, it is hard to find successful climate change governance in Asia because many Asian countries are trapped in the high-carbon economy model. Green politics on the basis of strong liberal democracy has not developed in Asia. Instead, the developmentalism model based on nationalism is a common characteristic in Asian countries. Nonetheless, environmentalists and local populations in Asia concerned about climate change have made a great effort to overcome climate change, and they have been working at the local, national, and global scales. It is also important to understand that the issue of climate change is constructed by various social forces. Social groups work together or fight each other over how climate change should be addressed. Climate change issues also cut across the traditional regime of governance.

However, new sustainability transition governance in which strong ecological modernization is successful can be constructed, if strong social solidarity movements for sustainability are sufficiently powerful. As it is, the cosmopolitan mindset has gained traction among peoples in Asia, though cosmopolitan governments have not yet taken root. In this book, we will therefore analyze how climate change governance in Asia is constructed and how it works on the national and local scales. Ultimately, climate change governance in Asia can be successfully constructed and implemented if governments are willing to work together not only with the business sector but also with civil society. In fact, there are already people in different parts of Asia who have mobilized themselves toward trying to achieve a sustainable society, thus the question is how we can take it to the next level.

Following this introduction, Chapters 2–5 discuss the Intended Nationally Determined Contributions (INDCs) (under the United Nations Framework Convention on Climate Change) which exhibit characteristics of cosmopolitan governance, as well as the carbon emissions policies, regulations, and timelines of each country, in addition to the transformational challenges faced by these high-carbon societies with regard to their country systems, decision-making processes, social pressures, and social conflict discourses, among other things. Superficially, these countries have responded to the Climate Change Convention by pledging to commit to the INDCs or establishing carbon pricing (such as a regional carbon trading platform). However, in reality, they are trapped in their domestic high-carbon economic structure thus resulting in transitional difficulties. The domestic path dependence has locked these countries into the brown economy, which has dominated the development of their industries and energy use. While this research orientation is aligned to the perspective of institutionalized cosmopolitanism, in order to carry out system intervention and norm setting within a country as part of the Climate Change Convention, it requires further discussion on the transitional challenges, especially on the attitude of the country toward its economic development pathway, such as on the discursive struggle between the low-carbon economy and the brown economy between the government, industry, and society, whilst facing constraints imposed by the political and economic power of industry players and the capability of civil society.

As compared to Part I of the book which is focused on the policy and structural analysis, Part II deals with climate knowledge on a micro level: the risks, environmental frames, and sustainability. Climate knowledge and the governance structure should not be seen only as an issue to be dealt with at the international level, and while it is the basis for government decision-making and regulations, at the same time it is also relevant at the level of local knowledge, and neither level should be neglected. Chapter 6 on the example of Thailand shows that in the actual management of climate policy, if the discussion and understanding of local knowledge were to be neglected, this would lead to a decision-making gap, in that the national interpretation of climate information at the global level is relative to the interpretation by local communities of climate information at the local level, which could result in differences in governance. Chapter 7 compares the risk perceptions and attitudes of Japanese and European citizens toward energy and climate policies and highlights the differences in their perceptions. We can therefore observe the attitudes of people in these countries toward climate and energy policies, and the differences in public opinion in relation to cosmopolitan governance. Chapter 8 adopts the systemic risk perspective to study the opportunities for transition and the structural challenges faced in Taiwan regarding its climate policy. The author analyzes the existing system of climate decision-making and real-time climate and energy landscape from the perspective of the transitional management of society and technology, and discusses the possibility of policy innovation. Chapter 9 adopts the perspective of ecological modernization to look at the issue of carbon capture and storage; it explores the struggles of environmental movements and the social impact of carbon capture and storage (CCS) framing.

The third part of this book is focused on the urban sustainability, climate change adaptation, disaster management, and social network orientation of four countries: participatory knowledge lies at the heart of all the cases depicted. Chapter 10 discusses three case studies in Seoul in South Korea, to explore how grassroots participation and institutional innovation were able to successfully develop a pathway for sustainability transition. Chapter 11 discusses coffee cultivation in India, and how the participation of local farmers and their knowledge was used to develop the adaptation strategies to mitigate climate change. It therefore captures the viewpoints of the traditional knowledge of local communities toward climate issues, which in turn lead to various planting strategies. Chapter 12 analyzes the framework for climate disaster management in the Philippines and includes various case studies to highlight how local government units train local communities how to respond to and form a strategy to deal with climate disasters, and in the process construct relevant actions, highlighting the way in which the local governance of disaster reductions can be imbued with local knowledge. Chapter 13 discusses the climate strategies and the citizen participation network in Tainan, Taiwan. The authors adopted the social network perspective to analyze community participation, stakeholder perception, and local knowledge actions developed by the local flood control groups, so as to showcase the climate governance as developed by the urban social ecosystem. From these examples, it is possible to observe how different societies and their various contexts are able to produce the multitudes of actors and social networks that bring about very rich and diverse forms of governance.

Although the chapters do not directly address trans- and cross-national research, the authors in Chapters 2, 3, 4, and 5 review the carbon reduction timeline and targets of each country under the global norms of the INDC, and discuss the corresponding challenges faced under the framework of a high-carbon economy. Chapter 6 (in Part II) and Chapters 11 and 13 (in Part III) discuss climate change knowledge and its interconnection with national climate management and local knowledge, the climate knowledge of local coffee cultivation, and local flood control knowledge and networks, and how these interactions exemplify the diversity of social construction in cosmopolitan climate governance. Other chapters also include indirect discussion of cosmopolitanism in climate change issues in various social contexts, such as in the public's climate perception (Chapter 7), social transformation toward a low-carbon system (Chapter 9), and the innovative means of sustainability transition promoted in Seoul.

The reality is that the simultaneous climate change risks that Asian countries are facing already constitutes the compulsory cosmopolitanism that Beck (2008) detailed, and further research therefore needs to be conducted using this framework. This book is a preliminary attempt at using the approach of 'embedding the national' to discuss climate governance, climate norms, and climate knowledge at a global level, and their actual practice in various countries; these experiences highlight the diversity of approaches at the local and national level, reflecting the diverse meanings of cosmopolitan climate governance.

References

Beck, U., & Grande, E. (2004). *Das kosmopolitische Europa. Gesellschaft und Politik in der Zweiten Moderne*. Frankfurt a.M.: Suhrkamp.

Beck, U. (1996). World risk society as cosmopolitan society? Ecological questions in a framework of manufactured uncertainties. *Theory, Culture & Society, 13*(4), 1–32.

Beck, U. (2002). The cosmopolitan society and its enemies. *Theory, Culture & Society, 19*(1–2), 17–44.

Beck, U. (2008). *World at Risk*. London: Polity Press.

Beck, U. (2009). Critical theory of world risk society: a cosmopolitan vision. *Constellations, 16*(1), 3–22.

Beck, U. (2014). On methodological cosmopolitanism. Public lecture at the Seoul Conference 2014 with Professor Ulrich Beck Seoul.

Beck, U., & Grande, E. (2010). Varieties of second modernity: the cosmopolitan turn in social and political theory and research. *British Journal of Sociology*, 61(3), 409–443. London: London School of Economics.

Beck, U., & Levy, D. (2013). Cosmopolitanized nations: re-imagining collectivity in world risk society. *Theory, Culture & Society*, 30(2), 3–31.

Beck, U., & Sznaider, N. (2006). Unpacking cosmopolitanism for the social sciences: a research agenda. *British Journal of Sociology*, 57(1), 1–23.

Bulkeley, H. (2005). Reconfiguring environmental governance: towards a politics of scales and networks. *Political Geography*, 24(8), 875–902.

Chang, K. S. (2010). The second modern condition? Compressed modernity as internalized reflexive cosmopolitanization. *British Journal of Sociology*, 61(3), 444–464.

Chou, K. T., & Liou, H. M. (2012). Cosmopolitan reform in science and technology governance. *Journal of State and Society*, 12, 101–198.

Chou, K. T. (2018). The cosmopolitan governance of energy transition. In K.-T. Chou (Eds.), *Energy Transition in East Asia* (pp.1–5). New York: Routledge.

Delanty, G. (2006). The cosmopolitan imagination: critical cosmopolitanism and social theory. *British Journal of Sociology*, 57(1), 26–47.

Global Carbon Project. (2018). Global carbon atlas. Retrieved September 9, 2019, from www.globalcarbonatlas.org/en/CO2-emissions.

Grande, E. (2006). Cosmopolitan political science. *British Journal of Sociology*, 57(1), 87–111.

Han, S. J., & Shim, Y. H. (2010). Redefining second modernity for East Asia: a critical assessment. *British Journal of Sociology*, 61(3), 465–488.

Hannigan, J. (1995). *Environmental Sociology*. New York: Routledge.

Heinrichs, H., & Gross, M. (2010). Introduction: new trends and interdisciplinary challenges in environmental sociology. In Matthias Gross and Harald Heinrichs (Eds.), *Environmental Sociology: European Perspectives and Interdisciplinary Challenges* (pp.1–16). Dordrecht, Heidelberg, London, and New York: Springer.

Hulme, M. (2010). Cosmopolitan climates hybridity, foresight and meaning. *Theory, Culture & Society*, 27(2–3), 267–276.

Jasanoff, S. (2010). A new climate society. *Theory Culture Society*, 27, 233–253.

Reusswig, F. (2010). The new climate change discourse: a challenge for environmental sociology. In Matthias Gross and Harald Heinrichs (Eds.), *Environmental Sociology: European Perspectives and Interdisciplinary Challenges* (pp.39–57). Dordrecht, Heidelberg, London, and New York: Springer.

Zhang, J. Y. (2015). Cosmopolitan risk community and China's climate governance. *European Journal of Social Theory*, 18(3), 327–342.

Zurn, M. (2016). Survey article: four models of a global order with cosmopolitan intent: an empirical assessment. *Journal of Political Philosophy*, 24(1), 88–119.

Part I

Framework of climate change governance in Asian countries

2 Climate change governance in Japan

Critical review on Japan's INDC and its energy policy

Koichi Hasegawa

Climate change governance in Japan

On greenhouse gas (GHG) emissions, Japan is the fifth largest emitter with a 3.5 percent world share following China with a 28 percent share, the US with 15 percent, India with 6.5 percent, and Russia with 4.5 percent in 2016. Germany with 2.3 percent and South Korea with 1.8 percent lesser emitters. Per capita, Japan is the fourth largest emitter with 9.0 tons following the US with 14.9 tons, South Korea with 11.5 tons, and Russia with 10.0 tons in 2016. Japan's GDP is the third largest following the US and China. Japan's responsibility to reduce GHG emissions is critical.

What are the major characteristics of climate change governance and climate change policy in Japan (Hasegawa and Shinada 2016)? On climate change policy, it has been criticized for being very tardy in taking active measures like introducing an aggressive carbon tax and carbon pricing system. The central government is reluctant to promote renewable energy resources, whereas it has been highlighting the role of coal-fired and nuclear plants. Although scholars and environmental non-governmental organizations (NGOs) sought "energy transition" prior to the 2011 Fukushima nuclear accident, government and mainstream economic sectors like Keidanren (Federation of Economic Organizations) have been negative about such a transition in order to protect their interests. The reform of energy policy is still very superficial. Even after the Fukushima accident, there are very few policy changes as described in detail later.

Why did Japan's energy policy remain almost the same in spite of the Fukushima accident? What are the political barriers to energy transition in Japan? The political opportunity structure on climate change policy and energy policy has been very closed. The Japanese government does not understand the real meaning of the system of climate change *governance*. Though the word of governance is ambiguous and has numerous connotations, in the context of environmental governance such as climate change governance, participatory governance has been focused. It stresses democratic engagement through the participation of multiple stakeholders, including NGOs and citizens in the processes of decision-making.

In Japan, the political leadership for tackling climate change issues has been unclear. A former Prime Minister Ryutaro Hashimoto and Minister of Environment Hiroshi Oki at the Kyoto Conference, called COP3, were exceptional. Both devoted their efforts to leading the conference successfully as the political leaders of the host country. Unfortunately, climate change issues do not have a high priority in Japanese politics. Most Ministers of Environment only hold the post for about a year. Prior to taking up the post they are not familiar with any environmental issues including climate change issues. In simple terms, they were non-professional, they could be called *amateur ministers*. Remember, the current German Chancellor Merkel was Minister of Environment from 1994 to 1998 under the Kohl administration. She has shown outstanding political leadership in climate change politics in her administration since 2005.

Historical Paris Agreement

On December 12, 2015, the Paris Agreement was adopted at the Paris Conference of the United Nations Framework Convention on Climate Change, called COP21. At the venue I was watching the historical scene when Chair, French Foreign Minister, Fabius announced he was adopting the document saying "this small wooden hammer adopts the large document."

At the 1997 Kyoto Conference only 38 advanced countries set a target of decreasing GHG emissions by 2012. These countries are listed in Annex I of the Kyoto Protocol, including 15 EU Member States which had increased to 27 states by 2012. Among the countries listed the US didn't ratify and Canada withdrew from the Kyoto Protocol in 2012.

In the Paris Agreement, all parties including developing countries agreed that the long-term temperature goal was to keep the increase in global average temperature well below 2 degrees Celsius above pre-industrial levels and that they should make every effort to limit the increase to 1.5 degrees Celsius, recognizing that this would substantially reduce the risks and impact of climate change. This should be done by peaking emissions as soon as possible, in order to "achieve a balance between anthropogenic emissions by sources and removals by sinks of greenhouse gases" in the second half of the 21st century. It also aimed to increase the ability of parties to adapt to the negative impacts of climate change, and give financial support to developing countries.

Under the Paris Agreement, each country had to determine, plan, and regularly report on the contribution that it had made to mitigate global warming. Each country's target should go beyond previously set targets. The target was not legally binding, whereas it had been under the Kyoto Protocol.

The world came to reach an agreement to move toward a decarbonizing society. It would be achieved by ending the burning of fossil fuels, developing the renewable energy market, and introducing a carbon pricing mechanism.

Leaving from the second commitment of the Kyoto Protocol

One of the most impressive sights at the Paris Conference was the sign of *oideyasu* at the airport in Paris. *Oideyasu* means welcome in Kyoto dialect. "Welcome" was translated in ten languages on the sign, including *bienvenue* in French, *bienvenido* in Spanish, and *willkommen* in German. In Japanese, welcome was translated as *oideyasu*, in Kyoto dialect, not *yokoso* as usual. Why? In my opinion, for the same reason that the office of the United Nations Framework Convention for Climate Change (UNFCCC) gave the special thanks to Kyoto where the 1997 conference was held and the Kyoto Protocol was adopted. The Paris Agreement was based on the Kyoto Protocol. Without this protocol, it might not have been possible to get the Paris Agreement.

However Japanese leaders didn't refer to the Kyoto Protocol at this conference. At the opening address of the first day, Prime Minister, Shinzo Abe only mentioned the fact that 18 years ago, the Kyoto Protocol was adopted as the first step toward protecting the world from global warming. In the second week, at the meeting of the Ministers, the Minister of the Environment, Tamayo Marukawa never mentioned the Kyoto Protocol. Although having been the host country of the 1997 Conference Japan should have reminded the other countries of the role and meanings of the Protocol, neither was mentioned. The chair of the Kyoto Conference and the Minister of the Environment at that time, Hiroshi Oki, passed away on November 13, 2015 at the age of 88 prior to the Paris Conference. Why did both Abe and Marukawa missed the opportunity to pay tribute to Oki? If they had mentioned his enormous contribution to the Kyoto Conference and his dying hope of the success of the Paris Conference, it might have made a deep impression on all of the delegates attending the conference. The Japanese government missed a wonderful opportunity to remind everyone of Japan's deep commitment to and its political leadership on the climate change negotiation process. In the Japan pavilion of the conference venue, there was no reference anywhere to the Kyoto Conference and the former minister, Oki. Why did they miss this golden opportunity? Had they forgotten the Kyoto Conference, the scenes there, and Japan's initiative at that time? Why did Japan's government disregard them although the office of the UNFCCC paid special respects to Kyoto?

The reason is that the Japanese government didn't positively evaluate the Kyoto Protocol and announced that it would not be signing up to the new targets in the second commitment period, along with New Zealand and Russia at the COP17 in Durban, South Africa, December, 2011. See Figure 2.1 which shows the relationship between the major signatories of the Kyoto Protocol, the first and second engagement, and the Paris Agreement. In my opinion not adopting the second commitment was one of the major misjudgments and failures of the Democratic Party of Japan (DPJ) administration at that time.

One of the major manifesto pledges by DPJ in the 2009 general election by was the ambitious target for a 25 percent decrease of GHG emissions including buying foreign credits from the 1990 level. It was one of the main reasons for

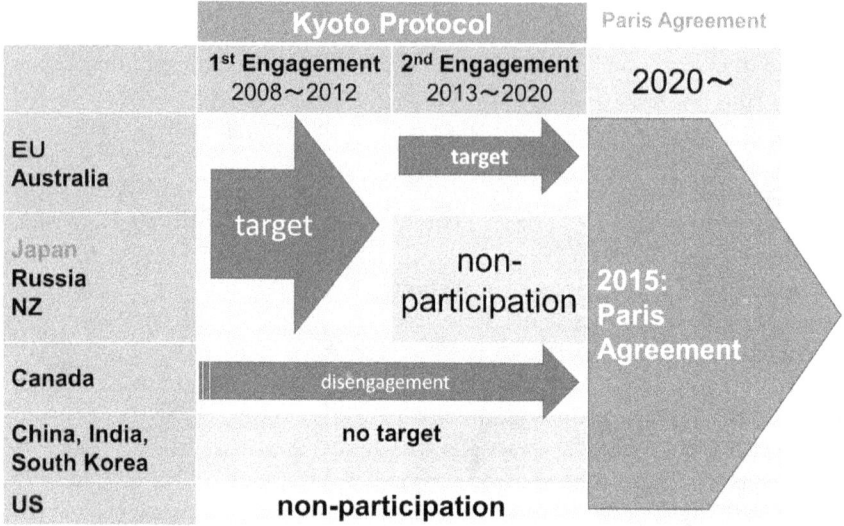

Figure 2.1 Kyoto Protocol and Paris Agreement

winning the election. However, immediately after the election the DPJ administration suddenly adopted a negative attitude toward protecting the world from climate change. Moreover, there were no political achievements in relation to climate change issues. The government revealed serious budget cuts for climate change protection and didn't sign up to the second commitment period of the Kyoto Protocol. The current second Abe cabinet is also reluctant to adopt a positive policy toward climate change protection.

Why did Japan's government adopt a negative attitude and leave the second commitment? The government claimed that it was unfair for advanced countries to take on the Kyoto target. China, the largest emitter of GHG at 25.5 percent, and the US, which was the second largest emitter at 16.9 percent in 2012, didn't ratify the Kyoto Protocol in 2001, nor did South Korea and the BRICS (Brazil, Russia, India, China and South Africa) countries adopt the target as Figure 2.1 shows. The government criticized the Kyoto Protocol for only covering 22.6 percent of the total amount of the world GHG emissions as Figure 2.2 shows. The industrial sector, for example Keidanren (the Japanese Business Federation), also claimed that due to being the host country of the Kyoto Conference, Japan had to concede and accept huge disadvantages. The Kyoto target of a 6 percent decrease was too tough for the Japanese economy.

The Kyoto Protocol was the first UN treaty to bear the name of the Japanese city, Kyoto. It was the symbol of Japan's commitment to climate change

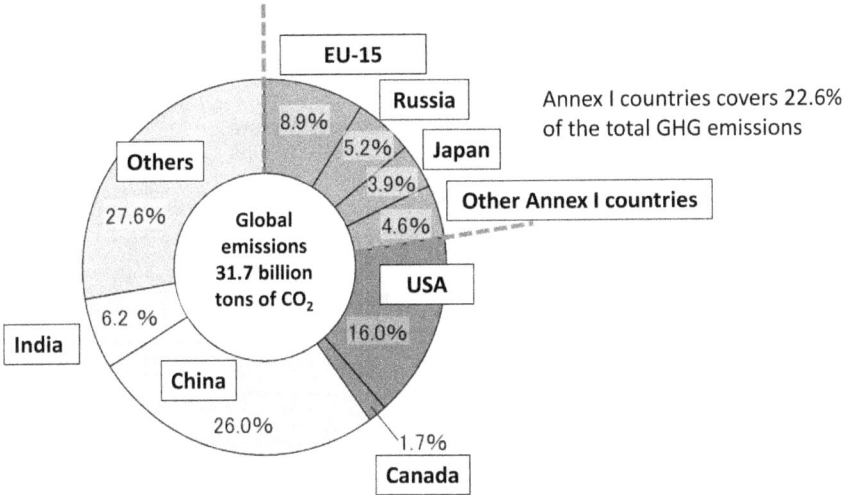

Figure 2.2 The Kyoto Protocol only covers 22.6 percent of the total GHG emissions

Source: The author revised and translated the figure from the Japanese Ministry of Environment website, 2014, accessed October 18, 2019 at www.env.go.jp/earth/cop/co2_emission_2012.pdf.

Note: Canada withdrew from the Kyoto Protocol in December 2012.

diplomacy. However, since it refused to sign up to the second commitment, Japan has been considered to be one of the countries with a huge economy that has a negative attitude toward climate protection. Since then Japan has not given any sign of taking a positive role toward climate protection including at the 2015 Paris Conference and at the more recent 2019 UN Climate Action Summit.

At the Paris Conference, Japan was eager to appeal confirming the strict qualification of enactment of the Paris Agreement because the Japanese government was afraid of free riding and that major emitters, such as the US, would not ratify the Kyoto Protocol. Finally, as Japan's position was accepted and entered into force, the Paris Agreement requires 55 parties to convention accounting for at least 55 percent of total GHG emissions have deposited their instruments of ratification. This was the same condition as for the Kyoto Protocol.

Not signing up to the second commitment meant the loss of the legal base of the reduction target. After the first commitment period ended in FY 2012, Japan no longer had a target. The mid-term goal was finally set in November 2013. But it was a very small target of a 3.8 percent decrease from the FY 2005 level by FY 2020, that is equivalent to allowing a *3.1 percent increase* from the

FY 1990 level. This was 30 percent lower than the target recommended by the EU Commission which proposed Japan should make a 24 percent decrease from the FY 1990 level by FY 2020 (Asuka 2015: 87).

Declining social interest on climate change issue after the Fukushima accident

Figure 2.3 shows the change in the number of news articles including the phrases "climate change" or "global warming" in major Japanese newspapers and the *New York Times* (*NYT*). The peak year with more than 3,000 articles was 2008 when the G8 Toyako Summit was held in Hokkaido at which one of the major topics was climate change. In 2007, former US vice-president Al Gore's film *Inconvenient Truth* was extremely popular all over the world. In 2009, the media hoped that an agreement on the new framework after 2012 – when the Kyoto target ended – would be reached at COP15 in Copenhagen, Denmark. However, the number of articles in Japanese newspapers as well as the *NYT* dropped dramatically thereafter. Japanese papers remained at about the 700 level and the general interest moved to nuclear issues or electricity supply. After the Paris Agreement, it continued to decline. The drop in the number of articles in the *NYT* in 2010 reflects the disappointment at the failure of the Copenhagen

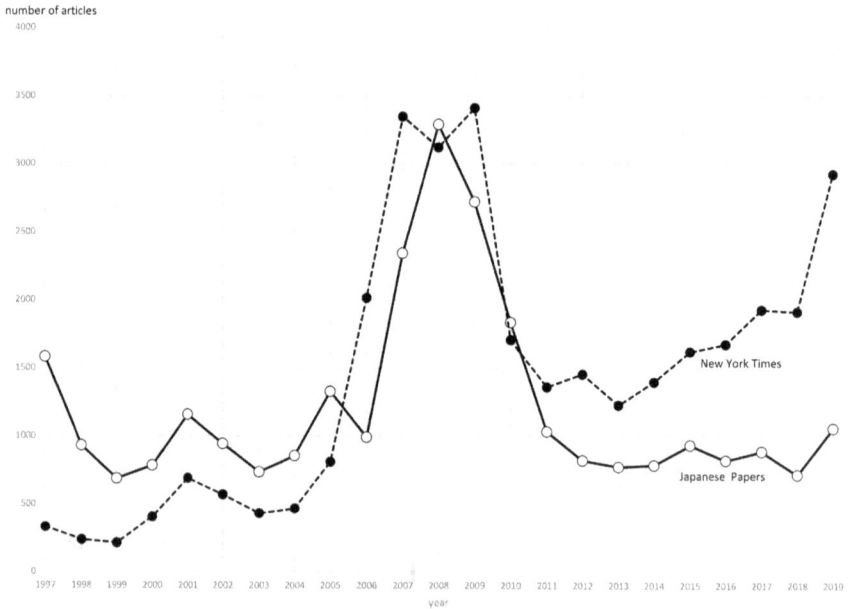

Figure 2.3 Number of newspaper articles including the phrases "climate change" or "global warming" from 1997 to 2019

Note: The solid line is the average number of articles in the Japanese papers, *Asahi*, *Yomiuri*, and *Nikkei*; the dotted line is the number of articles in the *New York Times*.

Conference. But since 2013 the number of articles in the *NYT* has been increasing constantly. This is a sharp contrast between Japan and the US.

Prior to 2007 and 2008, the number of articles in Japanese papers had three peaks in1997, 2001, and 2005: 1997 was the year of the COP3 Kyoto Conference; 2001 was the year the Bush administration declared it would not be ratifying the Protocol; and 2005 was the year the Protocol finally entered into force.

Meeting the Kyoto Target

In February 2015, the UNFCCC announced that many advanced countries had met the Kyoto target and a 23 percent decrease in GHG emissions had been reached from the 1990 level. The original goal was a 5 percent decrease. Internationally, the Kyoto target was recognized to be working quite successfully. This is why the office of the UNFCCC praised the Kyoto Protocol.

But let us look at the case of Japan more closely. Japan also met the original target of a 6 percent decrease in GHG emissions from the 1990 level. This was officially announced by the Ministry of the Environment in April 2014. But the ministry and media were reluctant to publicize this fact and most citizens were unaware of it. Why were they reluctant to publicize it?

Until FY 2007, Japan's GHG emissions increased gradually. Central government estimated meeting the Kyoto target would be impossible or too tough. Indeed, the FY 2007 emissions increased 8.6 percent, in Figure 2.4 the FY 2005 emissions increased 7.1 percent from the 1990 level. How was the target reached?

Figure 2.4 shows that during the target years from FY 2008 to FY 2012, Japan produced an annual average 1,278 million tons of carbon emissions, 1.4 percent higher than the FY 1990 level. However, based on the protocol, Japan was recognized as nominally having achieved a 9.8 percent decrease, comprised of a 3.9 percent decrease from sink of forests, called LULUCF, land use, and land-use change and forestry, and a 5.9 percent decrease from buying foreign carbon credit. Subtract a 1.4 percent increase from a 9.8 percent decrease, then you get 8.4 percent decrease. It means the target of a 6 percent decrease was achieved. This is the secret of Japan's success in meeting the Kyoto target.

In Figure 2.4, the standard of the FY 1990 level was 1 261 million tons. In only two cases, FY 2009 and FY 2010, was the emissions level lower than that of FY 1990. The decline from FY 2008 to FY 2010 was brought about by the economic recession triggered by the bankruptcy of Lehman Brothers in the US. The tsunami disaster on March 11, 2011 affected the emissions in FY 2011. The tsunami disaster affected the emissions in FY 2011. In summing up, Figure 2.4 reveals the economic recession from FY 2008 to FY 2012 was the most critical factor in reducing GHG emissions rather than any measures that were taken. The nominal 3.9 percent decrease from sink of forests and 5.9 percent decrease from buying foreign carbon credit were also critical factors. This was not the most impressive way for the Ministry of Environment to meet the Kyoto target. In my opinion, this may be the reason why the government was reluctant to publicize it.

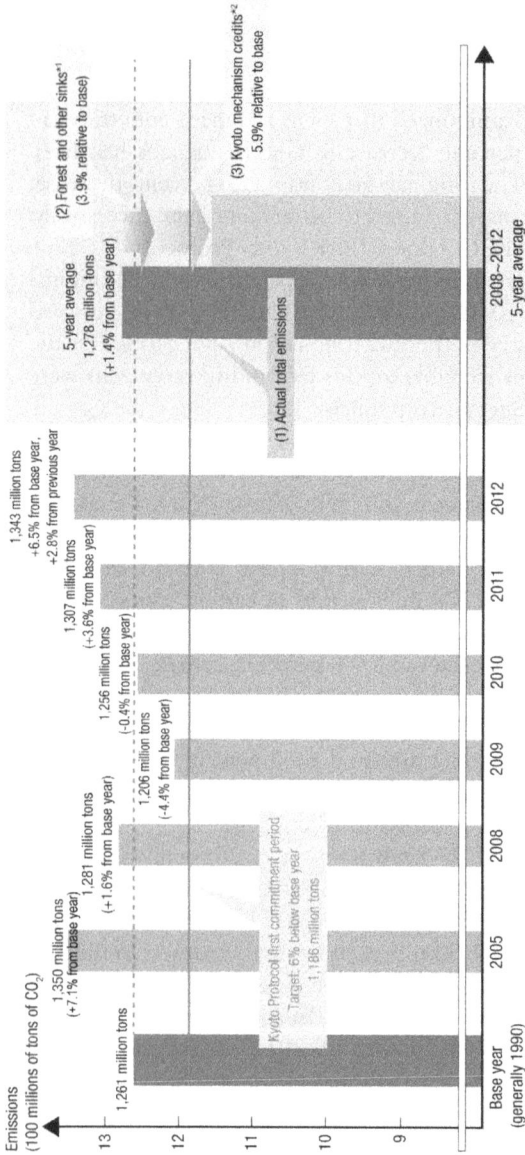

Figure 2.4 How Japan met the Kyoto target

Source: Japanese Ministry of Environment, 2014, accessed October 18, 2019 at https://www.env.go.jp/en/wpaper/2014/pdf/02.pdf.

Note: A 1.4 percent increase minus a 3.9 percent of LULUCF minus a 5.9 percent of credits is a 8.4 percent decrease beyond the Kyoto target, a 6.0 percent decrease from the 1990 level.

The effects of the 2011 Fukushima nuclear disaster

The 2011 Fukushima nuclear disaster is the second largest nuclear accident in the world after the Chernobyl disaster in 1986 (see Hasegawa 2015). Four reactors were simultaneously endangered for the first time in world nuclear history. People were suddenly forced to face the real nuclear risks and radiation exposure. They became extremely angry with, distrustful of, and disappointed with the national government, the Tokyo Electric Power Company (TEPCO), and media and nuclear experts. The most tragic and distressing cases are the 160 000 evacuees, the peak number in June 2012, who had been living near the melted-down nuclear station. A zone of up to 20 km from the nuclear site was designated an 'Access-restricted Area' and no entry was allowed unless authorized until the end of March 2017. Some areas, like Iitate Village, outside the 20-km zone were also designated 'Deliberate Evacuation Areas.' Even in September 2019, eight and half years after the accident, the prefectural government was officially reporting that around 42 000 people were still evacuated from their homes, 11 000 people within Fukushima prefecture, and 31 000 outside it. More than an estimated 24 000 other people were officially uncounted evacuees who were living in new house but still wanted to return to their own homeland in the near future.

Curious political stability within Japan

Remember what had happened after the 1986 Chernobyl accident. Mikhail Gorbachev, the last president of the Soviet Union, confessed "the accident at the Chernobyl nuclear power station was graphic evidence, not only of how obsolete our technology was, but also of the failure of the old system" (Gorbachev 1996: 189). There were only five and a half years between the accident and the collapse of the Soviet Union in December 1991 following the fall of the Berlin Wall in November 1989.

In contrast with the Soviet Union, politically, the current Japanese government has curiously enjoyed stable public support and won all six national elections in both the lower and upper house since the change of government in December 2012. The right-wing Shinzo Abe administration is one of front runners of the recent "We-firstism" governments (like Donald Trump's "America First") and is fortunate enough to be the longest seven-year cabinet in Japan's history. Before the second Abe cabinet, each of the six former prime ministers, including Abe himself during his first cabinet from September 2006 to September 2007, had to resign within approximately a year or less because of lack of public support. It would appear that the Fukushima accident did greater political damage to the Democratic Party of Japan (DPJ), which had been the ruling party from September 2009 to December 2012, than to the Liberal Democratic Party (LDP), then an opposition party. The DPJ cabinet of Prime Minister Naoto Kan misled the public and created a lot of confusion, it also deliberately concealed information and delayed disclosing information (Hasegawa 2015: 11–13). In fact, the LDP as the dominant ruling party for more than five decades bears most of the responsibility for promoting nuclear power and the policies that led to the Fukushima accident.

Drastic transition to denuclearization in Germany, Taiwan, and South Korea

What changes did the Fukushima nuclear disaster bring about in Japan energy policies in Japan and throughout the globe?

Internationally the Fukushima nuclear disaster brought about a drastic transition to denuclearization. In July 2011, Germany adopted an energy shift policy that included abandoning all 19 of its nuclear reactors by the end of 2022. Switzerland also decided to abandon new reactor construction and close all five reactors after 50 years of operation, which means all its reactors will be shut down by 2034. In East Asia, in January 2017, the new government of Taiwan decided to abandon all six of its reactors by 2025, and to this end a new law was passed (Chou 2017). In June 2017, the new Korean President, Moon Jae-in, also declared denuclearization would be implemented over the next 40 years. Both South Korea and Taiwan are dependent on oil imports, like Japan. In South Korea, 30 percent of electricity is provided by nuclear power, which is a similar situation to that of Japan before the Fukushima accident. In Taiwan, 19 percent of its electricity is provided by nuclear power. In both cases changes of government and the role of strong political leadership seem to have played a critical role in bringing about policy shifts toward denuclearization. Thus, denuclearization and promotion of renewables and energy efficiency have been the new mainstream policies worldwide since the Fukushima accident. But, what about in Japan?

Very few policy changes after the Fukushima accident

Immediately after the accident, some scholars, journalists, and environmental NGOs hoped to move to energy transition as a result of the Fukushima nuclear accident. The majority of public opinion shifted from supporting nuclear energy to closing down nuclear power plants in the near future (Iwai and Shishido 2015). However, the conservative government supported by economic sectors and the Ministry of Economy, Trade and Industry (METI) managed to return to the administration and keep the nuclear energy and climate change protection policy. The government is not interested in energy transition. It doesn't appear to have learnt any lesson from the Fukushima accident.

There have only been a few policy changes in the country in which the Fukushima accident happened. First, a new law to promote renewable energy with a feed-in tariff system, called FIT, was enacted in July 2012. Second, a new strict and independent regulatory system of nuclear energy started in September 2012, but LDP and the pro-nuclear groups have been exerting strong political pressure to allow the reopening of nuclear reactors one by one. Third, the DPJ's new energy policy, which appeared in September 2012, was abandoned by the LDP-Komei government after the general election in December 2012. The government revised the Strategic Energy Plan of 2014 in July 2018 and continued to allow the basic role of nuclear reactors. As of the end of December 2019, nine reactors in the western part of Japan are reopening but none of the reactors have been operating in the eastern part of Japan where the tsunami hit the coastal area. Fourth, in December 2016, the Japanese government finally decided to abandon the multi-decade troubled and highly costly fast breeder reactor, Monjyu. The nuclear fuel cycle program has thus lost one of its essential parts.

After FIT was introduced in July 2012, the installed capability of solar photovoltaics (PVs) increased six times but the number of wind turbines still only increased slowly. The Japanese government, heavy industrial companies, and power companies prefer nuclear energy and coal-fired power stations, and are still reluctant to promote renewable energy.

Why do power companies want to keep nuclear power plants after the Fukushima accident? Why can't Japan change its energy policy even after the Fukushima accident? And why is Japan's position on climate change protection so lacking in ambition? The unchanged energy policy and climate change policy and the stable political situation may be closely linked.

Electric power companies are afraid that if they decide to abandon nuclear energy, they will suffer an extraordinary loss as they will lose asset values in terms of their nuclear power generation plants, equipment, and nuclear fuel. If the extraordinary loss exceeds a company's net asset value, the company technically becomes insolvent and goes into liquidation. As an extraordinary loss cannot be included in the costs used to calculate electricity rates, the companies cannot cover the loss by way of higher electricity charges. Consequently, the electric power companies are against denuclearization because of the fear of corporate failure.

The electric power companies have been a critical support base for the ruling LDP. Especially in the local economy, an electric power company is the largest company and can therefore control the local media as the largest sponsor of advertisements. The labor union of the electric power company had been one of the major support bases for the former largest opposing party, DPJ, and this is still the case for the current Constitutional Democratic Party of Japan. Within the ruling and the major opposition parties, it is impossible to find any political leadership to break through the political barriers.

Chancellor Merkel in Germany, President Tsai in Taiwan, and President Moon in Korea have all shown great determination and political leadership to move energy transition forward to denuclearization and a stronger climate change protection policy including closing all or some of the coal-fired power plants. Such strong political will and leadership is indispensable if effective energy transition policies are to be introduced.

However, in Japan, the METI and companies like Hitachi, Mitsubishi, and major power companies are also barriers to energy transition because they work to protect their existing interests. The president of Hitachi, Nakanishi is also head of Keidanren.

Japan's INDC for the Paris Conference

The office of the UNFCCC requested each party to submit its Intended Nationally Determined Contributions (INDCs) by March 2015 if possible, or before COP 21. On July 17, 2015, the Japanese government submitted its INDC of a 26 percent decrease from the carbon emissions level in FY 2013. This means a 17.4 percent decrease from the FY 1990 level and a decrease of 25.4 percent from the FY 2005 level. As a base year, FY 2013 was selected rather than FY 1990. Domestic and international environmental NGOs criticized this target for being too small and passively. For the industry sector the target is only a 7 percent decrease, whereas for both the residential sector and the commercial and other sectors a 40 percent decrease is requested. This target looks unrealistic and too tough for both sectors.

Japan's INDC is consistent with its energy policy, that is METI's long-term outlook published in July 2015, which outlines the achievement of Japan's 2030 GHG emissions target for the UNFCCC Paris Conference (UNFCCC, COP21). METI said that total electricity generation of 1 065 billion kWh in the FY 2030 will be composed of 22–24 percent renewable energy, 22–20 percent nuclear power, 27 percent LNG-fired thermal energy, 26 percent coal-fired energy, and 3 percent of oil-fired energy as shown in Figure 2.5. The basic assumption is that

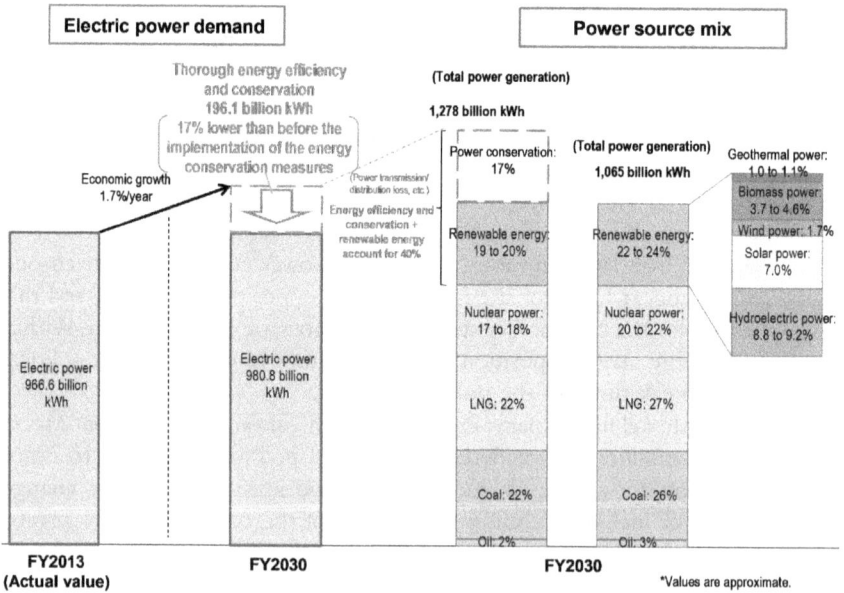

Figure 2.5 Long-term energy supply and demand outlook in July 2015

Source: Japanese Ministry of Economy, Trade and Industry (2015).

economic growth will continue at an average of 1.7 percent increase per year from FY 2013 to FY 2030. The required power generation will be 1 278 billion kWh in FY 2030, and it will be reduced to 1 065 billion kWh by 17 percent of power conservation (see Figure 2.5). What is wrong with this outlook?

Figure 2.6 shows that in reality, the peak of total power generation was 1 200 billion kWh in FY 2007 and it declined 15 percent to 1 020 billion kWh in FY 2015. The prediction of 1 278 billion kWh in FY 2030 was a huge overestimation, representing a 25 percent increase in the level of electricity supplied in FY 2013. It is important to bear in mind that Japan is already experiencing a shrink in population and has a low economic growth of an average of 0.7 percent in GDP per year between FY 2010 and FY 2015. The estimated share of 22–20 percent to be provided by nuclear power, 213–234 billion kWh in FY 2030, could be replaced by power conservation with no carbon emissions. Denuclearization by the 2030s is possible in Japan.

The decoupling of economic growth and power consumption, as is the case in Germany, has already started in Japan. Japan is decoupling economic growth and GHG emissions without relying on nuclear power. The peak of Japan's GHG emissions was reached in FY 2013, when Japan shut down all its reactors. Emissions then continued to decline by 3.4 percent, 6.1 percent, 7.3 percent, and 8.4 percent in FY 2014, FY 2015, FY 2016, and FY 2017 respectively

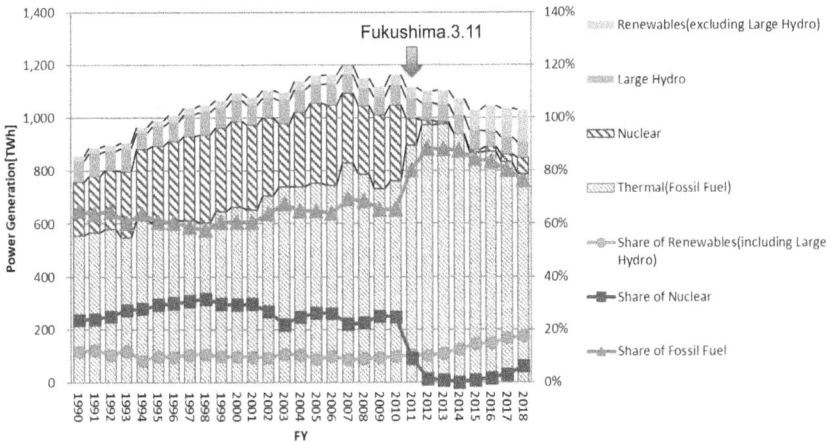

Institute for Sustainable Energy Policies

Figure 2.6 Power generation in Japan (FY 1990–2018)

Source: ISEP (2019).

Note: Ratio of renewable energy power generation to total power generation percent has remained unchanged for the past two decades at 10 percent. In FY 2018 it increased to 17.5 percent.

Emissions
(Billion t-CO$_2$ eq.)

1,360 Mt
(-1.6% from FY2005)

1,410 Mt
(+2.0% from FY2005)

1,308 Mt
<-7.3% from
FY2013>
(-5.4% from
FY2005)

1,244 Mt
[-3.6% from FY2017]
<-11.8% from FY2013>
(-10.0% from FY2005)

1,396 Mt
(+1.0% from FY2005)

1,399 Mt
(+1.2% from
FY2005)

1,362 Mt
<-3.4% from
FY2013>
(-1.4% from
FY2005)

1,382 Mt

1,324 Mt
(-4.2% from
FY2005)

1,356 Mt
(-1.9% from
FY2005)

1,324 Mt
FY2013>
<-6.1% from
FY2005)

1,291 Mt
<-8.5% from
FY2013>
(-6.6% from
FY2005)

1,305 Mt
(-5.6% from
FY2005)

(-4.2% from
FY2005)

1,275 Mt

1,251 Mt
(-9.5% from
FY2005)

14 — 13 — 12 — 11 — 0

1990 2005 2006 2007 2008 2009 2010 2011 2012 2013 2014 2015 2016 2017 2018

1. *These preliminary figures for FY2018 were estimated based on annual figures in various statistics. Some annual figures from FY2017 were temporarily used in place of FY2018 figures that have yet to be released. Moreover, some estimation methodologies are currently being reconsidered in order to make more accurate estimations of emissions. As such, the final figures to be released in April 2020 may differ from the preliminary figures in this summary. Removals by forest and other carbon sinks will also be estimated and announced at the time of the release of the final figures.*
2. *Total GHG emissions in each FY and percent changes from previous years (such as changes from FY2013) do not include removals by forest and other carbon sinks from activities under the Kyoto Protocol.*

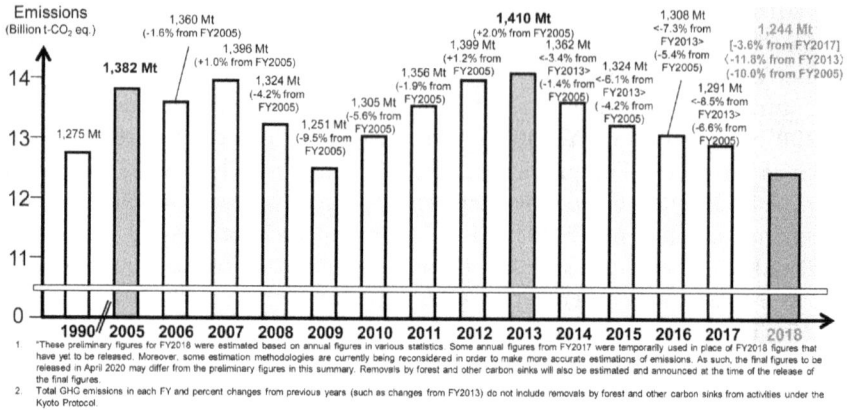

Figure 2.7 Trends of Japan's national GHG emissions FY 2005–2018
Source: Japanese Ministry of Environment, press release on November 29, 2019.

(see Figure 2.7) despite no operating nuclear reactors in 2014, and very few in 2015, 2016, and 2017. Efforts at energy conservation in industry were successful. Overestimation of economic growth and electricity consumption brought a kind of excuse for dependence on nuclear power and coal-fired plants. There was no suggestion of energy transition in METI's long-term outlook.

The Japanese government has been reluctant to introduce an effective climate change policy and has encouraged a "voluntary action plan" by each industry sector, such as the oil refining industry and the iron and steel industry, sectors of the economy it wants to keep (Hasegawa and Shinada eds. 2016). There is no carbon pricing system and no nationwide cap-and-trade program. Weak carbon tax with 289 Yen/CO$_2$ tons which means an extra payment of 100 Yen/per month for each household. There is no effective framing of the target goals of climate change protection in FY 2030 and FY 2050.

There is weak political leadership on climate issues for in both the ruling parties and the opposition parties. Mitigation policies for climate issues does not bring popularity, votes, or money for any politician. Few diet members who back climate issues receive any benefits for their efforts.

INDCs: EU and Germany

The World Resource Institute (2016) provides the latest situation of each party's INDCs in the *Paris Contributions Map*. Each target was submitted on a voluntary basis, was not legally binding, and the target years were different. The amount of the total targets will result in an increase of more than 2.0 degrees Celsius. The most aggressive target was Switzerland's with a 50 percent decrease by 2030 from the 1990 level including carbon credits. The target of the whole of the EU, with its 28 Member States, of a 40 percent decrease by 2030 from the 1990 level is also very aggressive.

The EU has already reached about a 19 percent decrease in carbon emissions from the 1990 level, but has simultaneously achieved a 44 percent increase in total GDPs. It is the best example of a decoupling of high GHG emissions and economic growth. The average emissions per capita in the EU declined to 9 tons in 2012 from 12 tons in 1990. It is estimated that it will be 6 tons in 2030. However, in Japan it is gradually increasing from the 1990 level. The EU shows a *peak out* of GHG emissions in 1979 which was the peak year for emissions.

Germany has the most aggressive targets of a 40 percent decrease by 2020 from the 1990 level, a 55 percent decrease by 2030, and a 80 or 95 percent decrease by 2050 (OECD/IEA 2013). At the UN Climate Action Summit, September 2019, Merkel promised German's GHG emissions would reach substantially zero by 2050. There is a sharp contrast between Japan and Germany. Germany decided to close all its nuclear plants by the end of 2022, and does not permit the construction of any new coal-fired plant. Additionally, in January 2019, a government committee on coal-fired plants recommended the government should close all of the existing 84 coal-fired plants by 2038 which provided 40 percent of Germany's electricity. In Germany, the coming of the post nuclear and post coal-fired society is a reality.

Rushed coal-fired power plant projects

Since the Fukushima nuclear accident, the use of coal-fired power plants has become a contested issue in Japan. In FY 2016, the electricity market was finally deregulated, which had been a major challenge for the government and power companies since the late 1990s. However, one result of the introduction of deregulation has been to stimulate a rushed construction of new coal-fired power plants across Japan. In 2012, prior to the deregulation, 50 new coal-fired power plants were constructed. They will bring approximately 43 GW of capacity thereby doubling the electricity output of all current coal power plants in Japan. If all these new plants start operation, Japan's estimated total GHG emissions in 2030 will reach 1,110 million tons beyond its target level of 1,042 million tons set by the Paris Agreement. This is 5 percent higher than the target. It means it will be impossible to meet the target of a 26 percent decrease from the level of FY 2013 emissions. This is a nightmare for climate change protection in the world which is trying to keep global temperatures from exceeding two degrees of warming under the Paris Agreement. Coal-fired plants also emit other kinds of air pollution, notably sulfur dioxide (So_x), nitrogen dioxide (NO_x), fine particles with a diameter of 2.5 μm or less ($PM_{2.5}$), mercury, and water pollution through sewage emissions.

Since the Paris Agreement was enacted, the international situation regarding coal-fired plants has changed dramatically. In European countries and the US, a total of 64 GW operating plants were retired in 2015 and 2016 as a result of the 2015 Paris Agreement. Even in China and India a total of 68 GW of coal-fired plant capacity under construction is now frozen at over 100 project sites. As of January 2017 the amount of coal-fired power capacity in pre-construction planning went down to 570 GW from 1 090 GW in January 2016 in China and India. In October 2017, the United Nations Environment Programme (UNEP) recommended an end to the construction of new coal power plants and an accelerated phasing out of existing plants as key steps toward achieving the goals of the Paris Agreement. Recently, the fossil fuel divestment movement has been very active in the US and Europe to reduce carbon emissions, applying public pressure to stigmatize "fossil fuel companies" that are currently involved in fossil fuel extraction and get them to move investments away from fossil fuels and toward renewable energy.

Nonetheless, METI has been supporting construction of coal-fired plants. Companies are eager to build coal-fired plants because their fuel price is cheaper than natural gas in the Japanese situation. It seems that power companies and the national government are strategically forcing citizens into a tough choice between "reopening nuclear plants" or "constructing coal-fired plants." METI and power companies have been reluctant to promote renewable energy as they seek to uphold the monopolized position of the current energy producers. The anti-nuclear sides prefer renewable energy resources, whereas the pro-nuclear factions prefer nuclear or coal energy. For many years, Japanese NGOs have protested against nuclear energy but ignored coal energy. The current Abe Cabinet, METI, and power companies want to export nuclear and coal energy to developing countries such as Indonesia, India, and Turkey.

As of the end of March 2020, 13 projects had been cancelled or converted to wood-based biofuel plants. But another 15 units have come online including the plant in Sendai. Most of the small projects of less than 112.5 MW have started operation without a legally environmental impacts assessment (EIA) being required.

In the city of Sendai, only a few years after the devastation of the 2011 tsunami, the municipality has permitted construction of a new coal-fired power plant of 112 MW. The Sendai Power Station Ltd cleverly exploited a legal loophole to avoid making an environmental assessment which would have delayed the project by two years. This broke *Sendai*'s proud coal-free history for the past ten years. Within a 5-km radius of the proposed plants, there are approximately 150 000 residents, and 32 schools including 17 primary schools, where children will be at risk from fine particulate and NOx pollution. Ignoring the global imperative to fight climate change and transition to renewables, Japanese corporations that have invested heavily in coal abroad need an outlet for their dirty fuel. The Sendai Power Station is preying on the economically-challenged tsunami regions of Sendai. Local citizens are complaining that the electricity produced will go to Tokyo, profits will go headquarters in large cities, while the air pollution will choke local residents, and the climate risk will burden today's young generation.

Three coal-fired power plant projects, including Sendai, Kobe, and Yoko-suka, have been sued for an injunction to stop operation or construction by local citizens. Kobe's project for two plants of 650 MW is under construction and Yokosuka's case which also has two plants of 650 MW is still at the planning stage.

The rushed construction of coal-fired power plants means there is little likeli-hood of a carbon price system being introduced in the near future in Japan. Generally coal-fired power plants produce electricity for more than 40 years.

International fossil-fuel divestment movements

However, the dirty image of a "coal-fired plant" has become stronger year by year. Powering Past Coal Alliance was established in November 2017 at COP23. Recently a total of 86 entities, including 30 countries such as the UK and Canada, states, 28 local governments, and 28 private sectors have joined the alliance. Its target is to ban the construction of domestic new coal-fired plants, to decommission existing coal-fired plants, and to stop financial support for the construction of coal-fired plants in developing countries by 2030.

The international fossil-fuel divestment movement have also criticized Mizuho Bank and Mitsubishi UFJ Bank for being the largest and the second largest lenders in the global banking industry to coal plant developers (https://www.banktrack.org/coaldevelopers accessed October 18, 2019). At every climate change conference, like COP21 and the UN Climate Action Summit, 350.org, Green Peace, and Friend of Earth have taken action to criticize the Japanese government's and major banks' policy of promoting coal-fired plants in develop-ing countries.

Who can break through the dead end situation?

Who can break through Japan's dead end situation of energy transition and climate change protection? Political leaders? Leaders of the economic sectors? The author estimates that the likelihood of such leaders appearing will be low as already described. Leadership of bureaucrats will also be low because generally speaking their attitudes tend not to be innovative. Additionally, the influence of the Ministry of the Environment are much smaller than METI in Japan.

What about environmental NGOs? Japan's environmental NGOs, like the Kiko network, the Citizen's Alliance for Saving Atmosphere and the Earth (CASA) and the Institute for Sustainable Energy Policies (ISEP) have all estab-lished a domestic base, and World Wildlife Fund (WWF) Japan, Friends of the Earth (FoE) Japan, and GreenPeace Japan, which are chapters of international NGOs, have been making efforts to promote climate change protection and energy transition through a variety of routes. Some members sit on government committees regarding these issues, but their influence is not very great because they are few in number on these committees. Bouts of anti-nuclear activism in 2011 and 2012 failed to bring activists victory in the following national elections

(Hasegawa 2018). Under the political backlash led by ethno-centrism and populism, supporters of nuclear energy and coal-fired power plants still represent large political barriers to the energy transition. They have failed to find an effective political route to change the Japanese energy policy and climate change policy due to the political weakness of current opposition parties. These civil society organizations lack effective political partners and strategies.

References

Asuka, Jyusen. (2015). *Kuraimeto jyasutisu: Ondanka taisaku to kokusai keizai no seiji·keizai·tetsugaku (Climate Justice: Climate Change Policies and Politics, Economy, and Philosophy of International Negotiation)*, Tokyo: Nihon Hyoron sya.

Chou, K. T. (2017). Tri-helix Energy Transition in Taiwan, in Kuei-Tien Chou, ed., *Energy Transition in East Asia: A Social Science Perspective*, London: Routledge, 9–27.

Gorbachev, Mikhail S. (1996). *Memoirs*, Garden City, NY: Doubleday.

Hasegawa, K. (2015). *Beyond Fukushima: Toward a Post-Nuclear Society*, Melbourne: Trans Pacific Press.

Hasegawa, K. (2017). Risk Culture, Risk Framing and Nuclear Energy Dispute in Japan Before and After the Fukushima Nuclear Accident, in Kuei-Tien Chou, ed., *Energy Transition in East Asia: A Social Science Perspective*, London: Routledge, 9–27.

Hasegawa, Koichi (2018). Continuities and Discontinuities of Japan's Political Activism Before and After the Fukushima Disaster, in David Chiavacci and Julia Obinger, eds., *Social Movements and Political Activism in Contemporary Japan: Re-emerging from Invisibility*, London: Routledge, 115–135.

Hasegawa, K. and T. Shinada, eds. (2016). *Kikohendo seisaku no syakaigaku: Nihon wa kawarerunoka (Sociology of Climate Change Policies: Can Japan Happen to Change?)*, Kyoto: Showado.

Institue for Sustainable Energy Policies (ISEP). (2019). *Deita de miru nihon no shizen enerugi no genjyo: 2018 nendo denryoku hen (Japan Status of Renewable Energy 2018)*. ISEP. Retrieved October 18, 2019, from https://www.isep.or.jp/wpdm-package/JapanStatus20190805.

Iwai, N. and Kuniaki, S. (2015). The Impact of the Great East Japan Earthquake and Fukushima Daiichi Nuclear Accident on People's Perception of Disaster Risks and Attitudes toward Nuclear Energy Policy, *Asian Journal for Public Opinion Research*, 2(3): 172–195. 10.15206/ajpor.2015.2.3.172.

Japanese Ministry of Economy, Trade and Industry. (2015). *Long-term Energy Supply and Demand Outlook*. Retrieved October 18, 2019, from https://www.meti.go.jp/english/press/2015/pdf/0716_01a.pdf.

OECD/IEA. (2013). *Energy Policies of IEA Countries: Germany 2013 Review*. Retrieved April 11, 2020, from https://www.iea.org/reports/energy-policies-of-iea-countries-germany-2013-review.

World Resource Institute. (2016). *CAIT Paris Contributions Map*. Retrieved October 18, 2019, from http://cait.wri.org/indc/#.

3 Climate change governance in Taiwan

The transitional gridlock by a high-carbon regime

Kuei-Tien Chou and Hwa-Meei Liou

Introduction

Modern society faces large-scale transformative societal change, especially since the engine of modernity of contemporary social operations is powered by fossil resources or coal-fire power which have dominantly molded the centralized system, resulting in the structure of political decision-making tending towards centralized, linear thinking, which has led to the control, distribution, and management of society's development (Loorbach & Rotmans 2006; Frantzeskaki, Loorbach & Meadowcroft 2012). However, this set of operating mechanisms has created the lock-in of negative externalities, and the long-term resistance to innovation and change has resulted in a high degree of systemic vulnerability (Beck 1986, 1992), which is the criticism from reflexive modernization of the self-development of industrial society that has led to the massive risk society, and even of the world risk society. Or, based on what Geels believes, is such that a socio-technical system based on fossil energy would result in contested modernization (Geels 2006; Rotmans & Loorbach 2009).

Urry (2011) criticized such development from a path dependency perspective, saying that the various social systems that we see today, including the energy systems, fuel inefficient automobile transportation, long-distance commuting, and energy inefficient technology, etc., have constructed today's modern capitalistic system, which has basically established a system of high-carbon operations. And also, the large-scale, socio-technical system and operating mechanism, through the accumulation by dispossessions, has further developed into the high-carbon capitalistic ideology today. In Urry's opinion, the most critical of them is how the high-carbon regime has achieved the legitimacy and ability to sustain its operations.

With such a sense of the problem, this chapter will mainly discuss the issue of Taiwan's climate governance, and provide an initial analysis of Taiwan in the context of Asia's political and economic development, and the structural dilemmas and challenges to governance.

Theoretical framework

Path dependency

The analytical path of reflexive governance needs to start from a self-critical and reflective evaluation of the country's decision-making and regulatory structure, which should include analysis at two levels of the institutions and agents (Grin 2006; Voß, Smith, & Grin 2009), of which the first could focus on the government's decision-making, policies, systems and regulation, while the second could be directed at the diverse actors, such as government personnel, industrial groups, civil society, and other stakeholders.[1]

Transition management, which possesses a reflexive governance importance, will focus on examining the path dependencies that cause social ecological transformation, including of the *cognitive, institutional, technical, and economic patterns*, and considers whether or not as society moves towards transformation these facets will impose strict restrictions and obstacles and lock-in social innovation and change onto a particular track (Voß, Kemp & Bauknecht 2006; Rip 2006). Further, a persistent problem is that complexities are deeply-rooted in the social structure, which has formed a structural path dependency that has locked-in social change and innovation to certain technological biases, dominant networks and administrative obstacles, etc. Overcoming these systemic failures requires the restructuring of the social system, and the reconfiguration of society's development and its values, in order to transform (Rotmans & Loorbach 2009, 2010).[2] In other words, transformation will be undertaken via the systems change and institutional innovation of the fundamental structure of society, using a long-term and systems way of thinking, and the participation of diverse fields and multiple stakeholders, to develop the economy, culture, technology, and ecology at the various institutional scales (Loorbach & Rotmans 2006).

Marshall and Alexandra (2016) highlighted the obstruction to policy innovation caused by institutional path dependency. They analyzed Australia's Murray-Darling Basin's study of the retardation and predicaments of the transformation of water resources policy as an example, and pointed out that researchers should use context-specific institutional analysis in their observations. Barnett et al. (2015) also concluded that institutional path dependency can also result in the obstruction of change, and affect the speed of change. They pointed out that path dependent institutions are resistant to change. When such resistance causes climate adaptation to become slow, these institutions will then become obstacles and limitations. In general, the barriers that are commonly agreed upon include a lack of or inconsistent leadership, insufficient knowledge of risks and responses, inadequate funding, difficulties in negotiating between competing values and goals, a lack of institutional support, and poor coordination across various levels of government. Van Buuren, Ellen and Warner (2016) emphasized that in the Netherlands, the dominant policy coalition created obstacles, which delayed the envisioning of a new paradigm for the Dutch flood risk management. Pierson (2000) pointed out very early on that four characteristics of politics were the main cause of path dependence: the conservative nature of institutions, the high density of institutions resulting in barriers to entry, power asymmetries which become self-reinforcing, and the complexity of policy environments that gets in the way of learning (cited by Buuren, Ellen, & Warner 2016). Obviously, the research on path dependency needs to highlight the decision-making system, knowledge, social cognition, and values of the framework used by decision makers.

Developmental state

In analyzing the climate and energy governance of some Asian countries, in addition to referencing the transition management theory, and using the transformation of institutional obstacles, market structures, technological systems, ideologies, policy discussions and values, etc., as pathways to the problem, it is most important to use an analytical perspective of reflexive governance as an entry point to understand the government's decision-making process and the social contexts in these Asian countries.

The research approach to reflexive governance is appropriate for many important Asian countries which are driven by high-carbon economies, especially since these countries also currently face the demands and challenges of a low-carbon economy and social transformation. Wang (2012) pointed out that the reflexive governance approach proposed by Beck (2006) could be used as a new method to reflect on the "development" of individual countries. In particular, developmental states emphasize economically-driven social change, and the quick imitation and catching up of advanced industrialized countries whose foundations were laid using fossil fuels, but which left behind numerous environmental and labor exploitations. Faced with major transformation today, it is necessary to conduct a more comprehensive and extensive study of the "developmental" implications of each country.

In the late 1980s, latecomer countries like South Korea had to play catch-up with the industrialized countries. In learning from and imitating the type of development of these industrialized countries, they had to deal with the fast-pace, accelerated, and compressed process of industrialization which resulted in the serious neglect of society's labor, environment, gender, etc. and their rights and related issues, which in turn spawned long-term problems of exploitation.

In terms of the reflexive criticism of political and social development, Chou (2000, 2002) explained that in addition to the fast-pace, accelerated, and compressed process of industrialization in the early 1980s, there was also the configuration of authoritarian politics and its transformation into democracy, but although there was considerable resistance to the labor, environmental, and political issues at that time, civil society still lacked the systematic and endogenous ability to construct risk knowledge, and was unable to respond quickly which had implications for its ability to be in control of the situation. On the other hand, the structural dominance and continuation of authoritarian expert politics and the developmental state logic of prioritizing economic development have led to various environmental and technological risks becoming hidden and neglected. In the long run, the result has been to create a delayed high-tech risk society with institutionalized hidden risks. South Korean scholar Chang (2010) consequently used the concept of compressed modernization to explain the development of his country.

But the delayed risk society poses massive challenges for a country which is rapidly undergoing the transformation of its climate and energy governance, and if it is unable to suitably construct transparent and participatory technology policies, it might evolve even deeper systemic risks, as underlined by the Organisation for Economic Co-operation and Development (OECD) (2003).

And we can see that although the newly-elected ruling Democracy Progress Party (DPP) has established social justice as its cornerstone in the Guidelines on Energy Development (under social risk communication and public participation) it has recently launched, and is proactively involving citizen participation in the Energy Transition White Paper that it has formulated, the nature of authoritarian expert politics in decision-making has not been completely eradicated, which has resulted in high vulnerabilities and systemic risks lurking in the development of Taiwan's recent climate and energy governance. As such, there needs to be greater transparency in transformation, and learning from and participating in the socio-technical system is therefore not something worth encouraging.

In fact, it can be initially said that the politics, economy, technology, and governance, etc. of the whole of Asia has been dominated by authoritarian expert politics as part of its historical formation, with economic development as the driving force, although such a governance model is being gradually challenged by society. And it is therefore even more necessary that in the context of authoritarian expert politics and delayed risk society in Asia the concept of accumulation by dispossession as emphasized by Harvey (2003) should be revised.

Methodology

As a developmental state dominated by authoritarian expert politics, Taiwan's government is similar to other Asian countries, in prioritizing economic development as a driver for social transformation, but as a result it has neglected environmental and labor risks over the long-term; and therefore now finds itself faced with difficulties in climate and low-carbon transformation. As the approaches to path dependency analysis is wide-ranging, this chapter will use the perspective of reflexive government as an entry point to analyze the question of social transformation, first, by limiting the analysis to the industrial and energy structures (economic models), electricity and water prices, and fossil fuel subsidies (regulations), to demonstrate the formation of the high-carbon, brown economic package.[3] Second, a brief social analysis will also be conducted on the struggle of environmental movements, and the perceptions and attitudes of the public towards climate and energy issues, in order to look for indications that society has already undergone a paradigm shift. Third, by analyzing social transformation from the perspective of governance, we explore the declarations of the carbon reduction policies, the lack of a clear roadmap, and the discussion of climate and energy issues, to understand how we are held back by the challenges of the discussions over the high-carbon and brown economy, which has resulted in a social transformation quandary.

For the data collection and chart depiction, the authors have largely used reports and statistical data from government sources to conduct the analysis. The data on carbon emissions from the industrial and energy sector and on energy subsidies comes from the Environmental Protection Agency and the Bureau of Energy; data for the industries' contribution to gross domestic product (GDP) is derived from the Directorate General of Budget, Accounting and Statistics, and data on subsidies is taken from from the International Energy Agency (IEA). Information on the government's major energy and industry conferences, policies, and announcements mainly come from the Executive Yuan, the Bureau of Energy, and the Environmental Protection Agency. The analysis of environmental movements and initiatives is taken from the cumulative research outcomes of one of the authors. Basing our brief analysis of the results of the surveys that the author conducted on climate change risk perception in 2012 and 2015, we point out that Taiwan's society is currently undergoing a paradigm shift. By connecting and amassing these sources of information, we developed a framework to understand Taiwan's high-carbon economy and high-carbon energy structure.

The transition lag of a high-carbon society

Path dependency – of a high-carbon industrial structure

Observing the trend curve of the growth of carbon emissions in Taiwan over a period of time will enable us to see the problem of the emissions composition clearly. As Taiwan's development is highly dependent on the foreign trade economic-orientation of the manufacturing industry, and in a situation where the development of indigenous and renewable energy has not been fully developed and imported energy largely comprises fossil fuels, the sustained development of an energy-consuming manufacturing industry necessarily directly leads to an increase in greenhouse gas (GHG) emissions. We can observe the cumulative trend of the emissions from the various sectors from 1990 to 2016 in Figure 3.1. In 1990, the total carbon emissions released in Taiwan were approximately 109.46 million metric tons. The total carbon emissions increased to 209.26 million metric tons in 2000, with 98.52 million metric tons (or 47.08 percent) coming from the industrial sector. By 2016, the emissions had increased to 262.66 million metric tons, of which 130.52 million metric tons (49.69 percent) came from the industrial sector (Bureau of Energy (BOE) 2018).

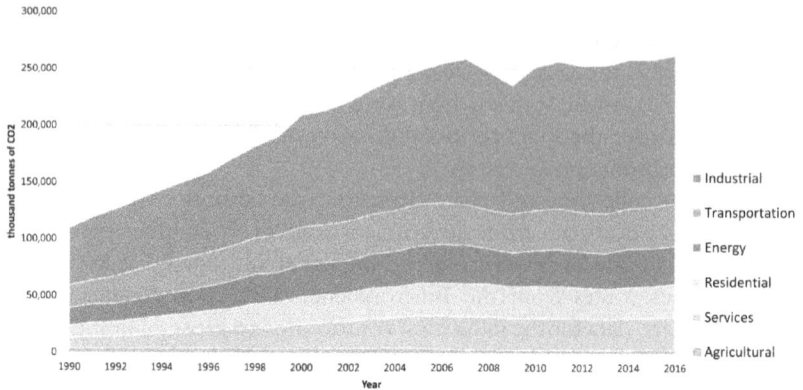

Figure 3.1 Carbon emissions from fuel combustion by sector, 1990–2016 (includes emissions from electricity consumption)

Source: BOE (2018).

If we further analyze the emissions situation among the main industries within the industrial sector, it can be seen that the emissions from the entire industrial sector have been consistently increasing with economic growth. However, simultaneously with industrial development, the proportion of emissions from the paper and textile industry has gradually decreased, while growth of emissions in the cement industry has remained flat, with a steady growth of emissions in the steel and petrochemical industry. In 2011, carbon emissions reached a peak in the petrochemical, cement, and steel industries, at 4,6511,000 metric tons, 1,1325,000 metric tons, and 2,288,500 metric tons, respectively, and since then they have led the carbon emissions in the industrial sector. The emissions in the electronics industry reached 20,524,000 metric tons in 2000 and continued to grow, reaching 25,655,000 metric tons in 2016 (BOE 2018). It is evident from Figure 3.2 that the electronics, steel, and petrochemical industries have been the main sources of emissions among Taiwan's industries over the years. From the statistical analysis done by the BOE (2018) it can be seen that in both the petrochemical and the steel industry, the carbon emissions accounted for more than half of the emissions from the industrial sector for nearly a decade prior to 2016, which shows that they are in fact the main sources of the high carbon emissions.

According to BOE's annual statistical report (BOE 2018), the total emissions of carbon dioxide from energy fuel combustion in 2016 was about 262.66 million tons, and the industrial sector, which made up 33.8 percent of the national GDP, accounted for 50 percent of the carbon emissions. Further, when sorting by industrial sector, the petrochemical industry accounted for 34 percent

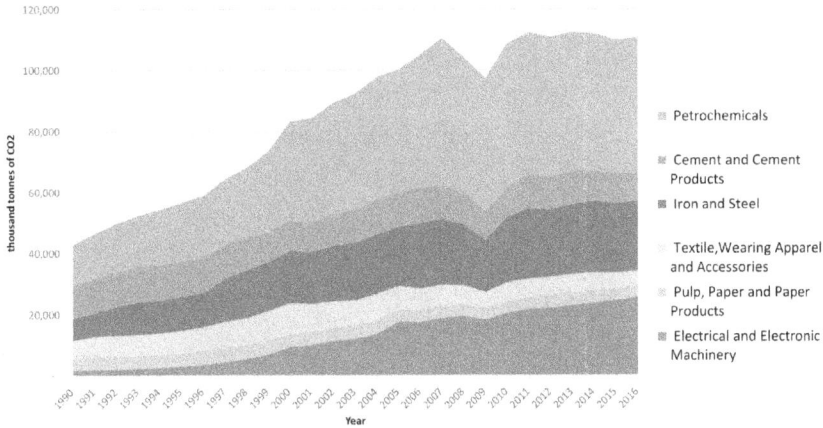

Figure 3.2 Cumulative trend of carbon emissions from the industrial sector allocated to major industries in Taiwan, 1990–2016

Source: BOE (2018).

of the carbon emissions from the industrial sector, but its output value only accounted for 11.37 percent of the industrial sector; the steel industry accounted for 18 percent of the carbon emissions from the industrial sector, while its output value accounted for 11.78 percent among the major industrial sectors. In contrast, although the electronics industry accounted for 20 percent of the country's carbon emissions, the output value accounted for 50 percent of the industry (Figures 3.3, 3.4, and 3.5).

These findings show that the major energy-consuming industries in Taiwan are weighted towards the steel and petrochemical industries, resulting in these two industries making up more than half of the carbon emissions in the industrial sector; however, their national output is relatively low and imbalanced.

Figure 3.6 shows that in the past 24 years (from 1992 to 2016), Taiwan's overall industrial energy consumption has only slightly decreased from 37.49 percent of national consumption to 31.78 percent, but the sector's contribution to GDP as a proportion of total GDP has increased from 27.89 percent to 34.26 percent. In 2013, the industrial sector's contribution to GDP as a proportion of GDP even exceeded its proportion of total national energy consumption. Of course, the contribution of energy-intensive industries is also increasing (9.83 percent to 18.57 percent), but their proportion of energy consumption is still significantly higher than the proportion of their contribution to GDP (31.9 percent to 27.46 percent). Among them, the petrochemical industry has increased its energy consumption but its contribution to GDP

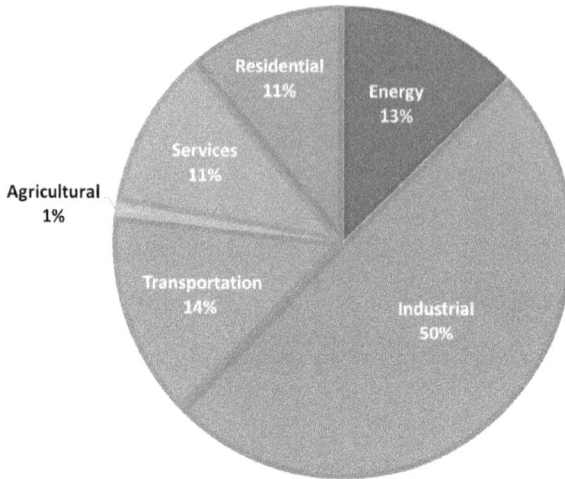

Figure 3.3 Share of carbon emissions from fuel combustion by sector in Taiwan, 2016 (includes emissions from electricity consumption)

Source: BOE (2018), author's chart illustration

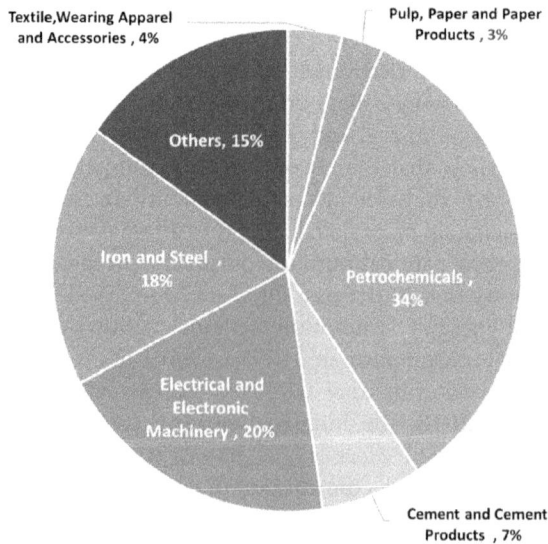

Figure 3.4 Share of carbon emissions from industrial sector allocated to major industries in Taiwan, 2016

Source: BOE (2018).

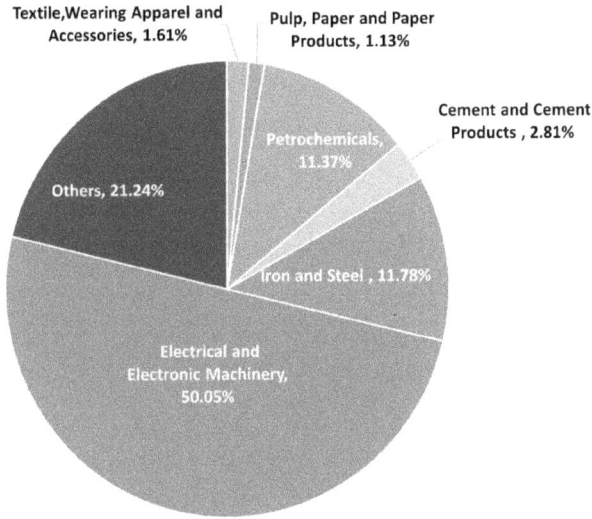

Figure 3.5 Share of real GDP from industrial sector allocated to major industries in Taiwan (as percentage of GDP in the industrial sector), 2016

Source: BOE (2018), author's chart illustration.

has shown a slight decline, with its energy consumption increasing slightly from 10.25 percent to 11.20 percent, but its GDP contribution decreasing from 3.94 to 3.45 percent.

Path dependency – of a high-carbon energy structure

If we look at Taiwan's power supply and examine the composition of Taiwan's power generation over the past decade, thermal power generation has always been amongst the biggest generators of carbon emissions. Figure 3.7 shows that thermal power generation accounted for 76.65 percent of Taiwan's power supply in 2003, and in 2008 and even in 2013, its share increased by nearly 78 percent, and surged past 80 percent in 2016 to reach 85.95 percent in 2017.

Further detailed analysis and comparisons showed that in terms of electricity generation, for thermal power generation, which released higher carbon emissions, coal-burning accounted for 48.06 percent of the electricity generated, and gas accounted for 27.55 percent; whereas nuclear energy accounted for 16.5 percent. Ostensibly, carbon-free renewable energy accounted for 4.28 percent, but after making allowances for conventional hydropower, it only accounted for 2.13 percent of the total amount of electricity generated. As a result of the strong promotion of energy transformation by the DPP government, in 2016 coal-burning accounted

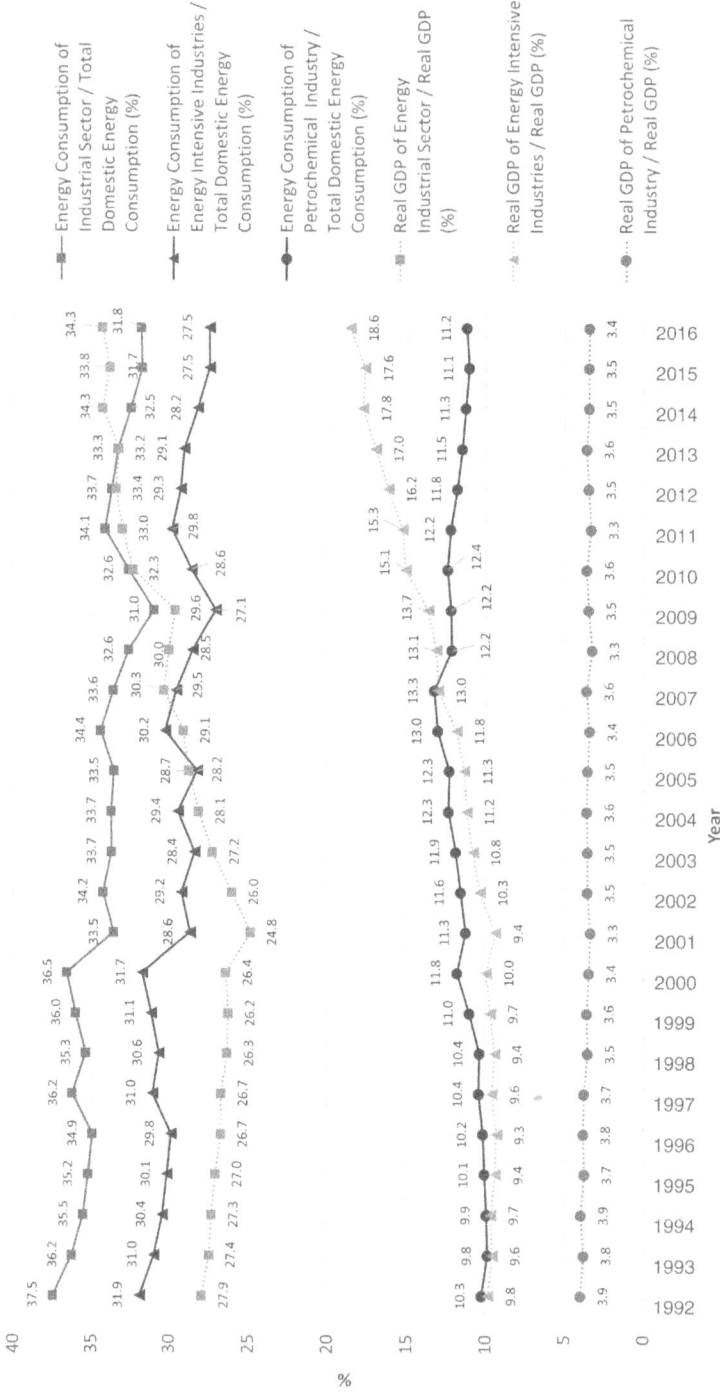

Figure 3.6 Comparison of energy consumption and GDP contribution at national-level and of energy-intensive and petrochemical industries in Taiwan, 1992–2016

Source: BOE (2018), Directorate General of Budget, Accounting and Statistics (2017), author's chart illustration

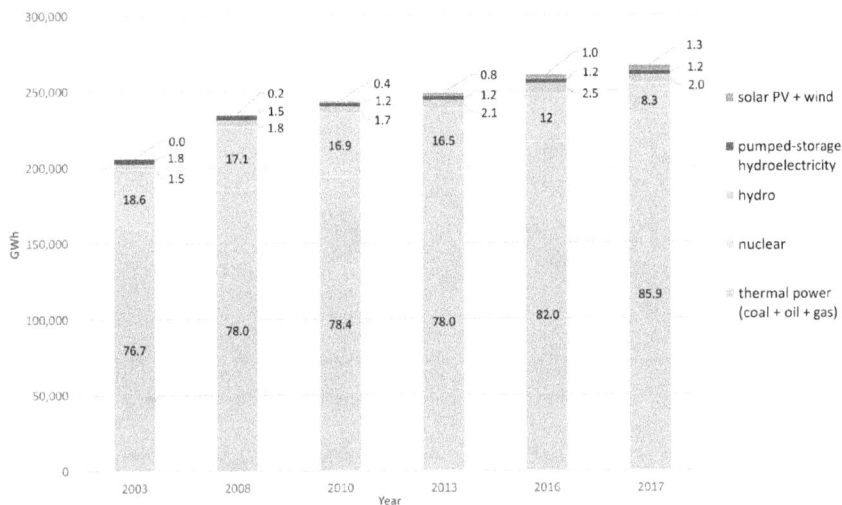

Figure 3.7 Electricity generation composition in Taiwan, 2003–2017
Source: BOE (2018).

for 45.44 percent of electricity generated, gas accounted for 32.41 percent; comparatively nuclear energy accounted for 11.99 percent, and renewable energy generation was still low, shares of renewables in electricity was about 3.5 percent in Taiwan in 2016 (BOE 2016). Clearly, the proportion of thermal power generation in the composition of Taiwan's power generation is still too high, and the proportion of electricity generated by coal, which releases higher carbon emissions, and gas, which releases lower carbon emissions, needs adjusting, and carbon emissions in general need reducing. It is also necessary to vigorously promote carbon-free renewable energy to replace thermal power generation.

Power sector's trends in carbon emissions

When the creation of GHG emissions in Taiwan is analyzed by sector, it becomes clear that the energy sector creates the most: the emissions from the power sector alone exceed half of the total emissions in Taiwan. According to the BOE (2018), in 1990 the emissions of the fuel input for the power generation industry (thermal power and cogeneration plants) was 39,628 thousand metric tons of CO_2. At that time, the total carbon discharge for the whole of Taiwan was 109,491 thousand metric tons of CO_2, carbon emissions of which the power generation industry accounted for only 36.19 percent of carbon emissions. By 2000, the emissions from the power generation industry were 108,944 metric

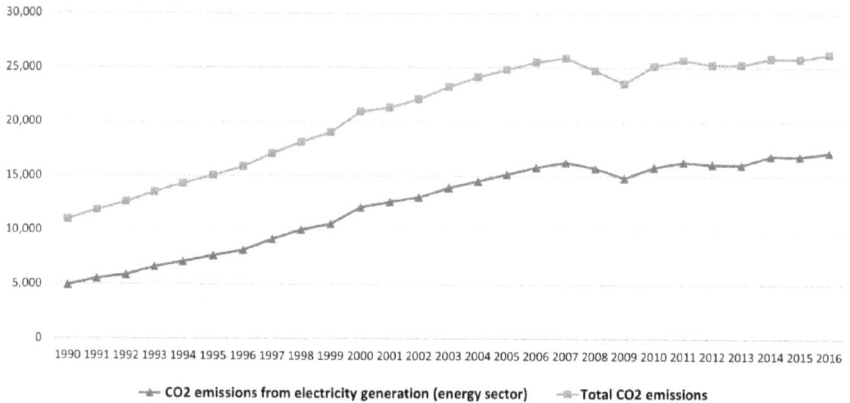

Figure 3.8 Carbon emissions from electricity generation in the energy sector and the total carbon emissions in Taiwan, 1990–2016
Source: BOE (2018).

tons of CO_2. Then, the total carbon discharge in Taiwan was 209,364 metric tons of CO_2, carbon emissions of which the power generation industry accounted for more than half. It then increased to 52.04 percent and 59 percent in 2012 and 2013 respectively, and then dropped back to 52.67 percent in 2015.

Figure 3.8 compares the power generation and national carbon emission trends between 1990 and 2016. It is clear that the emissions from the power generation sector have gradually increased since 1990, with the increase intensifying from 1998, reaching its peak in 2007; the ebb and flow of the trend curve evidently driving the curve of total national emissions, resulting in the total national emissions also increasing since 1998, and reaching its highest point in 2007. By 2015, the carbon emissions from the power generation sector had exceeded more than half of the total carbon emissions in Taiwan. In other words, the power sector is the main driver of carbon emissions in Taiwan, therefore the question of how carbon reduction measures should be promoted in the power sector will dictate whether or not Taiwan is able to achieve the targets set in the overall carbon reduction policy by 2025.[4]

Path dependency – low energy prices and fossil fuel subsidies

In addition to the above-mentioned high-carbon industrial and high-carbon energy structure, another contributory factor to Taiwan's high-carbon society is the low cost of electricity and water.

According to the latest 2017 statistics from the IEA, Taiwan's industrial electricity price was NT$2.4491 per kWh, which was seventh lowest in the world. By comparison, in South Kores it was NT$2.9911 per kWh, which was the fourteenth lowest in the world, and in Japan it was NT$5.1027 per kWh, which was the world's second highest. In terms of household electricity consumption, Taiwan's residential electricity consumption was the second lowest in the world at NT$2.5679 per kWh.

According to a 2014 survey by the International Water Association of the water price per 100 units in 160 cities, the average annual household water bill in Taiwan was US$39.53, or about NT$1,186, comprising 0.10 percent of GDP with the water burden ranking twenty-eighth. According to the statistics, the price of water in Europe, the US, and Asia is higher than that in Taiwan, while the average price of water in European countries is almost three times that of Taiwan (Taipei Water Department 2017).

According to the latest 2014 to 2016 statistics from the IEA (2017), the fossil fuel subsidies in Taiwan were US$159 million in 2014, US$20 million in 2015, and US$108.8 million in 2016 (see Table 3.1).

In addition, the 2016 statistics of various fossil fuel indicators from the IEA (2017) showed that in recent years Taiwan has been affected by the fluctuations in oil and gas prices . The per capita subsidy is US$4.6 (US$/person), which ranks as the fifth lowest among 41 countries, higher only than Sri Lanka (3.1 US$/person), South Korea (3 US$/person), Ghana (1.1 US$/person), and Vietnam (1.1 US$/person). Also, according to the statistics from the International Monetary Fund (IMF) (2015), Taiwan's total fossil fuel subsidies in 2013 were US$310 million, not including external costs. However, when factoring in air pollution, GHG pollution, and external costs such as traffic jams, then the subsidies would go up to US$26.6 billion, and it was estimated that this would increase to US$31.6 billion (or approximately NT$1.1 trillion) in 2015 (Figure 3.3), which would account for 5.43 percent of GDP (Chao 2017).

Table 3.1 Fossil fuel subsidies in Taiwan, 2014–2016

Unit: Real 2016 million US$

Country	Product	2014	2015	2016
Chinese Taipei	Oil	159.0	20.0	108.8
Chinese Taipei	Electricity	–	–	–
Chinese Taipei	Gas	–	–	–
Chinese Taipei	Coal	–	–	–
Chinese Taipei	Total	159.0	20.0	108.8

Source: IEA (2017).

Brown economic package

In other words, the structural problems that Taiwan faces in its social transformation are its high-carbon industries, high-carbon energy, low electricity and water prices, and long-term fossil fuel subsidies. To add to the problems, relative to the environmental costs of a high-carbon society, the cost of labor in Taiwan is also low relative to the world's major industrialized countries. as a consequence of the low electricity and water prices and the government's long-term subsidies for high-carbon industries and energy companies have indulged in rent-seeking behaviors and lacked the motivation to develop research and development innovation, resulting in them gradually losing their international competitiveness. This has led to the perception internationally that Taiwan has fallen into the middle-income trap (Liberty Times Net 2017).

In fact, such structural development has resulted in a brown economic package, leading to very tricky, long-term, and structural path dependency in the transformation of climate governance in Taiwan, which has even locked- in the developmental path of Taiwan's transformation.

Social resistance to path dependency

Contextual environmental movement

Other than the lag in the government's governance which requires our further analysis, there are two other areas of social developments which are worth mentioning, although they are not the current focus of this chapter's analysis. First, since the mid-1980s, there have been various anti-chemical or petrochemical movements in Taiwanese society, and these environmental movements have been directly and indirectly related to climate governance, whether in form or substance.

Chou (2017) divides these into anti-pollution movements and climate change risk movements, the former of which were protests led by residents due to concerns over their own health, and were mostly regional movements although a few which took on a national focus, and were centered mainly on air pollution. The latter can be exemplified by the anti-Pinnan industrial petrochemical protests in 1995, and the anti-Kuo Kuang petrochemical plant protests in 2010. These protests were to voice opposition to high energy consumption, high carbon emissions, high water consumption, and high pollution levels, and were well organized based on fairly systematic known-how. Such social movements have created a sustained change in the public's awareness of sustainability and environmental protection issues, and continue to exert pressure on the state and industries.

Public climate risk perception and paradigm shift

Second, the public's climate risk perception has become an important element in driving society towards a low-carbon transformation orientation. As a

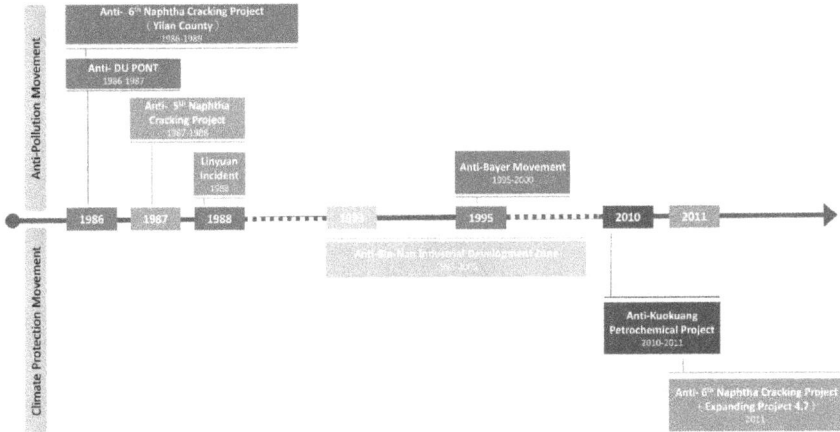

Figure 3.9 Evolution of anti-chemical movements since 1986

result of Chou's (2017) analysis of the two climate change risk perception surveys conducted in 2012 and 2015, the Taiwanese are able to support the general direction of energy and industrial transformation. No doubt, given the price that people are willing to pay for transformation, this needs to be meticulously looked into, but the basic disposition of the public is supportive of the government's developmental orientation towards a low-carbon society.

First, the Taiwanese already recognize the risks of climate change, and the need for change. As they accept a price has to be paid for transformation and that implementing an environmental protection agenda will not hinder Taiwan's economic growth, it means that Taiwan could develop environmentally-friendly industries, as the public would be supportive of such industrial development. A comparison between Figures 3.10 and 3.11, of the attitudinal surveys in both 2012 and 2015, shows that there is a slight increase in the willingness of the public to pay higher taxes for electricity, and there is also an increase in the number of the public who believe that environmental protection is not in conflict with economic development. In Figure 3.11, the two left columns show that 85.1 percent of the public are willing (very willing and willing) to pay higher electricity prices to enable renewable energy transformation. According to a 2018 Greenpeace Foundation survey, when provided with adequate information and when clearly informed of the price increase and purpose, the proportion of people who were willing for electricity prices to be raised increased from 52.5 percent to 71.2 percent. If electricity prices were to be increased to increase the speed at which renewable energy was developed, 25.2 percent of the respondents would be willing to pay 10 percent more for electricity, or NT$3.1 per kWh, 17.6 percent would be willing to pay 30 percent more, or NT$3.5 per kWh, 14 percent would be

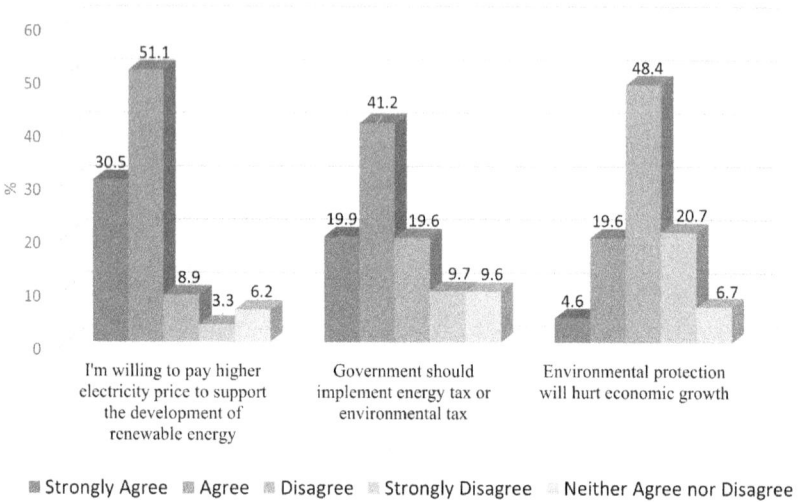

Figure 3.10 The willingness to pay the price for transformation, and the intention associated with environmental protection and economic growth, 2012

Source: Risk Society and Policy Research Center (RSPRC), National Taiwan University (2012)

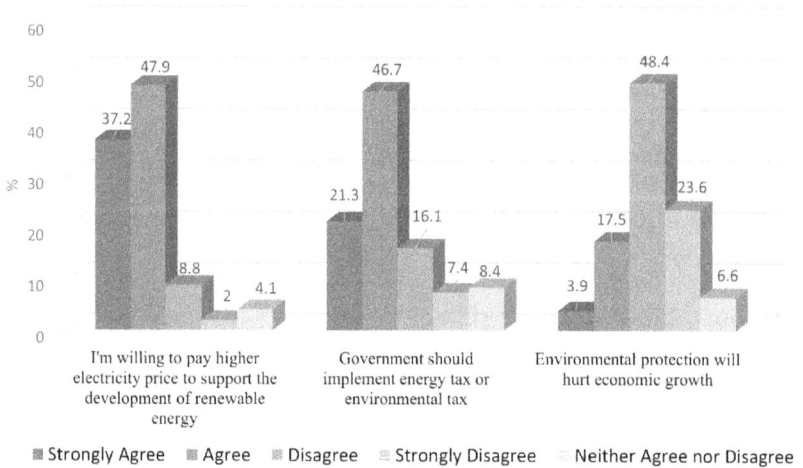

Figure 3.11 The willingness to pay the price for transformation, and the intention associated with environmental protection and economic growth, 2015

Source: RSPRC, National Taiwan University (2015).

willing to pay 50 percent more, or NT$4.2 per kWh, and 14.4 percent would be willing to pay more than 50 percent more. This demonstrates the validity of the surveys, and also reflects the willingness of the public to pay. As for a carbon or carbon-related tax which would directly contribute to energy and industrial transformation, Figures 3.10 and 3.11 shows that 68 percent of respondents would be willing to pay higher taxes to protect the environment. At the same time, the public has caught up with international trends, and understands that it is not a given that environmental protection will conflict with economic development, and the public has also recognized that a balance can be struck between economic development and environmental sustainability. As you can see in Figure 3.11, the right-hand column shows that 72 percent of the public does not think that environmental protection will hinder Taiwan's economic growth.

Based on the above-mentioned public surveys on climate change and related policies, the Taiwanese are aware and supportive of the government's active adjustment of the industrial and energy structure to deal with climate change risks. The public's acceptance level of the need to levy an energy tax or adjust electricity prices during the process of transformation is also high. Attention needs to be paid to the question of risk communication, and the purpose of transformation and the roadmap also needs to be clearly planned. The decision-making process and outcomes should be made public to help the public understand and accept these policies.

Both the contextual environment movement and the paradigm shift to a low-carbon society have brought about very important social transformations and symbols, and directly challenged the high-carbon society and brown economy. Society's low-carbon transformation drive has confronted the high-carbon economic system head-on, with the two pulling against each other, resulting in a rather tense relationship. But even as these two drivers of society move towards sustainable development, the transformation of the government's governance and industrial transformation is still dragging its feet.

The state's tardy transition

Important governmental conferences and policies, but without a clear roadmap

In response to global climate change conventions, carbon reduction, and sustainable development trends, we can trace Taiwan's government industrial policy transformation plans, which have been spread out across the important national energy conferences, sustainable economic development conferences, policy platforms, and guidance over the last two decades. In response to the signing of the Kyoto Protocol, Taiwan's government held the first-ever National Energy Conference in 1998, followed by important conferences, policy guidelines, and white papers (see Figure 3.12): the 2005 2nd National Energy Conference, the 2006 National Conference on Sustainable Development, the 2006 National

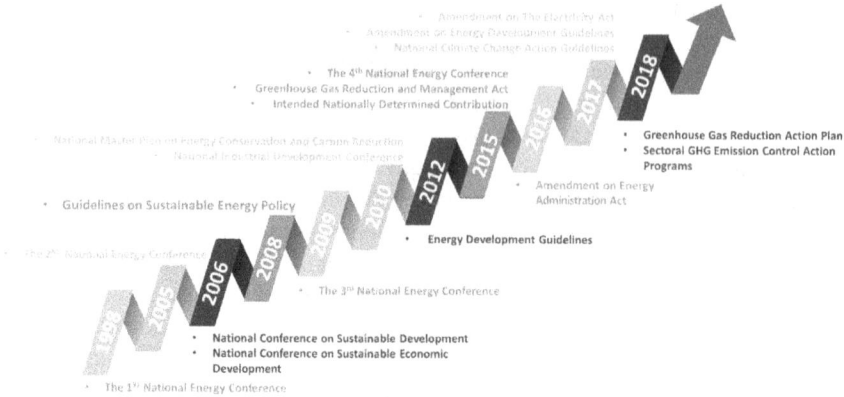

Figure 3.12 Key climate policies and conferences

Conference on Sustainable Economic Development, the 2008 Guidelines on Sustainable Energy Policy, the 2009 3rd National Energy Conference, the 2010 National Master Plan on Energy Conservation and Carbon Reduction, the 2010 National Industrial Development Conference, the 2012 Energy Development Guidelines, the 2015 4th National Energy Conference, the 2015 Greenhouse Gas Reduction and Management Act, and the 2015 Intended Nationally Determined Contribution (INDC), etc.

In all these various national-level energy and economic conferences and policy guidelines, there were two important directions on the transformation of energy, industries, and reducing carbon emissions. The first is to simulate carbon emission scenarios and set quality standards in response to the climate change convention, and the second is to develop a plan for industrial policy transformation. However, as Chou (2015, 2017) pointed out, the outcomes of these important meetings, policy announcements, and plans have not been genuinely fulfilled, revealing that in its policy implementation to counter global climate change, Taiwan's previous government had hit a roadblock in its promotion of industrial transformation. As in the case of the high-carbon industrial and high-carbon energy structures in the earlier analysis, it can be said that the government was faced with a conundrum in its transformation of climate governance, which delayed it undertaking social and economic transformation.

If we look at the different stages of policy announcements (see Figure 3.12), it can be seen that Taiwan's government has been setting targets for the last two decades to meet the requirements of international climate and carbon reduction conventions, and has attempted to carry out corresponding industrial, economic, and social transformations. The 1998 National Energy Conference set for the first time carbon emissions reduction targets, which aimed to reduce carbon emissions in 2020 to the 2000 benchmark level (1998 proposal); the 2008 Guidelines on Sustainable Energy Policy further set the carbon emissions reduction targets (2008 proposal), and the 2009 3rd National Energy Conference and the 2010 Legislative Yuan's National Master Plan on Energy Conservation and Carbon Reduction (2010 proposal) set the carbon emissions reduction targets from 2016 to 2020 to return to the level of 2008's emissions, to 2005's levels by 2020, and to 2000's levels by 2025 respectively, using 214 million tons as a benchmark. In September 2015, Taiwan's government announced to the world its INDC (2015 proposal), to reduce GHGs by 50 percent below Business As Usual (BAU) by 2030 (equivalent to 214 million metric tons of carbon dioxide). That is, that by 2030, GHG emissions will be reduced by 20 percent as compared to the 2005 baseline level (see Figure 3.13).

Before the COP21 in 2015, the symbolic promulgation of the Greenhouse Gas Reduction and Management Act also stipulated that by 2050 Taiwan's carbon emissions would be reduced to half (134 million metric tons) the 2005 benchmark levels (269 million metric tons) (see Figure 3.14).

The path dependency produced by a high-carbon social structure has hindered the transformation of climate governance, and this has become more obvious in recent years. The DPP government which came into power in August 2016 set up the Office of Energy and Carbon Reduction under the Executive Yuan to spearhead energy transition and carbon reduction initiatives, but so far it has had little success. The DPP government launched the 2016 Amendment on Energy

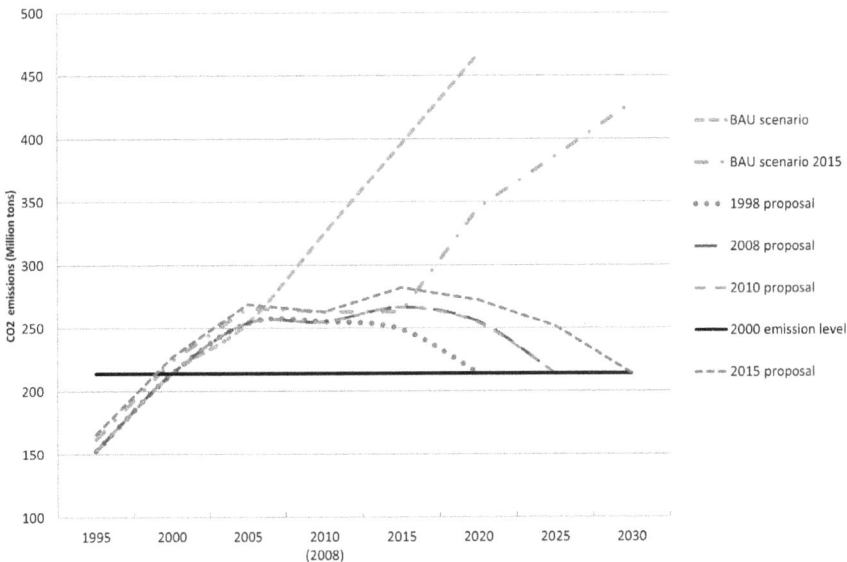

Figure 3.13 Carbon emissions in BAU scenarios and CO_2 reduction targets in Taiwan since 1995

Source: Amended version of Chou and Liou (2012); Chou (2017).

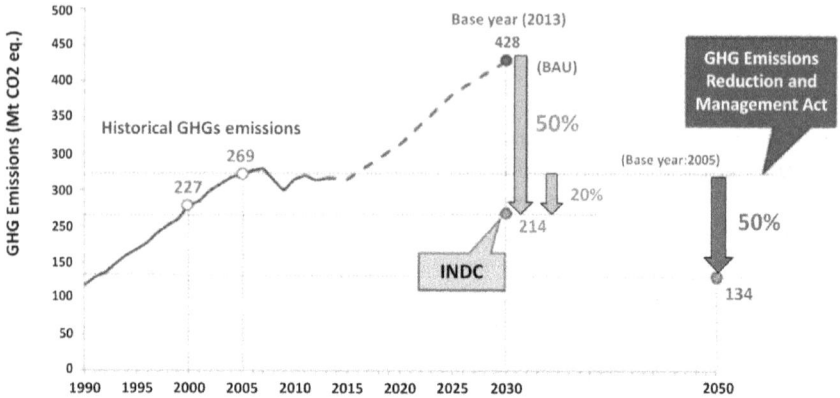

Figure 3.14 GHG emissions trajectory
Source: Ministry of Economic Affairs (2015).

Administration Act, the 2017 Amendment on the Electricity Act, the 2017 Energy Development Guidelines, the 2017 National Climate Change Action Guidelines, the 2018 Greenhouse Gas Reduction Action Plan, the 2018 Sectoral GHG Emission Control Action programs, and the 2018 energy transition white paper, but all these efforts have been met with formidable obstacles.

Transitional discursive struggle by brown economic regime

Whilst the government delays making the transformation of climate governance because of the obstacles to reform that have been brought about by the hardened high-carbon and brown economic structures, externally we also witness the robustness of civil society, that is, the strong and sustained environmental movements and thriving green energy initiatives, which have acted to monitor and put pressure on the government to take measures to make the transformation. However, the delay in transformation is not only due to a lack of a clear roadmap within government, and regulatory policies and tools, but most importantly, Taiwan's government still hasn't debated these aspects of the transformation and formed a clear plan of action. In contrast, the old high-carbon regime has continuously exerted its influence on academia, the media, and industries, and has been unceasing in its influence on discourses and interventions in policies under the guise of the brown economy package. The most prominent example of this is the 'five shortages' as proposed in the Chinese National Federation of Industries (CNFI) (2015) white paper: the lack of water, electricity, manpower, land, and skilled workers, which very clearly defended the brown industry and economy. Other than the last shortage of skilled talent, which has less relevance to climate and environmental governance, all the shortages are aimed at consolidating the rent-seeking for prioritizing the interest of these industries over water price, electricity price. and labor costs, even to the extent of externalizing environmental costs. This triggered environmental protests and thereby resulted in a lack of public trust, but by misconstruing cause with effect, and claiming that lack of land was a source of the problem, these industries hoped that the government would modify its program on environmental impact assessment, eliminate public protests, and provide land for factories.

However, the CNFI's five shortages were heavily criticized by research think tanks, academics, and citizen groups. Researcher Ming-Te Sun from the Chung-Hua Institution for Economic Research critically pointed out that Taiwan currently has the "seven lows" of low water prices, electricity prices, oil prices, prices of goods, interest rates, exchange rates, and wages –all cheap factors of production costs (Chiu 2016). "People want to come to Taiwan to set up their computer facilities precisely because electricity prices are so cheap", and also because the exchange rates and interest rates are not high. But such cheap factors of production costs have not been able to improve industrial competitiveness, primarily because of the low efficiency of businesses. In the 2017 survey of the Carbon Disclosure Project (CDP) conducted by the Risk Society and Policy Research Center (RSPRC) (2018), it was pointed out that the major energy-consuming industries include the Formosa Petrochemical Corporation, the Formosa Chemicals & Fibre Corporation, and the Formosa Plastics Corporation, etc., which all had a C grade rating, which showed that their energy consumption was overly high. The Taiwan Environmental Protection Union (2018) pointed out that the five shortages raised by CNFI were fake issues, and that the government needed to come up with a proper forward-looking plan, build proper basic infrastructure, be fully open and transparent with information, and have good management and maintenance skills, so as to prevent the monopolization and abuse of resources (such as land and water) and improper profiteering.

Even though it was roundly criticized from all sides, the concept of five shortages which were impeding the brown economy continued to be pervasive, as a result of getting exposure and discussion of it in the media the government has taken a conservative line and been restrained in its transformation and reform.. Even when the DPP returned to power in 2016, it was still preserving the interests of these industries. When the new premier first officially addressed the media (Executive Yuan 2017), he said that priority would be given to dealing with the five shortages in the industrial sector. That is, these discourse over development maintained the high-carbon economic structure, which aimed to sustain and replicate the rent-seeking model of the past under the CNFI – the representative of the interests of conservative businesses – by attempting to lock-in social and economic development to the existing high-carbon, brown, and low-value industrial energy and labor structure. Such a development discourse also amplifies the brown economy package, and works to strengthen the legitimacy of the high-carbon industry economic structure. From another perspective, the combination of the existing power of the brown economic structure and discourse, which produces a conservative and regressive vision, this puts a brake on and constrains the transformation of low-carbon governance in Taiwanese society; this is another of the dilemmas being faced by the current government as it tries to promote a low-carbon social transformation.

What's more, this battling in the discourse has resulted in the government tardiness in developing a clear roadmap, policy tools, and regulations. The DPP government isn't sufficiently well informed and doesn't have the ability to present its policy in such a way as to be able to lead the whole of society by having a clear guiding vision of its transformation, and by constructing a social learning curve to advance the whole of Taiwan towards a momentous climate and energy transition (Chou 2017).

Discussion and conclusion

This chapter analyzes the dilemmas and challenges that Taiwan faces in its transformation of climate governance from three perspectives: first, from quite a broad perspective which presents the formation of the high-carbon and brown economy in Taiwan, including the analysis of the industrial and energy structure (economic model), and electricity prices, water prices, and fossil fuel subsidies (regulations); Second, a brief analysis was conducted of the environmental movements starting in the late 1980s, as they evolved from anti-pollution protests to climate risk movements against high-carbon emissions, high energy consumption, and high water consumption. At the same time, this chapter also analyzes the significance of the paradigm shift that has occurred in Taiwanese society in relation to the public's attitude towards the climate and energy. Third, a brief secondary analysis was conducted of the government's pledges to reduce carbon emissions, which showed it lacked a clear roadmap of climate governance and transformation discourse. Consequently, when challenged by existing interest groups advocating for a high-carbon and brown economy, the government's ability to proceed with the transformation is poor and weak.

Such a transformative predicament is very similar to the analysis done by Pierson (2000), to construct the concept of the conservative nature of institutions and the high density of institutions that create barriers to entry, including that of interest groups which drive the high-carbon industrial and energy structure that constantly seek to maintain the existing brown economic package, and attempt to strengthen policy rent-seeking via policy discourse – and the CNFI is exactly one of these such representatives. Although the analysis of the social aspect of the transformation is not covered at length in this chapter, the momentum of the environmental movements and the public's climate perceptions symbolize the strong support from the public for low-carbon social transformation. From 1998 to 2018, the government's pledge to pursue a low-carbon policy has been ostensibly portrayed as the government's response to the international and domestic requirements, but on the whole Taiwan is still a high-carbon society and brown economic body, and the DPP government's climate governance has therefore met with difficulties both internally and externally.

In Table 3.2 we have divided these issues into different types based on the transformation structure to analyze the transformational dilemmas and challenges faced by the developmental state under authoritarian expert politics, mainly from the angle of industry, civil society, and the state, and divided them into five types: discursive package, discursive strength, structure, agents, and instruments. Of course, the media, academia, and industry, and the interest complex of the government should be included in any analysis of transition management, but these are not focus of this chapter.

Based on the analysis in Table 3.2, we can see that the agents of industry and CNFI have combined with the former ruling party Kuomintang's past high-carbon regime, to continue to shape the legitimacy and importance of energy-intensive industries, using the annual white paper as an important tool and then mobilizing the media to promote the five shortages, which forms quite a strong development discourse. The advocacy of the lack of water, electricity, manpower, land, and skilled labor continues to strengthen and justify the conditions that are required by the existing high-carbon industrial and energy structure; such an attempt will lock economic and social development into the brown economic package, while ignoring the external pressures of global climate and carbon-reduction governance. That paradoxically becomes quite a strong policy pressure.

Some government agencies have also decided to detach themselves from global carbon reduction governance. Even though DPP regained power in May 2016, the regime change cannot be taken to represent the smooth transition of Taiwan's overall transformation. On the one hand, the linear,

Table 3.2 Structural predicament of climate governance

	Industry	*Civil society*	*State*
Discursive package	Brown economic package	Focused on green energy transition A lack of clear and strong climate governance discourse	Internal struggle between high-carbon and low-carbon economic packages
Discursive strength	Strong	Weak	Weak and chaotic
Structure	High-carbon industry and energy	Multiple social movement objectives	Lack of a clear roadmap
Agents	Industry agents and an outmoded high-carbon regime	Multiple civil society actors	Multiple agents lacking good coordination
Instruments	Publication of white papers and media mobilization	Social mobilization and environmental movements	Requires innovative policies, regulations, and incentives

high-carbon economic model which authoritarian expert politics leaned towards in the past seemed to be due to the inertia in thought and behavior of the government agencies; on the other hand, even if there was a lack of large-scale innovation among the high-energy consumption, high pollution, and high-carbon emission industries for more than 20 years, and the contribution to GDP continued to be flat and limited, after taking into account the tremendous political and economic forces the DPP government has had misgivings about its transformation and reform, and decided to adopt only a fairly conservative development pathway. Moreover, because each government agent has a different perspective on development, this has resulted in conflicting goals within the government on low-carbon transformation. The hidden internal struggle between the old and the new reflects the lack of a clear roadmap across the whole government. It has become clear that the DPP government's policy discourse of climate carbon reduction governance is weak and has descended into chaos. The longer there is a lack of a clear discursive package and low-carbon transformation path, the easier it is to be locked into the existing gray economic track.

In fact, such a development is no different from the transformation of Taiwan's climate governance over the past 20 years. As we analyzed in this chapter, the government has over the years convened and promulgated important energy and industrial conferences and carbon reduction policies, but the strong lock-in effect of the high-carbon and gray economic structure often blocks the implementation of reform, and everything comes to naught, with even the government still promoting major industrial projects which result in high energy consumption, high carbon emissions, and high pollution, such as the 2010 Kuo Kuang petrochemical plant project (Chou 2013, 2016, 2017). And such development outcomes have resulted in systemic risks because of the delay in transformation, and the resultant constant vicious cycle which has severely slowed down the speed of Taiwan's low-carbon transformation.

As for Taiwan's civil society, even though it began to mature around 2010, it has shown social robustness in some major environmental and risk movements, and has been able to systematically organize as well as construct risk knowledge and shaped the discourse to supervise and confront the government (Chou 2013). At the same time, as shown earlier, the public already possesses the cultural appetite for the transformation to a low-carbon society, and is supportive of the government's energy, industrial, economic, and social reforms. But on the whole, civil society represents different agents in society and is concerned about multiple issues, and although civil society is proficient in using the media as a tool for mobilization, its focus is quite diverse; on the other hand, the focus of relevant citizen groups committed to participating in the energy transition efforts promoted by the government is on denuclearization and energy rationing, and the discourse of strengthening the development of climate change and carbon reduction initiatives is lacking.

Even if issues of energy transition are highly relevant to carbon reduction, generally speaking under the energy transition structure led by the government, there is not a clear carbon reduction policy and decarbonization pathway to supporting energy transition. Civil society has correspondingly raised criticisms, but there has been no systematic development, mobilization, and construction of a discourse on low-carbon knowledge. And the agents in society who champion green energy, such as the Taiwan Independent Power Plants and the Taiwan Environmental Protection Union, etc., are mostly focused on promoting green energy and are less systematic when it comes to carbon reduction issues. Compared with the symbiotic anti-coal and anti-warming initiatives among citizen groups in Thailand and Philippines (Francisco 2017; Jakkrit 2017), the anti-coal initiatives in Taiwan are linked to the issue of anti-air pollution more than climate change. Basically, other than the discourse and thought in relation to the anti-carbon emission and anti-high energy consumption initiatives which arose from the anti-Kuo Kuang protests in 2010 (Wu 2010), the carbon reduction awareness in the whole of society is weak, and there is not a clear low-carbon discursive package.[5]

The problems arising from this divergence lies in the fact that the characteristics and model of Taiwan's environmental movements need to change, even though Chou (2017) has pointed out that Taiwan's environmental movements have been gradually evolving from being stakeholders playing an opposing role and fighting over concerns about their own health, to being role models in the climate change risk movement, which are becoming aware of the threat of the high energy-consuming and high carbon-emitting industries to Taiwan, though in reality, the latter still requires the involvement of more people. There are two aspects here: first, whether systematic discussion among civil society would be able to grasp the complexity of climate and energy issues, and second, whether these ideas can be effectively mass transmitted to the public. The fact is, Taiwan's NGOs used to be mainly concerned with protests over local pollution issues, so there is still some way to go for them to transit to changing focus and having the necessary background knowledge about low-carbon and sustainable industries. Although the Anti-Kuokuang Petrochemical Project Movement brought about increased anti-carbon risk knowledge, but the focus on carbon emissions will gradually fade away unless there are sustained large-scale environmental movements to connect and reproduce the knowledge and discussions. At the same time, as the main axis of the issue is not clear, even though many NGO groups have turned their attention to energy transition movements to advocate for green energy and oppose nuclear energy, etc., these movements are not effectively mobilizing discussion around the topic, and have therefore been unable to clearly and powerfully transmit the message of reducing carbon emissions to the general public.

On the other hand, by using the perspectives of economic models (high-carbon industry and high-carbon energy), social cognition (public perception and the issue of five shortages), systems (water and electricity prices and fossil fuel subsidies), government policies and regulations (carbon reduction policy, NDC, relevant energy and carbon reduction regulations), administrative agencies (regime change and transition), social actors (citizen groups), and others, the author has conducted a preliminary contextual and reflexive analysis of the path dependency of Taiwan's transformation towards a low-carbon society. It can be seen that due to the linear, high-carbon, economic, industry-driven model dominated by authoritarian expert politics, this has profound and lasting influences which have configured Taiwan's development pathway. Although Taiwan has undergone the third wave of democratization, and is at the stage of democratic consolidation, the culture of control that has been shaped by authoritarian expert politics (of overvalued economic growth and undervalued risks), and the supporting linear economic development model has established the temporary success of Taiwan over the past 20 years. The success of this development model was built on the gray economic package, which used low electricity prices, water prices, labor costs, and a plentiful land supply and ignored environmental risks, thereby enabling the Taiwan economic development miracle of the past.

However, such a development model has resulted in the rent-seeking behavior of industrial interest groups and the expansion of the brown economic discourse, and at the same time has shaped the various environmental movements by citizen groups for the last 20 years, to resist chemical industries and petrochemicals, and the high-carbon and energy-consuming industries. What's more, state bureaucracy can superficially propagate its response to international expectations on carbon reduction and industrial policies, and claim that the high-carbon economic model is a thing of the past. In general, the whole of Taiwan's society is caught between the debate and the push-pull of the gray economic package and the delayed and hidden industrial, government, and social transformations (Chou 2013). From the process of social transformation research, we can observe that the whole of society has been using an outmoded thinking, industrial method and governance framework to operate and resist social innovation or change of any kind; this has been locked into a fixed path.

Such a development pattern has not only resulted in a delay to transformation, but also constitutes systematic gaps and risks. Taiwan's society therefore needs to undergo a complete reversal and paradigm shift. Social transformation will require the restructuring of the social system and the reconfiguration of society's development and values, in order to counteract the systemic failures (Rotmans & Loorbach 2009). It is most important to construct a *Leitbild* (guiding vision) to lead the change, that is, it requires a clear and fresh discussion to bring about the transformation (Voß, Kemp & Bauknecht 2006; Grin 2006; Kemp & Loorbach 2006). It is therefore imperative for Taiwan to come up with a new definition of sustainable industrial, social, energy, environmental, and labor development, to construct a fresh vision, and develop a systematic inventory of the barriers that can hinder the change from a high-carbon to a low-carbon society. In particular, the state, industry, civil society, media, and even academia would need to navigate a new social learning curve. But this is probably a problem common to many of the emerging industrial countries in Asia that are advancing and catching up.

Notes

1 This chapter will largely focus on the first, and will explore the historical path of the transitional difficulties caused by the long-term impact of Taiwan's high-carbon industry and high-carbon energy structure, and how the persistent supporting policies, and the component subsidies and low electricity prices, have led to the neglect by industries of investment in innovation, and research and development, which has thereby resulted in a delay in transition.

2 From the theoretical elements of reflexive modernization (Beck 1986, 1992), the proponents of transition management had attempted to propose reversing the industrial risk society model, to move towards a green and sustainable new social development model. The form of sustainability foresight analysis requires self-critical and self-corrective reflexive governance (Voβ, Smith & Grin, 2009; Stirling 2007). Such a process is non-linear, and after an initial period of gradual change, structural change brought forth by breakthroughs would produce social, economic, cultural, ecological, and systemic co-evolution (Rip 2006) and rapid change (Kemp & Loorbach 2006).

3 This chapter defines the brown economic package as a society whereby the exploitation of the environment and labor forms the core of economic activity and capital accumulation, in particular among the high-carbon industries, and the resultant problems they bring about in terms of high energy consumption, high pollution levels and high water consumption. It can also be referred to as an exploitation system based on the externalization of environmental costs while at the same time depressing the value of labor (see Chou & Walther 2016). In addition, the brown economic package not only refers to brown industries, but also to brown energy such as nuclear power, coal-fired power generation, and other low-cost electricity production and electric power systems, etc., which have delayed industrial innovation, as well as research and development, and slowed down transition.

4 On November 2011, the government announced its vision for energy development, with the policy goal set at installing renewable energy capacity to reach 9,952 MW by 2025, to account for about 14.8 percent of the total installed power generation capacity. Please refer to Jerry J. R. Ou, *Creating a Green Energy Low-Carbon Environment – To Move Towards a Nuclear-Free Home*, Energy Report, December 2011. Energy Report, Taipower Company's Response and Strategy to the New Energy Policy, January 2012. Yahoo News, President Ma's Announcement of the Energy Policy: First, Second, and Third Nuclear Plants will not be Closed, Fourth Nuclear Plant to be Halted, November 3, 2011.

5 In the past, Taiwan's environmental groups have been mainly focused on protesting over environmental pollution, and the energy issues pertaining to carbon reduction were generally the domain of the anti-nuclear movement. There were only two major environmental movements – the anti-Pinnan Industrial Park Movement in 1995 and the Anti-Kuokuang Petrochemical Project Movement in 2010 – which were concerned with global warming and carbon reduction issues (relating to high energy consuming and high carbon emitting industries). Chou (2017) pointed out that this change in direction showed that the model of Taiwan's environmental movements has shifted from anti-pollution movements to climate change risk movements, in particular with the systematic discussion over carbon reduction and anti-global warming in the Anti-Kuokuang Petrochemical Project Movement, which exemplified how civil society possessed the robust risk knowledge to overturn government decisions. However, , the public's awareness of carbon reduction has not continued to increase. The important civil groups in Taiwan, like the Green Citizens' Action Alliance and Citizen of the Earth, Taiwan, have focused quite intensively on anti-nuclear movements, and even though established groups like

the Taiwan Environmental Protection Union and the Consumers Foundation Chinese Taipei have continued to provide criticism of carbon discharge at the United Nations Climate Change Conference (COP) held in December every year, it has been unsystematic. In recent years, the topics discussed have revolved around anti-pollution as an extension of the Anti-Kuokuang Petrochemical Project Movement, with the inclusion of low-carbon energy and carbon reduction issues.

References

Barnett, J., Evans, L. S., Gross, C., Kiem, A. S., Kingsford, R. T., Palutikof, J. P., Pickering, C. M., & Smithers, S. G. (2015). From Barriers to Limits to Climate Change Adaptation: Path Dependency and the Speed of Change. *Ecology and Society*, *20*(3), 5. Retrieved from http://dx.doi.org/10.5751/ES-07698-200305

Beck, U. (1986). Risikogesellschaft. In U. Beck (Ed.), *Auf dem Weg in einen andere Moderne*. München: Suhrkamp.

Beck, U. (1992). *Risk Society: Towards a New Modernity*. New Delhi: Sage.

Beck, U. (2006). Reflexive Governance: Politics in the Global Risk Society. In J.-P. Voß, D. Bauknecht, & R. Kemp (Eds.), *Reflexive Governance for Sustainable Development* (pp.3–28, pp.31–56). Cheltenham: Edward Elgar.

Bureau of Energy (BOE). (2016). *Ministry of Economic Affairs, Energy Statistics Handbook 2016*. Retrieved from http://web3.moeaboe.gov.tw/ECW/populace/content/wHandMenuFile.ashx?menu_id=682&file_id=1043

Bureau of Energy, Ministry of Economic Affairs. (2018). *Statistical Analysis of Taiwan's Carbon Emissions from Fuel Combustion*. Retrieved from https://www.moeaboe.gov.tw/ecw/populace/content/ContentDesc.aspx?menu_id=6998

Chang, K. S. (2010). The Second Modern Condition? Compressed Modernity as Internalized Reflexive Cosmopolitization. *The British Journal of Sociology*, *61*(3), 444–464.

Chao, C. W. (2017). Challenges and Solutions for the Governance of External Energy Costs. *How to be [in] Transition: Initiate the Key Transition of Energy in Taiwan* (pp.245–268). Taipei: Chu Liu Book Company.

Chinese National Federation of Industries (CNFI). (2015). *Chinese National Federation of Industries White Paper 2015*. Taipei. Retrieved from https://drive.google.com/file/d/0B7Srh1VrhBDIaTJlT3J3TjFfZGM/view

Chiu, P. S. (2016) Gordon Sun: Taiwan's 'Five Shortages' are Being Exacerbated by the 'Seven Lows' of production. Retrieved July 25, 2016 from https://www.chinatimes.com/realtimenews/20160725003840-260410?chdtv

Chou, K. T. (2000). Risk Society and Social Praxis. Contemporary Monthly. *Special Issue: From Risk Society to the Second Modernity*, *154*, 36–49.

Chou, K. T. (2002). The Theoretical and Practical Gap of Glocalizational Risk Delayed High-tech Risk Society. *Taiwan: A Radical Quarterly in Social Studies*, *45*, 69–122.

Chou, K. T. (2013). Governance Innovation of Developmental State under the Globalized Risk Challenges—Burgeoning Civil Knowledge on Risk Policy Supervision in Taiwan. *A Journal for Philosophical Study of Public Affairs*, *44*, 65–148.

Chou, K. T. (2016). Beyond High Carbon Society. *AIMS Energy*, *4*(2), 313–330.

Chou, K. T. (2017). *Sociology of Climate Change: High Carbon Society and Its Transformation Challenge*. National Taiwan University Press.

Chou, K. T., & Liou, H. M. (2012). Analysis on Energy Intensive Industries under Taiwan's Climate Change Policy. *Renewable and Sustainable Energy Reviews*, *16*, 2631–2642.

Chou, K. T., & Walther, D. (2016). Whose Turn Is It to Sacrifice? The Environmental, Ecological and Social Risks in Taiwan's Economic Developmental Process. In Shiuh-Shen Chien, & Jenn-Hwan Wang, (Eds.), *The Developmental Study of Contemporary Taiwan Society* (pp.409–438). Taipei: Chu Liu Book Company.

Directorate General of Budget, Accounting and Statistics. (2017). *National Statistics, Republic of China (Taiwan)*. National Statistics, (National Income and Economic Growth).

Energy Report. (2012). *Taipower Company's Response and Strategy to the New Energy Policy*. January 2012.

Executive Yuan, Republic of China (Taiwan). (2017). *Premier Lai Proposed Six Countermeasures to Solve the Industry's Manpower and Skilled-labor Shortages*. Executive Yuan News. Retrieved from https://www.ey.gov.tw/Page/5A8A0CB5B41DA11E/31d71971-d0d9-4a76-9862-af1922dea3ac

Francisco, M. (2017). *Environmental Movements in the Philippines*. International Conference on Environmental Movements in Southeast Asia in the New Century, Session 2 Environmental Movements in the Philippines, Malaysia, and Thailand. Asian Social Transformation Thematic Research Team (AST), Institute of Sociology, Academia Sinica, Taipei, Taiwan.

Frantzeskaki, N., Loorbach, D. & Meadowcroft, J. (2012). Governing Societal Transitions to Sustainability. *International Journal of Sustainable Development*, 15, 1–2: 19–36.

Geels, F. W. (2006). The Dynamics of Transitions in Socio-technical Systems: A Multilevel Analysis of the Transition Pathway from Horse-drawn Carriages to Automobiles (1860–1930), *Technology Analysis & Strategic Management, 17*(4), 445–476.

Grin, J. (2006). Reflexive Modernization as a Governance Issue or Designing and Shaping Re-structuration. In J. P. Voβ., B. Dierk, & K. René (Eds.), *Reflexive Governance for Sustainable Development* (pp.54–84). Northampton, MA: Edward Elgar.

Harvey, D. (2003). The New Imperialism: Accumulation by Dispossession. *Socialist Register*, 63–87.

International Energy Agency (IEA). (2017). IEA Fossil-fuel Subsidies Database. Retrieved from www.iea.org/weo2017

International Monetary Fund (IMF). (2015). *IMF Survey: Counting the Cost of Energy Subsidies*. Retrieved from www.imf. org/external/pubs/ft/survey/so/2015/NEW070215A.htm

Jakkrit, S. (2017). *Thailand's Environmental Movements amidst Political Turbulence since 2000*. International Conference on Environmental Movements in Southeast Asia in the New Century, Session 2 Environmental Movements in the Philippines, Malaysia, and Thailand. Asian Social Transformation Thematic Research Team (AST), Institute of Sociology, Academia Sinica, Taipei, Taiwan.

Kemp, R., & Loorbach, D. (2006). Transition Management: A Reflexive Governance Approach. *Reflexive Governance for Sustainable Development* (pp.103–130). Northampton, MA: Edward Elgar.

Liberty Times Net. (2017). Causes of an Aging Population, Etc. Nikkei: Taiwan and South Korea will Fall into the "High Income Trap". Retrieved from http://news.ltn.com.tw/news/world/breakingnews/2039158

Loorbach, D. & Rotmans, J. (2006). Managing Transitions for Sustainable Development. In Xander Olsthoorn & Anna J. Wieczorek (Eds.), *Understanding Industrial Transformation* Vol. 44 (pp.187–206). Netherlands: Springer.

Marshall, G. R., & Alexandra, J. (2016). Institutional Path Dependence and Environmental Water Recovery in Australia's Murray-Darling Basin. *Water Alternatives, 9*(3), 679–703.

Organisation for Economic Co-operation and Development (OECD). (2003). Emerging Systemic Risks. Organization of Economic Co-operation and Development. *Final Report to the OECD Futures Project.* Retrieved from http://www. oecd.org/dataoecd/23/56/19134071.pdf

Ou, Jerry J. R. (2011). Creating a Green Energy Low-Carbon Environment – To Move Towards a Nuclear-Free Home. *Energy Report*, December 2011.

Pierson, P. (2000). Increasing Returns, Path Dependence, and the Study of Politics. *The American Political Science Review, 94*(2), 251–267.

Rip, A. (2006). A Co-evolutionary Approach to Reflexive Governance – and Its Ironies. *Reflexive Governance for Sustainable Development* (pp.82–102). Northampton, MA: Edward Elgar.

Risk Society and Policy Research Center (RSPRC). (2018). *A Brief Analysis of the Emission Changes and Transformation Actions of the Top 10 Companies with the Highest Greenhouse Gas Emissions.* Retrieved from http://rsprc.ntu.edu.tw/zh-tw/ m01-3/en-trans/open-energy/925-180411-10ghg.html

Rotmans, J., & Loorbach, D. (2006). Managing Transitions for Sustainable Development. In X. Olsthoorn & A. J. Wieczorek (Eds.), *Understanding Industrial Transformation* (pp.187–206). Netherlands: Springer.

Rotmans, J., & Loorbach, D. (2009). Complexity and Transition Management. *Journal of Industrial Ecology, 13*(2), 184–196.

Rotmans, J., & Loorbach, D. (2010). The Practice of Transition Management: Examples and Lessons from Four Distinct Cases. *Futures, 42*(3), 237–246.

Stirling, A. (2007). Risk, Precaution and Science: Towards a More Constructive Policy Debate – Talking Point on the Precautionary Principle. *EMBO Reports, 8*(4), 309–315.

Taipei Water Department (2017). *What is the Average Water Price in Different Countries around the World?* Retrieved from https://www.water.gov.taipei/News_ Content.aspx?n=30E4EDA27F6D9953&sms=87415A8B9CE81B16&s=A555C5 E4B46B130F

Taiwan Environmental Protection Union. (2018) Media Release: Who Has Polluted Taiwan? A Rebuttal of the White Paper Produced by the Chinese National Federation of Industries Headed by Wang Wen-yuan. Retrieved from https://tepu. org.tw/?p=16151

Urry, J. (2011). *Climate Change and Society.* Cambridge: Polity Press.

Van Buuren, A., Ellen, G. J., & Warner, J. F. (2016). Path-dependency and Policy Learning in the Dutch Delta: Toward more Resilient Flood Risk Management in the Netherlands? *Ecology and Society, 21*(4), 43. Retrieved from https://doi. org/10.5751/ES-08765-210443

Voβ, J. P., Kemp, R., & Bauknecht, D. (2006). Reflexive Governance: A View on the Emerging Path. In J. P. Voβ, D. Bauknechtm, & R. Kemp (Eds.), *Reflexive Governance for Sustainable Development* (pp.419–437). Cheltenham: Edward Elgar.

Voβ, J. P., Smith, A., & Grin, J. (2009). Designing Long-term Policy: Rethinking Transition Management. *Policy Sci, 42*(2), 275–302.

Wang, J. H. (2012). Long Live Development Studies. *City and Planning, 39*(2), 1–18.

Wu, C. (2010). I Can Only Write a Poem for You. *Business Weekly*, No.279.

Yahoo News. (2011, November 3). President Ma's Announcement of the Energy Policy: First, Second and Third Nuclear Plants will not be Closed, Fourth Nuclear Plant to be Halted.

4 Climate change governance in Korea

Focusing on the process of the establishment of its NDC

Sun-Jin Yun

Introduction

Ever since the Kyoto Protocol was signed, the risk of climate change has got worse and, consequently, global society's awareness of it has increased. Annex I parties achieved their reduction target during the first Kyoto Protocol implementation period of 2008 to 2012, a 5.2 percent decrease from the 1990 emissions level. Nevertheless, global greenhouse gas (GHG) emissions increased 36.1 percent, from 38 $GtCO_2$ in 1990 to 49 $GtCO_2$ in 2010, because of a doubling in the developing world. Annex I parties' emissions decreased by 10.4 percent from 1990 without accounting for land use, land-use change, and forestry (LULUCF) and 16.2 percent with LULUCF. This was possible because of an economic recession in economies in transition. GHG emissions of the Organisation for Economic Co-operation and Development (OECD) members that were Annex I parties slightly increased, by 0.3 percent including LULUCF and by 1.9 percent excluding LULUCF.

On December 12, 2015, the Paris Agreement was adopted at the 21st Conference of Parties (COP 21) under the United Nations Framework Convention on Climate Change (UNFCCC). For the first time, world society agreed on all nations' participation in combating climate change and on developed countries' more active financial and technological support for developing countries that have less historical responsibility for triggering climate change and lower per capita emissions. The essential target of the Paris Agreement was to keep temperature rise in this century below 2°C compared with preindustrial levels and make more ambitious efforts to constrain the temperature rise even further, to 1.5°C.

The Paris Agreement took a voluntary, bottom-up approach in order to have all parties' cooperation on climate actions. Each party had been asked to submit its Intended Nationally Determined Contribution (INDC) by March 2015, if possible, or before the COP 21 if it needed more time. The UNFCCC issued a synthesis report about the probability of achieving the temperature goal based on the submitted INDCs on November 1, 2015. All parties must report their emissions and implementation efforts regularly based on the Paris Agreement. A global review will be undertaken every five years to assess the collective achievement. All parties agreed on no backsliding, or the principle of progression, meaning that Nationally Determined Contributions (NDCs) submitted every five years must be strengthened.

The Paris Agreement entered into force less than a year after its adoption. It opened for signature on Earth Day, April 22, 2016, and took effect on November 4, 2016, 30 days after the so-called "double threshold" (ratified by 55 countries that are composed of at least 55 percent of global GHG emissions). By the end of 2018, 184 out of the 197 parties at the Convention had ratified it. South Korea was among them: it submitted its INDC on 30 June 2015 and ratified the Paris Agreement on 3 November 2016. Both events happened under the Park Geun-hye government. South Korea's national GHG emissions reduction target was 37 percent less than the business-as-usual (BAU) level by 2030: 25.7 percent was planned to be reduced domestically and the remaining 11.3 percent was to be achieved by using carbon credits from international markets. In 2018, the Moon Jae-in government modified the basic roadmap for national GHG reduction. It kept the reduction target itself but reduced the share from international carbon markets from 11.3 percent to 4.5 percent (including absorption sinks); thus, the share of domestic reduction increased from 25.7 percent to 32.5 percent.

This chapter explores South Korea's governance on climate change, focusing on the process of establishing its NDC. Currently, Climate Action Tracker, an independent scientific analysis that is jointly produced by New Climate Institute for Climate Policy and Global Sustainability, ECOFYS, and Climate Analytics, evaluates South Korea's NDC as highly insufficient. According to Climate Action Tracker, if all government targets were similar to Korea's then global temperatures would increase by between 3°C and 4°C. According to German Watch, which compiles the Climate Change Performance Index (CCPI) for the 58 countries whose share of GHG emissions is over 1 percent of the global total, Korea's CCPI belongs to the last, poorest group. Countries are ranked from fourth to sixty-first, instead of first to fifty-eighth, because no country is eligible to be ranked first to third, according to German Watch. South Korea ranked fifty-seventh in the 2019 assessment. The OECD stated in *OECD Environmental Performance Reviews: Korea 2017* (OECD, 2017) that Korea's climate target would be hard to achieve based on current trends. This chapter addresses the following questions: Why did South Korea establish such a poor NDC level? What was the decision-making process? Who was involved in that process? What issues were raised by whom?

Theoretical background: governance and climate change policy

The paradigm of governance appeared after the two oil shocks and stagflation in the 1970s raised questions about the government's ability to manage state affairs (Pierr and Peters, 2000; Eun and Lee, 2009; Lee and Yun, 2011). It has changed over time, evolving from state-centric governance in the 1970s, market-centric in the 1980s, and civil-centric governance in the 1990s to the current recognition that network governance is required. In network governance, the state, market, and civil society interact in horizontal relationships (Namkoong, 2009). Additionally, governance has been viewed along diverse dimensions, including international, national, and regional, as well as by subject: it takes a variety of forms, not a consistent or fixed one.

The essence of governance is the participation of multiple actors other than the government in the decision-making process for public and/or social problem-solving. It can be understood as a cooperative problem-solving process that involves in-depth discussion and consensus among diverse actors from the government, market, and civil society, beyond the unilateral rule of the government. When a policy is decided through a governance-based approach in which various actors participate and collaborate, policy acceptability is enhanced by reflecting various opinions. Consequently, policy legitimacy and effectiveness are also improved (Yun, 2005).

The governance approach is important in climate policy decision-making processes because climate change has a much longer time lag between cause and effect than other environmental problems, is deeply linked to equity issues, and is a collective problem that society as a whole must take action on together (Yun, 2009; Lee and Yun, 2016). The nature of the climate change problem means it requires an active governance regime in which diverse members of society participate in decision-making processes. There is a high possibility of conflict among stakeholders during the process of developing and implementing policies to cope with climate change, especially sharing GHG reduction obligations, because of the different responsibilities and impacts.

When countries prepare NDCs, they must consider the international community's view that climate change is a global problem that requires cooperative participation by all countries. The establishment of national GHG emissions reduction targets is critical because global temperature rise is related to each country's targets. However, like climate change itself, this work influences the activities and daily life of members of society and stakeholders because GHGs are emitted from almost all sectors. Thus, various stakeholders are required to participate in decision-making processes while developing social consensus and distributing appropriate reduction responsibilities. The Paris Agreement declares that the participation of stakeholders other than governments is required to combat climate change.

In Korea, the industrial sector is very sensitive and critical because of its high energy consumption and large GHG emissions. Thus, there have been controversies and social conflicts over the proper distribution of GHG reduction obligations by sector. This means that a governance approach based on the participation of multiple stakeholders from diverse sectors in the INDC decision process is necessary in Korea.

The background and process of establishing the 2030 NDC Target

The status of GHG and combustion-based carbon dioxide emissions in South Korea

Korea's total GHG emissions in 2017 stood at 709.1 million tons CO_2eq. As shown in Table 4.1, they comprised 86.8 percent from energy, 7.9 percent from industrial processes, 2.9 percent from agriculture, and 2.4 percent from waste.

Table 4.1 Trend of GHG emissions in South Korea 1990–2017
(Unit: $MTCO_2$eq.)

Year	Total emissions	Net emissions	Energy	Industrial process	Agriculture	LULUCF	Waste
1990	292.2	254.4	240.4	20.4	21.0	−37.7	10.4
1991	315.5	282.0	258.2	24.4	21.3	−33.5	11.6
1992	343.5	311.0	279.1	30.0	21.7	−32.5	12.7
1993	378.7	348.0	308.6	34.4	22.1	−30.7	13.5
1994	404.0	371.2	328.0	39.1	22.5	−32.8	14.4
1995	435.9	405.0	352.2	45.2	22.8	−30.9	15.7
1996	471.5	437.2	385.8	45.9	23.3	−34.3	16.5
1997	502.9	464.0	411.9	50.6	23.3	−39.0	17.2
1998	431.8	384.3	351.6	41.1	23.0	−47.5	16.0
1999	469.7	414.5	382.3	48.8	21.8	−55.3	16.9
2000	503.1	444.8	411.8	51.3	21.2	−58.3	18.8
2001	516.7	458.9	426.2	50.1	20.7	−57.7	19.7
2002	538.7	494.2	445.5	54.0	20.5	−55.5	18.7
2003	549.1	490.6	453.2	56.9	20.3	−54.9	18.8
2004	557.4	501.3	460.3	59.0	20.3	−56.1	17.7
2005	561.8	507.7	468.9	55.7	20.5	−54.0	16.7
2006	567.0	511.9	475.1	54.3	20.6	−55.1	17.0
2007	580.7	524.9	492.8	51.5	20.8	−55.8	15.6
2008	592.2	535.5	506.4	49.5	20.8	−56.7	15.4
2009	598.0	541.6	513.1	48.2	21.3	−56.3	15.4
2010	657.6	603.8	566.1	54.7	21.7	−53.8	15.0
2011	684.2	630.7	595.0	53.1	20.7	−53.6	15.5
2012	687.5	638.9	596.3	54.2	21.3	−48.6	15.7
2013	697.0	652.8	605.1	54.8	21.2	−44.2	15.9
2014	691.5	649.3	597.5	57.3	21.3	−42.2	15.4
2015	692.3	649.9	600.8	54.4	20.8	−42.4	16.3
2016	692.6	648.7	602.7	52.8	20.5	−43.9	16.5
2017	709.1	667.6	615.8	56.0	20.4	−41.6	16.8

Source: National Greenhouse Gas Inventory Report of Korea, 2019

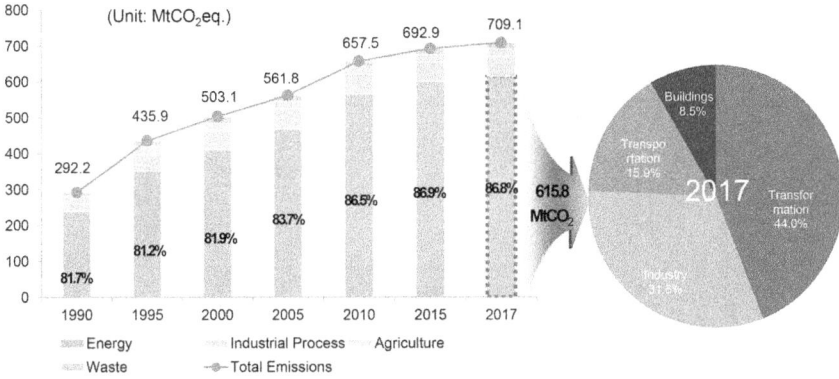

Figure 4.1 Trend in GHG emissions and the share by sector 1990–2017

Source: GIR, 2018, National GHG Inventory.

The emissions volume in 2017 had risen 2.4 percent year-on-year and 142.7 percent (2.43 times) since 1990. Looking at emissions trends, the annual average increase of 8 percent between 1990 and 1997 was sharply reduced by 14 percent year-on-year due to the 1998 financial crisis. It started to increase again from 1999 and reached a record high of 696.7 million tons in 2013. Since 2014, it has been showing a slight increase since a 0.8 percent decline in 2013 and then more increase in 2017 exceeding 700 million tons CO_2eq. Figure 4.1 shows the trend of South Korean GHG emissions from 1990 to 2017 and the share of GHG emissions from energy by sector in 2017.

Korea's total GHG emissions were the twelfth highest in the world in 2017, and sixth among OECD countries. Korea had the seventh highest carbon dioxide (CO_2) emissions from fuel combustion in 2017, 589.2MtCO$_2$; this was the fourth-highest among OECD members. This represented 1.8 percent of the world's CO_2 emissions, as shown in Table 4.2. Korea's population accounted for 0.7 percent of the world's total population, so it accounted for 2.6 times more CO_2 emissions than its share of the population. Per capita CO_2 emissions in 2017 amounted to 11.50 tons, 2.6 times higher than the global average of 4.35 tons. This was 1.27 times higher than the average of 9.02 tons per person in OECD countries and 3.48 times higher than the average of 3.30 tons per person in non-OECD countries. Despite a shorter industrialized period than advanced countries, Korea's cumulative GHG emission between 1900 and 2011 ranked sixteenth in the world. Moreover, the growth of emissions in Korea was the second-highest among OECD countries from 1990 to 2013, while CO_2 emissions per capita were the sixth highest among OECD countries. Korea has been responsible for causing climate change, and now it is responsible for responding to it.

One reason for South Korea's fast-growing emissions is the energy-intensive and export-oriented manufacturing industry (Yun, Ku and Han et al., 2014).

Table 4.2 Major statistics of world top 10 big CO$_2$ emitters 2017

Rank	Nation	CO$_2$ emissions from combustion (MtCO$_2$)		Per capita CO$_2$ emissions (tCO$_2$)	Population (millions)		GDP-PPP (2005 billion US$)		Total primary energy supply (MTOE)	
1	China	9,056.8	28.0	6.57	1,378.7	18.6	19,450.4	17.8	2,958.0	21.5
2	US	4,833.1	15.0	14.95	323.4	4.4	16,920.3	15.5	2,166.6	15.7
3	India	2,076.8	6.4	1.57	1,324.2	17.8	7,904.5	7.2	824.7	6.0
4	Russia	1,438.6	4.5	9.97	144.3	1.9	3,219.8	2.9	852.4	6.2
5	Japan	1,147.1	3.5	9.04	127.0	1.7	4,759.8	4.4	425.6	3.1
6	Germany	731.6	2.3	8.88	82.3	1.1	3,553.4	3.3	310.1	2.3
7	**S. Korea**	**589.2**	**1.8**	**11.50**	**51.2**	**0.7**	**1,796.1**	**1.6**	**268.4**	**2.0**
8	Iran	563.4	1.7	7.02	80.3	1.1	1,263.8	1.2	282.4	2.1
9	Canada	540.8	1.7	14.91	36.3	0.5	1,454.9	1.3	280.1	2.0
10	Saudi Arabia	527.2	1.6	16.34	32.3	0.4	1,595.6	1.5	210.4	1.5
	Sum*	21,391.8	66.5%	5.94	3,603.6	48.2%	56,112.0	56.7%	8,483.5	62.3%
	OECD	11,591	35.9%	9.02	1,284	17.3%	49,034	44.9%	5,275	38.3%
	World Total	32,316	100%	4.35	7,429	100%	109,231	100%	13,761	100%

Source: International Energy Agency, 2019.

Note: *Sum means subtotal of 10 big emitters and their share of the world total.

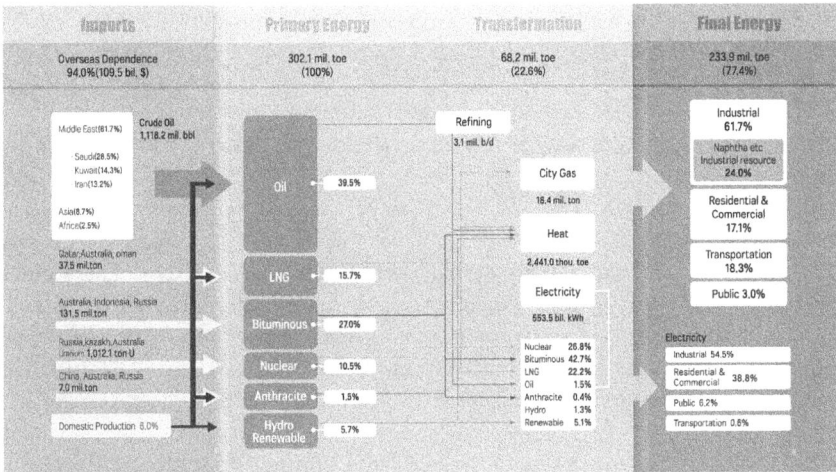

Figure 4.2 Energy balance flow in South Korea (2017)
Source: KEEI, 2018 Energy Info. Korea (downloaded from www.kesis.net/sub/sub_0003.jsp).

Unlike other OECD countries with relatively high per capita emissions (such as the United States, Australia, or Canada), whose energy consumption and CO_2 emissions per capita are projected to decline, those of South Korea are expected to continue to rise despite a declining population. In 2017, the industrial sector consumed 61.7 percent of final energy consumption and 54.5 percent of electricity (Korea Energy Economics Institute [KEEI], 2018), as shown in Figure 4.2.

Major policy measures to mitigate GHG emissions before INDC announcement

To reduce sectoral emissions, in 2012 South Korea introduced the GHG and Energy Target Management System (TMS), which was a precursor of the Emissions Trading System (ETS) and covered 60 percent of total national emissions. The TMS still covers emitters that consume significant amounts of energy and are not covered by the ETS. As a result of TMS operations, 65 companies collectively reduced their emissions by 0.74 MtCO$_2$e/year in 2015 compared to BAU (Republic of Korea, 2019). Introduced in 2015, the ETS manages 69 percent of national GHG emissions and covers more than 600 companies from 23 subsectors (Republic of Korea, 2017b) of the steel, cement, petrochemicals, refinery, power, buildings, waste, and aviation sectors. It includes all installations in the industrial and power sectors that have annual emissions higher than 25 $ktCO_2e$. The ETS system includes both direct and indirect emissions (emissions from electricity use).

In Phase I (2015–2017) of the ETS, all allowances were allocated gratis (IETA, 2016). Energy-intensive and export-oriented sectors received free allowances for all their emissions in all three phases. In Phase I, the plan was to decrease the absolute emissions cap from 573 $MtCO_2e$/year in 2015 to 562 MtCO2e/year in 2016 and 551 $MtCO_2e$/year in 2017 (Ministry of Environment, 2014). In addition to the overall cap, the ETS also set sectoral caps that reflected sectoral-based emissions reduction targets (ICAP, 2018). The sectors were selected based on their contribution to the country's overall emissions. They were expected to play an important role in meeting South Korea's previous target of reducing emissions to 30 percent below the baseline by 2020, with sector-wide reductions ranging from 17.5 percent for waste to 34.3 percent for transport.

As of June 2016, all but one of the 525 liable entities had successfully submitted permits for compliance. A total of 539 $MtCO_2e$ was allocated to the entities and they emitted a total of 543 $MtCO_2e$, demonstrating that the plan functioned quite well (Greenhouse Gas Inventory and Research Center [GIR], 2017). The cap for Phase II (2018–2020) of the ETS was announced in July 2018; it increased the total allowance by 5.2 percent compared to Phase I, the quotas for the first and second phases are not directly comparable due to differences in the number of companies that were given allocations, the scope of facilities subject to pre-allocation, and the emission calculation coefficient. In Phase II, 97 percent of the allowances were allocated free of charge and the remaining 3 percent were auctioned. For Phase III, 10 percent of allowances were set to be auctioned (Ministry of Environment, 2018).

In the 2030 GHG roadmap revised in 2018, South Korea provides details of sector-specific reduction targets as well as policy measures to be further encouraged. For the building sector, the plan mentions strengthening permit standards for new buildings, promoting green renovation, identifying new circular business models, and expanding the renewable energy supply. For the industrial sector, the plan focuses on energy efficiency and industrial process improvement measures and on the promotion of eco-friendly raw materials and fuel (Ministry of Environment, 2018). These planned measures are not seriously considered in current policy projections.

The process of 2030 INDC target establishment

When the timing of the INDC submission was agreed upon at the 2014 COP 20 in Lima, Peru, the South Korean government was obliged to establish and submit a post-2020 INDC well in advance of the 2015 COP 21, to be held in Paris. The Park Guen-hye government began work on establishing a long-term 2030 national reduction target in May 2014. The 2030 INDC establishment process was overseen by the Joint Climate Change Response Task Force, comprised of the relevant ministries and centered in the Office for Government Policy Coordination under the Prime Minister. Related ministries included the Ministry of Strategy and Finance; the Ministry of Foreign Affairs and Future Planning; the Ministry of Land, Infrastructure and Transport; the Ministry of Trade, Industry and Energy; the Ministry of Environment; the Ministry of Agriculture, Food and Rural Affairs; and the Ministry of Maritime Affairs and Fisheries.

The working-level draft of the 2030 INDC was created by a joint task force consisting of experts recommended by the relevant departments. The joint task force included researchers belonging to various national research institutes and government-affiliated organizations, including the Agricultural Science Institute, the Forest Research Institute, the Forestry and Fisheries Research Institute, the Korea Development Institute, the Korea Energy Economics Institute (KEEI), the Korea Energy Management Corporation, the Korea Energy Technology Institute, the Korea Environment Corporation, the Korea Fisheries Development Institute, the Korea Rural Economic Institute, the Korea Transportation Authority, the Korea Transportation Institute.

The reduction goal-setting procedures that were announced initially were as follows. First, the government's GHG emissions forecast would be derived by early 2015 from the forecast of activity (product output, etc.) and the forecast of energy demand based on major prerequisites. Second, it would analyze reduction scenarios and economic ripple effects during the first half of 2015. Third, the long-term goal for reducing GHG emissions would be derived by the end of 2015 by compiling the results of the emissions outlook, reduction scenarios, and analysis of ripple effects. Fourth, social consensus would be obtained through public hearings, and the final decision would be made through the Presidential Committee on Green Growth (PCGG) and the Cabinet. The timing of submissions to the international community would be decided by considering overall international trends.

The government published *Ways to Develop Long-Term Goals to Reduce GHG Emissions and Collect Opinions* on November 14, 2014. According to that report, the government would collect opinions from stakeholders, such as industries and civic groups, in the course of establishing long-term goals for reducing GHG emissions. To collect opinions, a joint task force with eight experts, four recommended by industries and four by civic groups, would be formed to review the details of the work and collect opinions. The plan to collect opinions during the process of establishing a long-term reduction target would be composed of four stages: Stage 1 was to be activity data (including preconditions such as population growth, GDP growth, international oil prices, and industrial structure) projection; Stage 2 was to be emissions prospects; Stage 3 was to be reduction potential analysis; and Stage 4 was to be setting reduction targets.

Data discussed during the review process were to be distributed only within the review team in order to maintain security. It was announced that data review and opinion collection would be conducted separately and that briefing sessions for economic and private sectors would be held as often as necessary. The government planned to ask for information about expansion plans, prospects for internal production (or energy demand), and means to reduce GHG emissions from each business sector, intending to collect opinions from each industry sector at the same time. In accordance with these procedures, the government established a long-term national reduction target. During this process, the number of participating experts was not sufficient for the joint review team, so the number was increased from eight to ten.

The government proposed four reduction targets on June 11, 2015, at a meeting of the public-private joint review team. The joint project team's calculation of the BAU GHG emissions forecast (based on major economic variables such as economic growth rates, oil prices, and the industrial structure) estimated that 782.5 million tons of CO_2eq would be emitted by 2020 and 85.6 million tons of CO_2eq would be emitted by 2030. Based on the joint team's analysis of potential reduction and reduction means, the government announced that it would choose a mid-term reduction target from four scenarios that would reduce the BAU by 14.7 percent, 19.2 percent, 25.7 percent, and 31.3 percent by 2030, as shown in Figure 4.3. After a public hearing on June 12, the day after the scenarios were announced, and a discussion at the National Assembly on June 18, the government decided to reduce BAU by 37 percent at the Cabinet meeting on June 30. This strengthened target was a result of accepting criticism based on the "no back-sliding principle" or "principle of progression" raised at home and abroad. However, the then Korean government (the Park

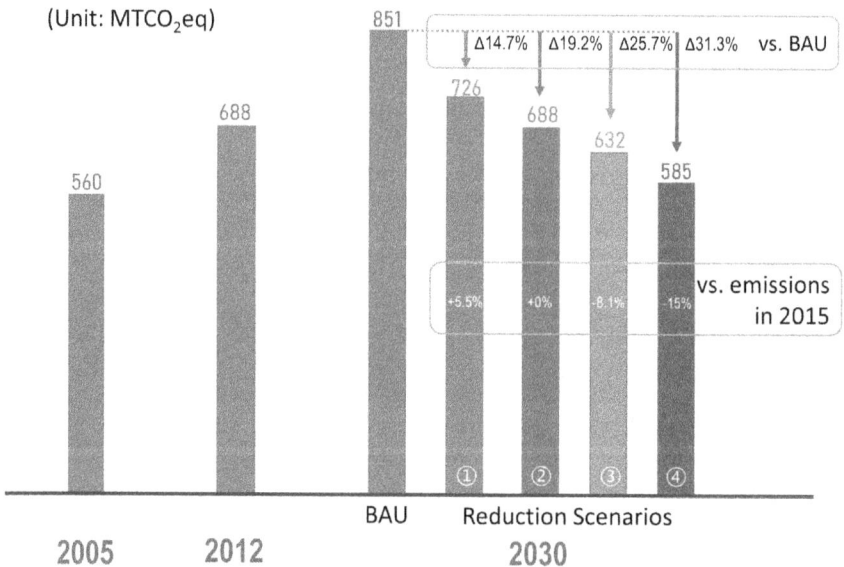

Figure 4.3 2030 reduction target scenarios suggested by the government

Source: Materials of public hearing for post-2020 national reduction target setting.

Table 4.3 The process of INDC establishment

Time		Contents
2014	April	Composition of Joint Climate Change Response Task Force of related departments
	May	Composition of a joint task force consisting of experts recommended by relevant departments (generally managed by the Greenhouse Gas Inventory and Research Center)
	Nov. 14	Meeting with five representative economic organizations and five representative civic organizations
	Dec. 17	Composition of a joint public-private review team
2015	May 20	A social forum organized by National Assembly Climate Change Forum (Submission date of the INDC was changed from the end of September to the end of June)
	June 11	Release of "Post-2020 Plan to Set Greenhouse Gas Reduction Goals" (4 scenarios presented)
	June 12	Post-2020 public hearings on national greenhouse gas reduction targets
	June.18	Discussion meeting hosted by the National Assembly Climate Change Forum
	June 30	INDC submitted to the UNFCCC Secretariat

Source: Based on Lee and Yun (2016, p. 275).

Geun-hye government) planned to reduce only 25.7 percent through domestic efforts and the remaining 11.3 percent through international carbon markets. On that same day, June 30, 2015, South Korea's INDC was submitted to the UNFCCC. The National Assembly passed a bill to ratify the Paris Agreement on November 3, 2016 and deposited it with the UNFCCC shortly before the Paris Agreement went into effect. Korea was the ninety-seventh country to ratify. Table 4.3 summarizes the process of INDC establishment in South Korea.

Restructuring the government's climate response entity and the establishment of the 2030 roadmap

In May 2016, the Park Geun-hye government amended relevant law to reflect a newly established national GHG reduction target and reorganized the climate response governmental system. Article 25 paragraph 1 of the Enforcement Decree of the Framework Act on Low-Carbon Green Growth was amended to replace the 2020 national reduction target (30 percent reduction from BAU) that had been announced in 2009 with the new 2030 target. In addition, the Park government changed the body responsible for inter-ministerial policies on climate change from the Ministry of Environment to the Office for Government Policy Coordination. The role of the Prime Minister and the Deputy Prime Minister for Economic Affairs to oversee and coordinate the response to climate change were strengthened and a ministry-responsibility system was adopted. The Ministry of Strategy and Finance was responsible for the emission trading system that had previously been handled by the Ministry of Environment. The relevant ministries were given responsibility for overseeing the ministry-related climate change policies.

In December 2016, the Park government announced a comprehensive roadmap and a basic plan to cope with climate change. The responsibility of relevant ministries and agencies for each sector was based on the overall coordinating functions of the Office for Government Policy Coordination. The Green Growth Committee reviewed the "First Basic Plan for Responding to Climate Change" and the "2030 National Greenhouse Gas Reduction Basic Roadmap" before they were finalized and announced at the Cabinet meeting. The "First Basic Plan for Responding to Climate Change" was established under the "Framework Act on Low Carbon Green Growth." It stipulated reducing GHG emissions, adapting to climate change, and international cooperation as the first comprehensive measures in the nation's mid- and long-term climate change strategy and included specific action plans for the new climate system. Along with the "Basic Plan" to respond to climate change, the "Basic Roadmap for National Greenhouse Gas Reduction" was finalized. It included systematic implementation measures to efficiently achieve the 2030 National Greenhouse Gas Reduction Target. Sub-targets of domestic GHG reduction (25.7 percent from 2030 BAU) were allocated by sector, as shown in Figure 4.4. Those sub-targets were drawn from studies about the reduction potential in each sector that were conducted by relevant national research institutes and considered the impact on the national economy. The government promised to keep the industrial sector emissions reduction rate from exceeding 12 percent when it announced the national reduction target.

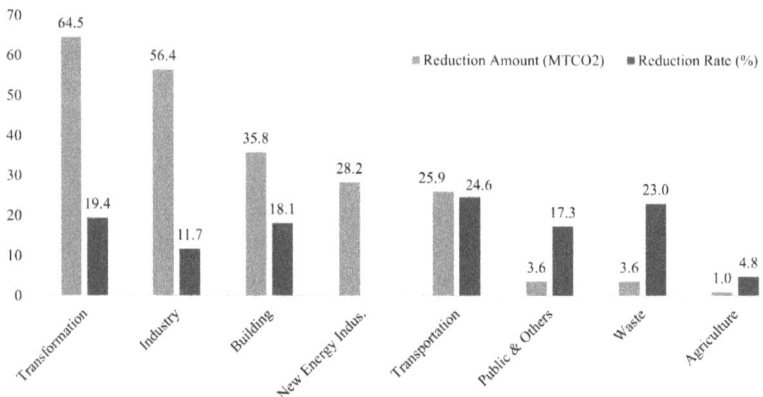

Figure 4.4 2030 GHG reduction targets by sector

Source: The Office of Prime Minister, 2016, "Presentation of National Mid- and Long-term Strategies and Policy Directions" (Press release, Dec. 6, 2016).

Modification of the road map to implement the 2030 NDC target

In 2018, the Enforcement Decree of the Framework Act on Low Carbon Green Growth and the Enforcement Decree of the Act on the Allocation and Trading of Greenhouse Gas Emission Permit were again amended after the Moon Jae-in government took office. The work of setting overall national GHG reduction goals was left with the Office for Government Policy Coordination, but the Ministry of Environment de facto took charge of practical operations, as it had in the past.

The GIR responsible for GHG statistics management, had belonged to the Ministry of Environment until the Park Geun-hye government transferred it to the Office for Government Policy Coordination in 2016. The Moon government returned it to the Ministry of Environment in 2018. Regarding the emissions trading scheme, the Ministry of Strategy and Finance was to be in charge of the basic plan along with the Ministry of Environment, while the allocation plan was also to be organized by the Ministry of Environment. Specific executive functions such as quota allocation, emission assessment and certification, market operation, and paid auctions were also returned to the Ministry of Environment. However, external project evaluation and certification and GHG reduction support projects were mainly carried out by the government agencies in charge of each sector, such as the Ministry of Trade, Industry, and Energy; the Ministry of Maritime Affairs and Fisheries; and the Ministry of Agriculture, Food, and Rural Affairs.

In July 2018, the South Korean government announced a revised roadmap that set an objective for national emissions to peak around 2020, despite no change in the national GHG emissions reduction target itself (37 percent reduction from the BAU). The revised roadmap reduced the scope of the reduction from international offsets with forest absorption to 4.5 percent (approximately 38.3 million tons) (see Figure 4.5). It increased the share of

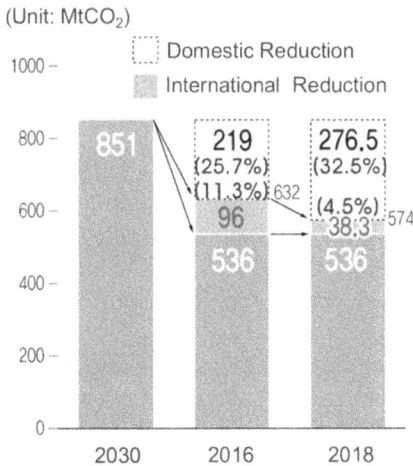

Figure 4.5 Revised roadmap for 2030

Table 4.4 2016 Roadmap vs. 2018 Roadmap

Sector		BAU	2016 Roadmap		2018 Roadmap	
			Emissions after reduction $(MtCO_2)$	Reduction rate vs. BAU (%)	Emissions after reduction $(MtCO_2)$	Reduction rate vs. BAU (%)
Reduction in emission sources	Industry	481.0	424.6	11.7	382.4	20.5
	Building	197.2	161.4	18.1	132.7	32.7
	Transport	105.2	79.3	24.6	74.4	29.3
	Waste	15.5	11.9	23.0	11.0	28.9
	Public & others	21.0	17.4	17.3	15.7	25.3
	Agriculture	20.7	19.7	4.8	19.0	7.9
	Leakage, etc.	10.3	10.3	0.0	7.2	30.5
Use of reduction measures	Transformation	(333.2)[1]	−64.5	–	(confirmed) −23.7 (additional potential) −34.1[2]	–
	New energy industry/ CCUS	–	−28.2	–	−10.3	–
	Forestry	–	–	–	−38.3	4.5%
	Abroad	–	−95.9	11.3%		
Existing domestic reduction			631.9	25.7%	574.3	32.5%
Total		850.8	536.0	37.0%	536.0	37.0%

Source: News release of the Ministry of Environment, July 24, 2018b.

Note
1) Emissions from the transformation sector (333.2 $MtCO_2$) were allocated to electricity and heat use in each sector other than the transformation sector and were excluded from the total.
2) 23.7 $MtCO_2$ was confirmed as the reduction amount from the transformation sector and additional reduction potential will be confirmed in 2020 when a long-term emission strategy for 2050 is submitted.

domestic mitigation necessary to reach the Paris Agreement target, with a domestic target of 32.5 percent (approximately 276.5 million tons) emissions reduction in 2030, up from the previous 25.7 percent. The sectoral reduction targets were also adjusted, as shown in Table 4.4. The path to achieve the 2030 reduction target is shown in Figure 4.6.

These revisions were developed by the Joint Task Force, beginning in September 2017. The Joint Task Force was composed of two groups. One group was a technical review team centered on national research institutes and the other was a roadmap team focused on related departments and civil

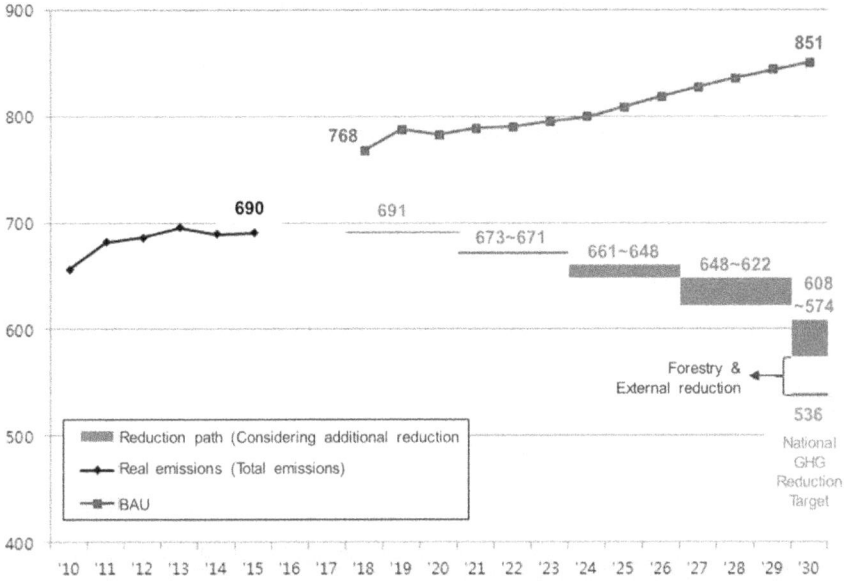

Figure 4.6 Korea's revised GHG reduction roadmap

Source: Joint Ministries, 2018, "Amendment of basic roadmap to achieve national GHG reduction targets by 2030., retrieved from http://www.me.go.kr/home/web/board/read.do? menuId=286&boardMasterId=1&boardCategoryId=39&boardId=886420.

experts. The former team worked on the technological prospects for reductions in each sector and technical tools and measures for GHG emissions reduction, while the latter discussed policy issues and worked on coordination and collaboration among sectors. Based on the revised draft, expert opinions were collected four times, through forums, panel discussions, and seminars. Opinions were gathered from stakeholders, especially actors from the business sector, through meetings and from the public by opening a dedicated homepage and social network service (SNS) channels with the Ministry of Environment. The Moon government then began working on a low emissions development strategy for 2050, to be submitted to the UNFCCC secretariat by 2020.

Critical examination of the INDC establishment in terms of governance

How appropriate are South Korea's reduction targets? There are many ways to evaluate the adequacy of each country's target. According to the Climate Action Tracker, which evaluates each country's GHG emissions reduction target, South

Korea's target is inadequate. It states that if all government targets were in line with South Korea's, temperature rise would reach between 3°C and 4°C. Although the Moon government strengthened the domestic responsibility for GHG reduction, South Korea still needs to substantially strengthen its target. Current and planned policy projections are far short of the Paris target level and much more so for the level of reduction needed to be considered as "1.5°C-compatible." More stringent policies are required, even for their current weak target.

This section examines the process of establishing the INDC. It is essential to have various actors participating in a process in which diverse stakeholders must discuss policy issues and reach a consensus. Who was invited to the process and how and what did they discuss? How did public debate come about?

The process of establishing the INDC

Initially, the government planned to submit the INDC at the end of September 2015. Although "The Plan to Establish Long-term Greenhouse Gas Reduction Goals and Collect Opinions," issued on November 14, 2014, stated that "the timing of submission to the international community should be decided in consideration of international trends, etc.," the press reported that the submission was scheduled for the end of September (*Chosun Biz*, May 20, 2015). In fact, there was no discussion about the timing of the submission even among the joint public-private review team. However, at the end of June, the government suddenly set the submission date and hurriedly held a public hearing. One reason for this was to follow the trend of major countries to submit the INDC ahead of the deadline in order to contribute to the launch of the new climate regime (Ministry of Environment, 2015). Another reason was the establishment of the 7th Basic Plan for Electricity Supply and Demand (2015–2029). This was scheduled for 2015 and had to reflect the GHG reduction plan.

Because of this sudden change in the timing of the submission, the public hearing was held only two days after being announced. Generally speaking, public hearings need to be held according to a proper time schedule. This was particularly the case in this instance as the climate change reduction targets would have a significant impact on the national economy and people's lives. Legally, Article 38 of the Administrative Procedure Act (notice to hold a public meeting) states that public hearings must be announced 14 days in advance and a pre-announcement period of 40 days is necessary to inform the public if legislation is very relevant to people's lives. However, the announcement of the INDC public hearing was made just two days before the event and the details were announced the day before. Thus, the public opinion collection process failed not only to meet the people's right to know, but also to observe the law of procedure. This legal failure was pointed out during a public hearing on June 12 and also during a parliamentary debate on June 18.[1] It would have been preferable to have set the goal after sufficient social discussion even if the deadline for submission was delayed, rather than pushing ahead with it and ignoring the domestic procedure law in order to meet the end-of-June deadline. Actors from the business sector complained that the government's proposal was unrealistically high, while actors from civil society criticized it as being too low.

The government said it collected opinions from experts through a joint public-private review meeting on June 11 before the public hearing, but the purpose of that meeting was unclear. It was held a day before the public hearing, on the very day that the government announced four INDC scenarios. Experts recommended by civic organizations for the joint public-private review team did not attend the meeting because they believed it was only for show. They judged that it was meaningless to participate in the meeting when there no change would be made to the contents that would be announced at the public hearings. Initially, the Park government said that it planned to divide the process of establishing the INDC into four stages: developing activity data (including the preconditions), forecasting emissions, analyzing reduction potential, and deriving a reduction target. It said it would listen to the opinion of the public-private joint review team at each stage. However, it failed to properly deal with the issues pointed out by representatives of civil society during the first and the second stages, and the procedure for collecting opinions was not well managed by the joint public-private review team during the third and the fourth stages. There was neither proper information sharing nor sufficient discussion.

After the public hearing, the government voted at a Cabinet meeting to change the reduction goal to 37 percent of BAU by adding an 11.3 percent reduction through the overseas emissions market to the strongest previous scenarios (a 25.7 percent reduction from the BAU in 2030), but there was no public discussion of the new plan. The sharply diverging opinions between industry and civil society, the fact that the establishment of the INDC for 2030 could have a decisive impact on the nation's climate change and energy policies, and legal flaws in the process are highly likely to be stumbling blocks to the future implementation of related policies.

Participants invited to the INDC establishment process

Who participated in the process of setting the INDC? In other words, who was invited as a participant? Identifying the participants helps to show which stake-holders' views could have been reflected. To collect the opinions of the non-governmental sector, the government initiated a joint public-private review team consisting of ten experts from industry and the same number from civil society. Their agencies are shown in Table 4.5.

The experts recommended by industry can be said to represent or have a direct understanding of existing energy-intensive industries, such as associations of energy-intensive industries, the Federation of Korean Industries, and the Korea Chamber of Commerce and Industry. These energy-intensive industries

Table 4.5 Participants in the public-private review team

The Industrial Sector	The Civil Sector
• Executive director of Lotte Chemical Corporation	• Professor of Economics at Chung-Ang University
• Korea Petroleum Association	• Chairman of the Steering Committee, Environmental Justice Forum
• Semi-conductor Industry Association	• Policy Committee member of Korea's NGO's Energy Network
• Executive director of the POSCO Management Institute	• Secretary General of Green Transport
• Professor of Economics at Yeonsei University	• Director of Institute for Climate Change Action
• Head of Industry, Federation of Korean Industries	• Director of Post-Nuclear Team, Korea Federation for Environmental Movements
• Display Association	• Deputy director of Environmental Justice Research Institute
• Chief of Environmental Climate Strategy Office of Business at Institute for Sustainable Development, Korea Chamber of Commerce & Industry	• Professor of Energy and Environmental Policy at Seoul National University
• East-West Power Company	• Director of Green Energy Strategy Institute
• Professor in Graduate School of Energy and Environment (Green School), Korea University	• Professor of Environmental Economics at Seoul National University

Note: The names of participants were not disclosed, but all participants are identified because the author participated as an expert recommended by the civil society.

needed to come to the table because they were likely to have conflict with the government: the reduction target would not only affect the nation's economy but also require significant cuts in emissions. However, it was difficult for these experts to represent the general interests of whole industries. In particular, there were no participants from industries such as the energy efficiency improvement and renewable energy industries, which would push for deeper cuts in emissions, or from agricultural, livestock, and fisheries industries, which are likely to suffer significant impact and damage from climate change. In other words, industries that would voice differing opinions and would be more likely to support an active approach were unable to present their views that a failure to make more aggressive efforts to tackle climate change at the national level could cause significant damage. In the case of civil society, it was difficult to have more diverse voices, such as consumer groups, religious groups, agricultural and fisheries organizations, and labor unions, as environmental groups were mainly involved. In addition, dialogue between stakeholders in energy-intensive industries and activists from environmental groups was difficult as they tended to draw parallel lines.

The composition of participants at events held to collect opinions, such as public hearings, parliamentary emergency debates, panel discussions, seminars, and forums, was similar to that of the public-private review team. There was considerable overlap in the interests of participants and speakers, making it difficult to reflect a wide variety of voices. More diverse voices should have been heard from a wider variety of sectors and areas. Furthermore, even though the national reduction goal-setting process needed to be a social learning process if it was to understand climate change and respond with policies and measures to address it, the scope of participants was too narrow to achieve such an outcome. It was also problematic that the role and function of the National Assembly, the representative body of the people, was minimal despite its input being of critical importance: target setting for the reductions is a crucial state affair.

Issues related to principles of INDC establishment

What issues emerged during the process of establishing INDC and by what criteria were INDCs established? Issues include the adequacy of the INDC-setting method, the appropriateness of the overall emissions forecast for the establishment of the INDC, the violation of the principle of "no backslide" in the government's reduction target scenarios, the adequacy of the final INDC itself, and the appropriateness of the means of reduction.

First, the adequacy of the INDC begins with whether it was appropriate to set a reduction level relative to BAU. When setting up INDCs by country, Annex I parties took absolute reductions compared to the base year, while developing countries and non-Annex I parties used a reduction level based on BAU or an energy intensity improvement target method compared to the base year. However, the question has been raised as to whether it was appropriate for Korea to adopt the latter method of setting reduction targets, as it is a member of the OECD and it is therefore difficult to classify Korea as a developing country. In particular, using the BAU scheme to revise the forecasts means that the reduction targets will be revised in line with the 2020 BAU target revisions. The question of the validity of using the BAU method was raised by academics and civic groups during public hearings and parliamentary debates.

Another issue was the appropriateness of the 2030 reduction target level. Actors from industry argued that even the lowest level among the four scenarios the government proposed was difficult. However, actors from civil society complained that even the highest scenario was lower than the 2020 mid-term reduction target announced in 2009, and thus ran counter to the principle of no backsliding agreed in Lima. Korea is not minimally responsible for global emissions, so it should set more aggressive reduction targets. What should the reduction level be?

Were the reduction measures appropriate? The inclusion of nuclear power was particularly controversial. The Park government proposed the construction of additional nuclear power plants as a major mitigation measure to realize the INDC. The government scenario supposed that three or four nuclear power plants would be added. Therefore, the 37 percent reduction target included nuclear power plant construction as a major policy measure. In fact, the energy demand forecast used to estimate the GHG emissions already included the construction of additional nuclear power plants. As the social acceptability of nuclear power had decreased after the Fukushima nuclear power plant accident, it was not easy to obtain national consensus for this construction plan. It was subject to social controversy.

In order to alleviate the burden on industry, the government promised to cap the burden on the industrial sector at 12 percent. In addition, the government decided to make a reduction of only 25.7 percent through the domestic market, with the remaining 11.3 percent coming from purchasing and cancelling the emission rights in overseas markets, effectively paying those expenses for industry. This was also likely to become a subject of controversy. Even with the revised INDC roadmap, it was not clear who was responsible for this payment.

Public climate awareness in South Korea

International evaluation of individual country's NDCs, such as that which Climate Action Tracker provides, is important. However, NDCs +must also be legitimized and accepted domestically. Therefore, public awareness of the urgent need for attention to climate change is critical. According to Yun et al. (2014), two allied networks are concerned with climate policy in Korea: the growth network and the environmental network. These networks are divided and compete against each other. It is important that the public be aware of these two networks.

In South Korea, public awareness of environmental conservation has been surveyed by an external agency hired by the Ministry of Environment every three to five years since 1995. Climate change was included as an issue for the first time in 2008. Since 2012, the state-run Korea Environment Institute (KEI) has conducted surveys annually. In 2013, it included climate change as a special issue, and since 2014, public perception of climate change has been continuously investigated as an important issue. Trends in public perceptions about climate change can be derived from these surveys.

In 2008, 90.9 percent of the public surveyed indicated that global warming/climate change was a serious (including responses of "very serious") environmental problem; this was the highest proportion of any issue, followed by household waste (89.7 percent) and industrial waste (88.7 percent). However, the issue of global warming/climate change as a top priority for the government to deal with. It ranked third (13.5 percent), after household waste (26.3 percent) and protection of the natural environment and ecosystem (19.7 percent). In 2013, global warming/climate change was considered a serious environmental problem by 94.0 percent of those surveyed, second to industrial waste (95.8 percent). Yet again, the issue of global warming/climate change was not regarded as the top governmental priority. Once more it ranked as the third priority, following protection of the natural environment and ecosystem (27.8 percent) and water management (15.8 percent). The most recent survey was carried out in 2018. In that survey, global warming/climate change was selected as a serious environmental problem by 85.1 percent, and it was again ranked as the third most important issue for the government to tackle. These survey results in Table 4.6 imply that the Korean public seemed to be seriously aware of the issue of global warming/climate change but did not regard it as the most urgent policy item for the government to tackle.

KEI's survey results tell a different story. When members of the public were asked to express how serious climate change was, they said it was very serious. However, when they were asked to select the most serious environmental problem, climate change ranked sixth or seventh, as shown in Table 4.7. The average ranking for climate change from 2012 to 2017 was sixth. The public were more concerned about visible and perceivable domestic environmental problems. Furthermore, climate change was not prioritized by the general public over other political and social problems, as shown in Table 4.8. Climate change averaged as the seventh-highest priority from 2012 to 2017. The policy importance given to climate change was significantly lower than other sociopolitical issues. On average, 41.2 percent of respondents indicated unemployment as the top priority, while only 2.6 percent chose climate change.

In the 2018 survey, KEI included more questions about climate change.[2] Concerning climate change-related terminology, 58.0 percent had heard of the Paris Agreement and 26.6 percent knew what it meant. This implies relatively lower recognition of climate change than other environmental terminology, such as green algae (97.2 percent), ecosystem disturbance (94.0 percent), and radon (89.2 percent). When asked to choose three images that came to mind when they heard the term climate change, the damage caused by abnormal weather was the highest, followed by average temperature rise, sea level rise, and GHG emissions rise.

With respect to the seriousness of climate change, two questions were asked: "How serious do you think climate change is at this point?" and "How serious do you think climate change is *for you* at this point in time?" The first question was about the social dimension and the latter was about the personal dimension. The results were 64.3 percent and 88.8 percent, respectively, implying more concern about climate impact on the social dimension than on the personal dimension. Regarding the timing of climate change's negative impact on society, 67 percent said they had already been affected and 18.6 percent were aware it would have a negative impact "within 10 years." This shows that climate change is not viewed only from a long-term perspective.

Concerning actions in response to climate change, positive answers were given. On a five-point scale with one point indicating very little agreement, the sentences "The nation's efforts to reduce greenhouse gas emissions are useless

Table 4.6 Seriousness of environmental issues vs. policy priorities (Unit: %)

	Serious-concern environmental issue						Most urgent policy issue					
	1995	2000	2003	2008	2013	2018	1995	2000	2003	2008	2013	2018
National environment & ecosystem	57.3	93.7	90.6	79.9	92.3	72.6	8.3	8.6	10.5	19.7	27.8	22.6
Air	80.9	97.1	93.4	70.8	92.8	78.6	27.9	25.7	17.6	6.0	10.2	18.6
Water (with underground water)	80.0	97.4	94.2	68.4	86.6	55.3	33.8	25.1	21.2	12.0	15.8	11.1
Soil	–	91.3	88.1	63.2	78.9	50.9	5.3	3.1	4.0	1.2	1.3	3.2
Household waste	68.0	95.4	92.6	89.7	93.5	77.1	24.5	16.9	22.9	26.3	14.4	9.1
Industrial waste	–	94.6	94.1	88.7	95.8	79.6	–	10.2	11.6	10.3	9.1	7.7
Noise & vibration	63.4	83.4	80.6	67.9	80.1	57.3	–	1.3	4.9	1.7	1.1	3.8
Odor	50.3	72.9	69.4	63.3	63.1	49.6	–	0.9	2.3	1.0	0.1	1.4
Toxic chemicals	–	90.3	88.2	83.5	90.1	77.1	–	–	5.0	8.7	4.7	4.1
Global warming/climate change	–	92.0	–	90.9	94.0	85.1	–	3.9	–	13.5	15.7	18.3

Source: The Ministry of Environment, 2019, *Report on the Survey Results of the Public Awareness on Environmental Conservation in 2018.*

Note: In each survey, samples of 1,500 or 2,000 men and 2,000 men and women were selected by region, sex, and age (by age 15 and over), and interviewed using structured questionnaires. Sampling error is ±2.2~2.53 percent, confidence level 95 percent.

Table 4.7 Most serious environmental issues

(Unit: %)

	2012	2013	2014	2015	2016	2017	Ave.
Depletion of natural resources	24.3	24.3	23.5	20.0	16.1	20.1	21.4
Waste increase	22.8	16.4	14.8	22.5	16.2	14.1	17.8
Consumption habit	9.9	9.6	7.6	8.4	3.3	4.8	7.3
Water pollution	9.8	10.1	8.7	9.8	14.7	12.8	11.0
Natural disaster	8.6	8.3	3.5	5.9	13.8	9.3	8.2
Air pollution	5.4	7.9	7.0	10.5	11.9	17.1	10.0
Climate change	**5.0**	**10.5**	**4.4**	**6.5**	**11.3**	**8.2**	**7.7**
Biodiversity loss	4.4	2.8	4.8	4.8	3.2	3.4	3.9
Accidents & health risks from chemicals	3.5	3.3	1.6	2.2	2.4	3.3	2.7
Human-made disaster	2.4	1.9	0.8	1.8	0.5	0.3	1.3
Agricultural pollution	0.9	0.5	0.0	0.5	0.2	1.1	0.5
Noise	0.9	1.9	21.8	1.7	2.9	1.3	5.1
Deterioration of the urban environment	0.8	0.7	0.8	2.1	2.5	2.9	1.6
Expanded use of GMOs	0.7	1.1	0.6	2.2	0.7	0.8	1.0
Car-oriented transportation system	0.7	0.7	0.1	1.3	0.3	0.6	0.6

Source: KEI, *Survey Reports of the Public Awareness on Environmental Conservation from 2012 to 2017.*

Note: For each year's survey, a total of 1,000 men and women were sampled by nationwide adult (age 19-69) by region, gender and age, and then interviewed using structured questionnaires. Sampling error is ±3.1 percent, confidence level 95 percent.

Table 4.8 Top government priorities

(Unit: %)

	2012	2013	2014	2015	2016	2017	Ave.
Unemployment	38.1	39.4	44.0	43.9	44.0	37.9	41.2
Prices	28.0	28.2	21.5	21.3	21.0	24.1	24.0
Welfare and health	8.4	10.4	9.0	11.3	7.7	9.4	9.4
Security	8.0	3.5	4.9	7.1	7.2	6.7	6.2
Environment and pollution	7.6	7.8	10.6	7.8	8.8	7.7	8.4
Climate change	**3.3**	**2.6**	**2.5**	**1.7**	**4.1**	**1.6**	**2.6**
Energy and electricity	2.9	2.9	3.1	1.8	2.1	1.7	2.4
National defense	2.0	1.0	2.2	3.0	1.9	7.1	2.9
Housing	1.0	2.4	0.9	1.1	2.4	2.4	1.7
Transportation	0.7	0.9	0.8	0.8	0.6	0.9	0.8
Foreign workers	0.1	0.6	0.3	0.2	0.2	0.5	0.3
Others	–	0.3	0.2	0.1	–	–	0.1

Source: KEI, *Survey Reports of the Public Awareness on Environmental Conservation from 2012 to 2017*

Note: For each year's survey, a total of 1,000 men and women were sampled by nationwide adult (age 19–69) by region, gender, and age, and then interviewed using structured questionnaires. Sampling error is ±3.1 percent confidence level 95 percent.

because other countries act the other way around," "I don't think my everyday behavior or lifestyle causes climate change," and "If the government makes more efforts to cope with climate change, I will also try" got 2.83, 2.55, and 3.95 points respectively. The most urgent environmental problem to solve was air quality, with a share of 33.6 percent. Solving damage from climate change and climate response was a long way behind as the second highest choice (14.3 percent). Nowadays, fine dust is the prominent issue in South Korea because high concentrations of fine dust last longer. This suggests that environmental issues felt in everyday living conditions have much higher public interest and policy support.

Conclusion

Establishment of the NDC is a critical climate action in every country. Keeping the temperature rise below 2°C, or even further to 1.5°C, is the international community's promise under the Paris Agreement. Because climate policy itself has a big impact on each country's economy and people's lives, there are diverse stakeholders in society whose participation in the decision-making process is crucial.

South Korea's 2030 NDC was established based on public and private actors' co-engagement in the decision-making process. The composition of the public-private review team and its work over a relatively long period (seven months) can be seen as a meaningful attempt at engagement. However, there were many limitations. The range of participating agencies was not diverse. In particular, groups that could benefit from an aggressive response were not invited to be part of the review team. Most participants in the review team and opinion collection processes represented energy-intensive companies and associations with negative attitudes towards climate response, particularly towards establishing a strong NDC. Civil actors mostly came from environmental groups and also did not present diverse voices. The one public hearing was not enough. There was little social dialogue, even though climate change influences all members of society and requires comprehensive social transformation.

Above all, there was little paradigm shift. Climate change is an alert from nature, informing us that the fossil-fuel civilization maintained since the Industrial Revolution is not sustainable. Thus, the answer is a paradigm shift from a system based on fossil fuels to an energy system based on energy conservation, efficiency improvement, and expansion of renewables. Before fossil fuel leaves us, we have to leave fossil fuel. The need for an energy transition that would meet the 2°C imperative can be satisfied with a backcasting approach, in which normative targets are set and ways found to achieve them, rather than with forecasting. However, there has been no urgency in the process of establishing an NDC. Sticking to a short-term perspective, the 2030 NDC process paid most attention to protecting energy-intensive industries and economic growth.

The Moon Jae-in government has adopted the spirit of participatory democracy that had been previously pursued by the Roh Moo-hyun government. Because it took power after the candlelight demonstration that resulted in the impeachment of the President Park Guen-hye, the Moon government takes governance seriously. Also, the Moon government has made post-nuclear energy transition a national task, in which climate change is an underlying driving force. However, there is still tension and conflict between the two allied networks, the growth network and the environmental network (Yun et al., 2014). The general public is another force for deciding future NDCs. According to KEI surveys carried out over a period of several years, the public usually get information about climate change from portal sites and TVs. This implies that the role of media is crucial in delivering the positions and opinions of the aforementioned two networks.

Currently, the Moon government is preparing the 2050 Low-Emission Development Strategies (LEDS), scheduled to be submitted to the UNFCCC by the end of 2020. As part of this process, the Moon government has organized a Low Carbon Society Vision Forum, which is composed of one general team and six individual teams from transformation, industry, transportation, building, waste, and agriculture, livestock, and fisheries. Each individual team has its own technical working group to assess reduction potential and investigate reduction measures. Government officials and actors from business and civic organizations work together in all the teams and working groups. Unlike the previous 2030 NDC process, a youth group has been added to the individual groups because the year 2050 requires reflecting future generations' perspective and wishes. By the end of 2019, the Forum will complete recommendations for the 2050 LEDS and issue it for public comment. Before the submission of recommendations, it plans to hold discussions that engage the public. The youth team will hold a Low Carbon Vision Seminar titled "2050: Future Generations' Expectations." The government plans to develop a final draft after public discussion to collect opinions and consulting with related ministries. This 2050 LEDS is expected to be propelled by a different visions and principles, with contents including emissions target presentation, reduction scenario setting, and emission target determination.

Currently, however, public support for climate policy is not strong enough, despite the perception that climate change is serious. The general public gets information about climate change mainly through internet portal sites (51.4 percent) and TV (36.8 percent), followed by social media (4.0 percent) and printed newspapers and journals (3.8 percent) (KEI, 2018). This means that the role of journalists is as crucial as that of scientists and policy makers. According to Yun's study (2016), based on in-depth interviews about journalists' climate awareness and reporting attitudes, South Korean journalists are generally aware that the climate change issue is "realistic" and feel the need to encourage public interest in it. Those journalists pointed out that most citizens say the climate change issue is important, but they are not active in implementing measures to solve it. Thus, most journalists saw a need for more consistent and in-depth reports. However, they saw lots of barriers ahead. One of the barriers most often pointed out was that the proportion of reports on climate change was not as high as their importance or necessity. This is also linked to the media's tendency to choose more shocking and dramatic elements to enhance commercial appeal, resulting in single-step coverage centered on weather disasters and damage. This tendency is linked to the fact that most respondents in the poll recognized climate change with images of average temperature rise, greenhouse gas increase, damage from abnormal weather, seasonal change, sea level rise, and biodiversity loss. The image of meteorological change and damage exists, but it does not connect well with the effects of climate change on social and economic activities or changes in social structures. Active resolution is needed, but the problem is that climate change is not addressed as a very serious problem that requires changes in the political economy and social and cultural paradigms in the near future. The tendency to perceive it as an out-of-date problem through long periods of single-step reporting has not led to broader social discourse. Therefore, in order for climate change to continue to be publicly debated through the media, newer and more meaningful research needs to be published.

Notes

1 The person who pointed out the problem was the author of this chapter, Sun-Jin Yun. When the government proposal was made public on June 28, the Korea Sustainable Development Solutions Network also said, "The government's 2030 greenhouse gas reduction plan has been set up wrong," adding, "The government should put off submitting its greenhouse gas reduction target to the U.N. until September, when the deadline is set."
2 KEI's 2018 survey method was different from the previous ones. Thus, it is not appropriate to compare 2018 results with others. It was a web survey with a sample of 3,081, not a face-to-face survey with 1,000 samples.

References

Chosun Biz, May 20, 2015, The US 28%, Japan 26%, EU 40% ... Countries Plan to Submit Their Reduction Plan by September. Retrieved from https://biz.chosun.com/site/data/html_dir/2015/05/20/2015052000116.html (in Korean).
Eun, J. H., and Lee, K. H. (2009). *A Study on National Governance*. Paju, S. Korea: Bubmunsa.
Greenhouse Gas Inventory and Research Center (GIR). (2018). *National Greenhouse Gas Inventory Report of Korea 2018*. Retrieved from http://www.gir.go.kr/home/board/read.do?pagerOffset=0&maxPageItems=10&maxIndexPages=10&searchKey=&searchValue=&menuId=36&boardId=49&boardMasterId=2&boardCategoryId=
German Watch. (2019). *Climate Change Performance Index: Results 2019*. Retrieved from https://www.climate-change-performance-index.org/sites/default/files/documents/ccpi-2019-results-190614-web-a4.pdf
International Carbon Action Partnership (ICAP). 2018. *Korea Emissions Trading Scheme*. Retrieved from https://icapcarbonaction.com/en/?option=com_etsmap&task=export&format=pdf&layout=list&systems%5B%5D=47

International Energy Agency (IEA). (2019). *Key World Energy Statistics 2019.* Retrieved from https://webstore.iea.org/key-world-energy-statistics-2019

International Emission Trading Association (IETA). (2016). *Republic of Korea: An Emissions Trading Case Study.* Retrieved from https://www.ieta.org/resources/ Resources/Case_Studies_Worlds_Carbon_Markets/2016/Korean_Case_ Study_2016.pdf

Korea Energy Economics Institute (KEEI). (2018). *Energy Info. Korea 2018.* Retrieved from http://www.keei.re.kr/web_keei/d_results.nsf/0/F8EA7698FE 2397D5492583C50001BBA2/$file/ENERGYINFO2018.PDF

Korea Environment Institute (KEI). (2019). *A Comprehensive Study on the Environmental Value for the Integration of Environment and Economy: 2018 Survey on People's Environmental Awareness.* KEI Project Report 2018-06-03. Retrieved from www.kesis.net/sub/sub_0003.jsp

Korea Environment Institute (KEI). (2017). *A Comprehensive Study on the Environmental Value for the Integration of Environment and Economy: 2017 Survey on People's Environmental Awareness.* KEI Project Report 2017-05-03.

Korea Environment Institute (KEI). (2016). *Public Attitude towards the Environment: 2016 Survey.* KEI Working Paper 2016-19.

Korea Environment Institute (KEI). (2015). *Public Attitude towards the Environment: 2015 Survey.* KEI Working Paper 2015-19.

Korea Environment Institute (KEI). (2014). *Public Attitude towards the Environment: 2014 Survey.* KEI Working Paper 2014-12.

Korea Environment Institute (KEI). (2013). *Public Attitude towards the Environment: 2013 Survey.* KEI Working Paper 2013-22.

Korea Environment Institute (KEI). (2012). *Survey Results Report of Public Attitude towards the Environment (Supplementary Appendix of Project Report of 2012 Comprehensive Study on Green Growth).* KEI Working Paper 2012-15-01

Lee, D. G., and Yun, S. J. (2016). A Comparative Analysis of the Formation of INDCs in Korea and New Zealand. *Korean Society and Administration Study, 27*(2), 261–294.

Lee, J. H., and Yun, S. J. (2011). A Comparative Study on Governance in State Management: Focusing on the Roh Moo-hyun Government and the Lee Myung-bak Government. *Development & Society, 40*(2), 289–318.

Ministry of Environment. (2019). *Report on the Survey Results of the Public Awareness on Environmental Conservation in 2018*

Ministry of Environment. (2018). *Understanding Revised 2030 National Greenhouse Gas Reduction Roadmap.*

Ministry of Environment. (2016). *Press Release: Mid-to-long term strategy and policy direction for effective climate change response at the national level in accordance with the launch of the new climate regime.* December 06, 2016.

Ministry of Environment. (2015). Korea's 2030 Greenhouse Gas Reduction Target was Confirmed to be 37% from BAU. Press release, June 30, 2015. Retrieved from www.me.go.kr/home/web/board/read.do?boardMasterId=1&boardId=53 4080&menuId=286

Ministry of Environment. (2014). *2013 Survey Reports of the Public Awareness on Environmental Conservation.*

Ministry of Environment. (2009). *2008 Survey Reports of the Public Awareness on Environmental Conservation.*

Ministry of Environment. (2003). *2003 Survey Reports of the Public Awareness on Environmental Conservation.*

Namkoong, Keun. (2009). *Policy Studies*, Paju: Byeopmunsa (in Korean).

Organisation for Economic Co-operation and Development (OECD). (2017). *OECD Environmental Performance Reviews: Korea 2017.*

Pierr, Jon, and Peters, Guy. 2000. *Governance, Politics and the State.* Basingstoke: Macmillan Education UK.

Yun, S. J. (2005). "A Study on a Way of Public Participating Energy Governance: Based on an Evaluation of the Citizens' Consensus Conference on Electricity Policies in Korea. Korean Society and Public Administration, 15(4),121–153 (in Korean).

Yun, S. J. (2009). The Ideological Basis and the Reality of "Low Carbon Green Growth, *ECO, 13*(1), 219–266 (in Korean).

Yun, S. J. (2016). The Climate Change Awareness of Korean Journalists and Their Reporting Attitudes. *ECO, 20*(1), 7–61.

Yun, S. J., Ku, D., and Han, J. Y. (2014). Climate Policy Networks in South Korea: Alliances and Conflicts. *Climate Policy, 14*(2), 283–301.

Part II
Risks and transition

5 Climate change governance in China

The role of international organisations in the Guangdong emission trading scheme

Kang Chen and Alex Y. Lo

Introduction

Supported by the reforms and opening-up policy initiated in 1978, China has experienced rapid economic development with average double-digit gross domestic product (GDP) growth over the past three decades (World Economic Outlook Database, 2013). Nevertheless, this remarkable economic growth has incurred serious environmental costs. From 1990 to 2015, Chinese total energy supply increased almost fourfold and reached 119.9 EJ in 2015, accounting for 21.7 percent of the total world energy supply (United Nations Statistics Division, 2018, p. 1). Moreover, the massive consumption of fossil fuels has caused severe air pollution. Less than 1 percent of the 500 largest Chinese cities meet the air quality standards recommended by the World Health Organisation (WHO), and seven cities are ranked amongst the ten most polluted cities in the world (Zhang and Crooks, 2015).

As the world's largest emitter of greenhouse gases (GHGs)[1] since 2007, China is a critical component of human contributions to climate problems. In 2014, China produced 28 percent of the world's total annual emissions (Porter, 2014). Given the international community has exerted increasing pressure on the Chinese government to take more responsibility for climate change mitigation (Kuhn, 2015), establishing an effective climate governing regime is of great significance for the country as well as the Asian region. However, most existing literature on China's climate change governance has concentrated on the state centric approach and government policies (Morton, 2005; Hübler, Voigt and Löschel, 2014; Bäckstrand and Kronsell, 2015). Although some scholars began to pay attention to the participation of non-state actors and explore the collaboration of climate governance in China (Jing, 2015; Lo et al., 2018), the complexity and persistence of climate issues require more diverse and multilevel solutions (Wang, Liu and Wu, 2018).

This chapter aims to explore the latest development of China's climate governing regime, with special focus on the engagement of international organisations. The climate change conference held in Paris (COP 21) was regarded as a historic moment in addressing global climate threats, China's intended nationally determined contribution (INDC) outlined its ambitious commitment to post-2020 global GHG reductions. Specifically, the main targets of China's INDC include achieving peak carbon emissions by around 2030, reducing carbon intensity by 60–65 percent compared to 2005 levels and increasing the share of non-fossil fuel sources in the primary energy mix to approximately 20 percent (Fu, Zou and Liu, 2015). However, as a developing country, China lacks the ability and resources to solve the climate problems alone. However, in order to compensate for their historical GHG emissions, industrial countries have an obligation to support the climate measures of developing countries with technical and financial assistance. The following sections look into the policy experiment of emission trading schemes (ETSs) in Guangdong, China and seek to provide a multi-dimensional analysis of the governance collaboration between Chinese institutions and international organisations.

Climate change and governance in China

Political commitments

Over the past few decades, China's system of environmental governance has grown, developed and transformed at the subnational, national and international levels (Mol and Carter, 2006). Wu (2009) indicated that few studies on China's environmental politics before its economic reforms are available. In Mao's China, although the problems of environmental pollution and material consumption were highlighted, environmental governance was mainly ignored (Zang, 2009). Before the 1990s, economic growth was a national political priority, and the abuse of natural resources was viewed as an acceptable side-effect (Shapiro, 2001). Environmental protection was finally placed on the political agenda as China's environmental problems worsened with the acceleration of industrialisation and urbanisation (Schröder, 2012). China began to build its environmental regulatory system in 1979 when the state Environmental Protection Law was issued. Since the 1990s, an increasing number of the rhetorical expressions of environmental protection have been found in China's five-year plans and environmental regulations. From this perspective, environmental governance has attained equal status with the previous political priority of economic growth (Zhang and Vertinsky, 1999).

China's attitude towards climate change governance has also transformed over recent decades. The Chinese government was initially sceptical about climate issues and condemned attempts to 'intervene in sovereign affairs' (Ross, 1998). China's attitude became cooperative in the 1990s, and the establishment of the National Climate Change Coordination Group in 1993 indicated that the Chinese government had begun to attach importance to climate change issues. In the following year, the concept of climate change adaptation was first proposed in the national China Agenda 21 (Hu and Guan, 2017). China later became one of the first countries to sign the United Nations Framework Convention on Climate Change (UNFCCC) and the Kyoto Protocol (Schröder, 2012). Since 2007, China has been under considerable international pressure to increase its participation in climate change mitigation and to establish ambitious GHG reduction targets given its rapidly increasing emissions (Qi, 2013). Then, President Hu Jintao explicitly proposed to 'strengthen capacity to address climate change and make new contributions to protecting global climate' during the 17th Communist Party of China National Congress in 2007 (Xinhua, 2007). In November 2014, Chinese President Xi Jinping announced that China would endeavour to maximise CO_2 emissions by 2030. The targets of China's INDC include decreasing energy intensity to 60–65 percent of 2005 levels and increasing the share of non-fossil energy to 20 percent of the total primary energy supply (NDRC, 2015). Although China's annual CO_2 emissions more than quadrupled between 1990 and 2016 with the economic boom, the country's CO_2 emissions per unit of GDP dropped sharply with each passing year (Figure 5.1). Indeed, the relative decoupling of economic growth from environmental pressures and resource use has started to happen in China, particularly in the more prosperous eastern coastal regions.

From government control to market mechanism

Morton (2005, p. 3) defined the Chinese government's environmental system as the 'state-driven environmental management.' Governmental actors at the central level are policy makers. They propose guiding principles and abstract

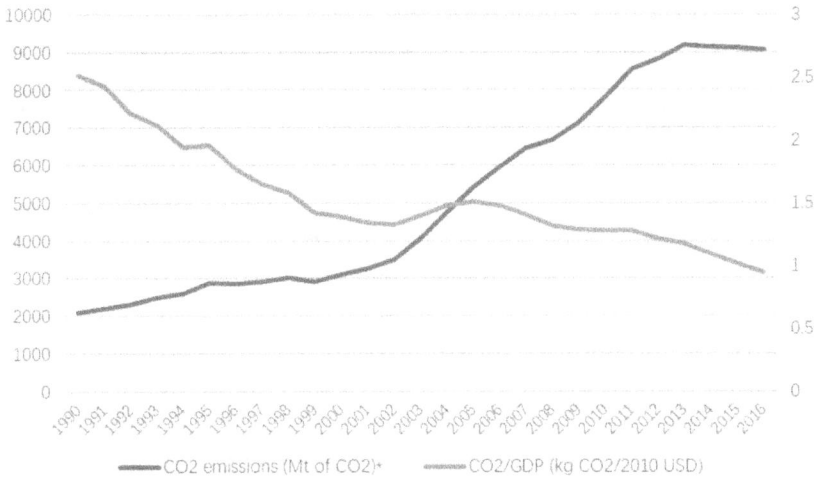

Figure 5.1 CO_2 emissions and CO_2 emissions per unit of GDP P.R. China, 1990–2016

Source: International Energy Agency (IEA), *Key World Energy Statistics 2018*, https://www.iea.org/statistics.

* CO_2 emissions from fuel combustion only. Emissions are calculated using the IEA's energy balances and the 2006 IPCC Guidelines

concepts to instruct environmental law-making and policy implementation. Moreover, the central authority controls local personnel appointment because it has the right to appoint, promote or remove local officials. By contrast, local officials are policy implementers who are motivated to act in ways consistent with meeting the goals set by the central authority (Zhou et al., 2013). This command-and-control approach has been widely employed in addressing environmental hazards in China through energy conservation, pollution monitoring, waste disposal, pollution discharge permit systems and environmental impact assessment (Zhu and Zhou, 2017). However, given the lack of sufficient resources, conflicting interests and inadequate incentive measures, the local implementation of national environmental policies can be problematic (Schröder, 2012; Tang et al., 1997; Ma, 2017; Ma and Ortolano, 2000), and local bureaucratic behaviours, such as 'bargaining' (Wu and Zheng, 1995) and 'muddling through' (Zhou et al., 2013), have emerged.

Economic reforms have stimulated the increasing use of the market mechanism to set prices and allocate goods and services, which also profoundly influence environmental and climate governance in China (Dickson and Chao, 2001). China currently utilises a mixture of government control and market-oriented instruments to address the increasing damage to its environment (Shuwen, 2004). As summarised by Economy (2007), China has adopted several market-based instruments for environmental governance. These instruments include subsidies, environmental funds, price regulation, taxes on natural resource consumption and emission charges and standards. Service contracting is one of the most commonly used approaches. Given that governmental organisations may lack the capacity and resources to cope with local issues, problems can be effectively solved by purchasing services from external collaborators (Jing, 2015). One example is the financial subsidy provided by the Chinese government. In accordance with related regulations, most commercial users can receive a subsidy of 30 percent on each highly efficient lighting product, whereas residential users receive a subsidy of 50 percent (West et al., 2013). The government has also made progress towards environmental tax reform. In 2006, taxation on consumption in China was modified by imposing high taxes on oversized and energy-inefficient vehicles (Zhou, Levine and Price, 2010).

Domestic and foreign participants

In addition to governmental actors, nonstate institutes, such as academics, think tanks, business actors, consultancies, nongovernmental organisations (NGOs), social groups and individuals, contribute their efforts into climate governance in China. The increasing impact of China's nonstate actors on regional, national and international climate governance has become a core interest in contemporary studies on domestic participants in global environmental politics (Koehn, 2016; Keck and Sikkink, 1998; Wang, Liu and Wu, 2018). Since 1994, environmental nongovernmental organisations (ENGOs) have been the most active NGOs with a profound social influence in China (Schwartz, 2004). Apart from NGOs, two of the most important types of civil society actors with pivotal roles in climate governance are student associations and academic organisations (Liu, Wang and Wu, 2017). The actors are closely associated with the interests of government bodies and have been defined as semi-official institutes or hybrid actors, such as public service agencies (*shiye danwei*), government-organised nongovernmental organisations (GONGOs), state-owned enterprises (SOEs) and other government-sponsored institutes. As acknowledged by Francesch-Huidobro and Mai (2012), some regional GONGOs that were established to promote a low-carbon economy have made valuable contributions to local climate governance.

Climate governance in China also involves foreign participants. Zhang and Wen (2008) indicated that the Chinese government welcomes innovative political instruments and technical means to facilitate collaboration between domestic stakeholders and international partners because China lacks sufficient experience and advanced technology to address climate issues alone. International NGOs, like the Environmental Defense Fund, with abundant funds, rich experience and well-connected experts, play an active role in strengthening China's climate governance. Apart from NGOs, other international bodies, including government actors, research institutes, private consultancies and business enterprises, also participate in the activities of climate change mitigation and adaptation in China. By holding events for global climate governance, introducing environmental management experience, providing low-carbon training and organising overseas field trips, international organisations have built close relationships with Chinese partners (Chen, 2010). Considering the limited attention that has been given to international entities, their impact on the climate governing regime in China warrants further elaboration.

Policy experimentation

Policy experimentation has been applied by the Chinese government for many decades (Saich, 2004). As a large nation experiencing rapid transition, China follows the ideology of gradualism, that is, the adoption of a step-by-step reform approach without revolution (Naughton, 1995). This developmental path was characterised by the great reformer Deng Xiaoping as 'crossing the river by feeling for stones'. Consequently, local experiments with pilot policies have become important and necessary to determine the future direction of national policy decisions (Qi et al., 2008). Any innovative policy will first be implemented in local communities; once the central leadership recognises a positive outcome, it will be enforced throughout the country (Saich, 2004). Along with decentralisation, subnational units receive increased authority and discretion from the central government to allow for effective local experimentation (Lau, Qian and Roland, 2000).

Policy experimentation and pilot projects have been initiated in many domains. For example, market mechanisms for coping with climate change have been introduced in recent years, and most of these mechanisms are implemented on regional or local levels as experimental programmes. O'Connor (1996) stated that China began to experiment with economic instruments to impose pollution charges as early as 1982. This approach, however, was ineffective. A later experience with sulphur dioxide (SO_2) emission trading was also unsatisfactory (Chang and Wang, 2010). The expected market-oriented instrument became a state-led 'pseudo market' because of governmental interventions (Tao and Mah, 2009, p. 175). By contrast, the policy experiment of establishing provincial clean development mechanism (CDM) centres to control GHG emissions through a market mechanism has achieved considerable success in 27 Chinese provinces (Schröder, 2012).

More recently, Chinese policy makers set up a domestic carbon trading system. In 2011, the Chinese government initiated seven pilot carbon ETS programmes in two provinces (Guangdong and Hubei) and five cities (Beijing, Tianjin, Shanghai, Chongqing and Shenzhen). Based on the productive experience of pilot ETSs, Fujian Province started its own ETS in 2016 and became the first carbon market in a non-pilot region. The following section will elaborate and reflect on the development of China's ETS experimentation.

Guangdong case introduction

Carbon emissions trading

Carbon markets are where allowances to emit GHGs or credits earned by avoiding or sinking GHGs are traded. Each allowance and credit is measured on the basis of 1 metric tonne of CO_2 equivalent (tCO_2e). Two mechanisms are generally available for emission trading. The first one is the allowance trading mechanism, which is also known as the cap-and-trade regime and is widely applied globally. Another is the crediting mechanism, also known as the baseline-and-credit system, which is mainly used in the CDM and Joint Implementation under the Kyoto Protocol (Gao et al., 2016). Kossoy and Guigon (2012) stated that in 2011, the carbon market traded US$176 billion of emission allowances and credits worldwide. Prior to the Kyoto Protocol the European Union Emissions Trading System (EU-ETS) was the most famous trading system that drove the global carbon market. The European Community issued the directive to create the EU-ETS in July 2003, and the European carbon market was launched in January 2005. Outside Europe, ETSs are currently operating in New Zealand, South Korea, Kazakhstan, California and ten northeastern states in the United States, Tokyo (Japan), Quebec (Canada) and several Chinese cities and provinces (Lo, 2016).

China first entered the carbon trading community as a primary producer of compliance credits under the CDM of the Kyoto Protocol. On the basis of the advantages of enormous GHG emission rights, technical strength, low-risk and relatively easy access to project financing, China became the world's number one CDM host country in 2007 and continues to account for the majority of CDM projects and certificates in the global market (UNEP Risoe, 2008). After the success of CDM, policy makers in China decided to embark on another climate market experiment on an even larger scale. In October 2011, the National Development and Reform Commission (NDRC, 2011) initiated seven pilot ETS programmes in Guangdong and Hubei Provinces and in the cities of Beijing, Tianjin, Shanghai, Chongqing and Shenzhen. The short-term goal of ETS pilots was to establish transprovincial and transregional ETSs that would be integrated into a national scheme by 2015–2016. After several years of development, the implementation experience at the regional level has been extracted by the central government. On December 19, 2017, the China NDRC announced the establishment of China's national ETS and released the National ETS Construction Programme (Deng et al., 2018). Table 5.1 gives some background information on the seven pilot ETSs as well as the new-born national carbon market (power generation industry only). As one of the two provincial carbon trading experiments, the Guangdong ETS is characterised by a stable market environment and sound regulatory system. The following paragraphs provide a brief introduction to Guangdong Province and an explication of the construction of the Guangdong ETS.

Introduction to Guangdong

Located near the South China Sea and encompassing the Pearl River Delta, Guangdong (formerly Canton) Province is one of three vast urban agglomerations in China. It is a highly industrialised and urbanised province and is home to over 113 million people. As the largest single economy in China, Guangdong's total GDP reached US$1414.8 billion in 2018 and accounted for over 10 percent of the country's total annual GDP (Table 5.2). Guangdong Province has been

Table 5.1 Background information on China's carbon markets in 2017

	Total market of China's seven pilot ETSs	China's national carbon market
Population (million people)	260	1390
GDP (trillion Chinese Yuan)	20.30	82.71
Annual allowance auction price (Chinese Yuan per ton CO_2e in China)	1–123	–
Amount of carbon emissions covered by carbon market (million tons CO_2e)	≈1280	≈3500 (it only covers power generation industry at present)
Carbon emissions covered by carbon market as a percentage of total carbon emissions	≈50 percent	>30 percent

Source: Environmental Defense Fund (2018). *The Progress of China's Carbon Market 2017.* Beijing. Available at https://www.edf.org/sites/default/files/documents/The_Progress_of_Chinas_Carbon_Market_Development_English_Version.pdf. Accessed 26 September 2018.

Table 5.2 Key features of Guangdong Province

Guangdong Province	
Total population (2018) (percentage of country)[a]	113 million (8.1%)
GDP (2018) (percentage of country)[a]	US$ 1414.8 billion (10.8%)
GDP per capita[a]	US$ 12,568
Overall GHG emissions (excluding LULUCF) (2012) (percentage of country)[b]	610.5 MtCO$_2$e (5.6%)

Sources:

a National Bureau of Statistics of China (data.stats.gov.cn). The GDP estimates originally in Chinese yuan were converted to US dollars at the rate of 6.8755 (USD/CNY) based on the Federal Reverse's exchange rate records (31 December 2018).

b International Carbon Action Partnership (ICAP) (icapcarbonaction.com). Accessed 27 August 2019.

known as the heartland of marketisation experiments since the opening-up policy was launched in the early 1980s. Relying on its superior location, resources and policies, Guangdong has constructed its vanguard status, and numerous pioneering reformist policies were first implemented in Guangdong Province prior to their national extension (Chen et al., 2017). Therefore, Guangdong is usually regarded as a showcase for China's economic reform achievements and is seen as having the highest level of marketisation of all the provinces in the country.

Nevertheless, Guangdong Province faces tremendous challenges in restraining its soaring energy consumption, air pollution and GHG emission rates. After three decades of rapid development, Guangdong has become home to nearly all kinds of industries, including high-energy industrial sectors such as petrochemicals, metallurgy and electronics. The province is also one of the largest emitter provinces of GHG in China. In 2012, Guangdong Province generated 610.5 million tonnes of carbon-equivalent emissions, which accounted for approximately 5.6 percent of the country's total emissions (Table 5.2). In contrast to bottom-up economic reforms, Guangdong's environmental governance follows the traditional command-and-control system. Although the regulatory mandate could receive immediate effect, eliciting local interest is difficult, and enforceability is weak (Kostka, 2016). In 2011, the NDRC selected Guangdong as one of the seven pilots to adopt the ETS policy. However, officers in Guangdong did not treat ETS adoption as a mere political task allocated by higher government levels. Given that the establishment of the carbon trading market also meets Guangdong's need for low-carbon transformation, the local authority attaches considerable importance to the construction of ETS (Chen et al., 2017).

Guangdong ETS

Guangdong officially launched its ETS in December 2013. In the first compliance year, four industries covering electricity generation, iron and steel, cement

Table 5.3 Key features of Guangdong ETS

Guangdong ETS	
Launch date	19 December 2013
Percentage of GHG emissions covered by the ETS[a]	60%
Cap on GHG emissions (2017)[a]	422 MtCO$_2$e
Sectors covered[b]	Electricity generation, Iron and steel, Cement, Petrochemicals, Pulp and paper and Aviation
Number of compliance enterprises[c]	Electricity generation, Iron and steel, Cement, Petrochemicals, Pulp and paper and Aviation Total (2017): 296 Existing entities (2017): 246 New entrants (2017): 50
Inclusion thresholds[b]	20,000t CO$_2$/year or energy consumption 10,000 tons coal equivalent (tce)/year
Trading mode	Spot trading, block trades, negotiated transfers
Banking and borrowing of allowances[d]	Banking is allowed during the pilot phase Borrowing not allowed
Offsets and credits[d]	Domestic project-based carbon offset credits – China Certified Emission Reduction (CCER) – are allowed. The use of CCER credits is limited to 10% of the annual compliance obligation. At least 70% of CCERs need to come from Guangdong. Pre CDM credits are not eligible, especially the credits that come from hydropower and most fossil fuel projects.
Total trading amount (percentage of seven ETS pilots and ranking)[e]	80.91 million tons CO$_2$ (33.7%, 1st)
Total turnover (percentage of seven ETS pilots and ranking)[e]	US$ 248.70 million (32.1%, 1st)
Compliance rate[e]	100% (three consecutive years from 2014)

Sources:

a International Carbon Action Partnership (ICAP) (icapcarbonaction.com). Accessed 23 August 2017.

b Guangdong Development and Reform Commission (DRC) (2016). Notice on the Allocation of Carbon Emission Allowances in Guangdong Province 2016. Guangzhou: People's Government of Guangdong Province, People's Republic of China. Available at http://dtfz.ccchina.gov.cn/archiver/LowCD/UpFile/Files/Default/20160714093558793864.pdf [in Chinese]. Accessed 23 August 2017.

c Guangdong Development and Reform Commission (DRC) (2017). Notice on the Allocation of Carbon Emission Allowances of Pulp and Paper, Aviation and White Cement in Guangdong Province 2016. Guangzhou: People's Government of Guangdong Province, People's Republic of China. Available at www.cnemission.com/article/news/ssdt/201701/20170100001201.shtml [in Chinese]. Accessed 23 August 2017.

d People's Government of Guangdong Province (2014). Trial Methods of Emission Management of Guangdong (No. 197, Decree of Guangdong Provincial Government). Guangzhou: People's Government of Guangdong Province, People's Republic of China. Available at http://zwgk.gd.gov.cn/006939748/201401/t20140117_462131.html [in Chinese]. Accessed 23 August 2017.

e Guangdong Research Center for Climate Change (GDRCCC), 2017, *Guangdong Pilot Emissions Trading Scheme Report (2017–2018)*. Guangzhou: Guangdong Provincial Development and Reform Commission.

and petrochemicals were included in the trial ETS programme. All enterprises with annual emissions exceeding 20,000 tons of CO_2 in the four industries were listed as compliance enterprises and accounted for approximately 55 percent of the total emissions in Guangdong Province (Guangdong DRC, 2013). A conservative approach was employed by Guangdong Development and Reform Commission (DRC) to measure and examine industrial emissions with convenient methodologies. Therefore, only core companies were included in the pilot ETS, and the estimated emission cap was acceptable for most participants; this approach proved to be crucial to the operational stability of ETS at the initial stage (Chen et al., 2017). A combination of free allocation and an auctioning mechanism for allowance allocation is adopted in the Guangdong ETS. Paid allowance is allocated by auction, and compliance enterprises can decide whether or not to buy; this approach contributes to the theoretical identification of the proper price for carbon emissions (Environomist, 2017).

Regarding the institutional setting, a multitiered management system was established to direct the Guangdong ETS (see Figure 5.2). On the top level, a leading group was set up in 2012 to accelerate the establishment of the Guangdong ETS. This leading group controls low-carbon development and emissions trading in Guangdong, and the governor of Guangdong Province serves as the group head. The joint meeting mechanism aims to oversee Guangdong's construction as a low-carbon pilot province, and the executive vice governor serves as the principal convener. Below is Guangdong Provincial DRC, which is the authority department of the pilot ETS. It also established the climate change division, a new branch that addresses climate change and is responsible for the actual

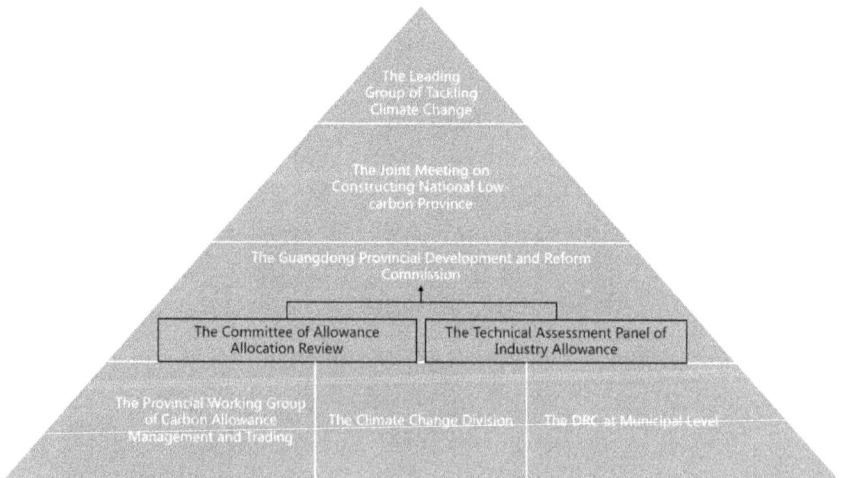

Figure 5.2 Emissions trading scheme management system of Guangdong

Source: Adapted from Guangdong Research Centre for Climate Change, *Guangdong Pilot Emissions Trading Scheme Report (2014–2015)*, p. 13.

management of the Guangdong trial carbon market. In addition, low-carbon experts and technicians were convened by Guangdong DRC to set up the provincial working group of carbon allowance management and trading. The working group is responsible for the study and design of the ETS mechanism, the operation of the trial carbon market and the supervision of compliance enterprises. Moreover, as the subordinate agencies of provincial DRC, municipal DRCs take charge of the local implementation of ETS policy and regulation and also facilitate carbon emission assessment, monitoring, reporting and verification.

Two organisations were created by Guangdong DRC to support industry allowance allocation. One is called the Technical Assessment Panel of Industry Allowance, which consists of industry experts and company representatives who are tasked with collecting industry feedback on allowance allocation and creating the corresponding modifications. The panel members submit assessment reports to Guangdong DRC and provide suggestions for allowance adjustment for different industries. Another assisting organisation is the Committee of Allowance Allocation Review, which is responsible for the review of allowance allocation proposals. The committee members include competent authority officials, low-carbon experts and representatives from different industry associations and enterprises, with no less than two-thirds of experts. The two mechanisms provide a communication channel between industrial actors and government officials and strive to ensure the scientific and fair distribution of carbon emission allowances (GDRCCC, 2015).

Governance in the Guangdong ETS

As demonstrated in Figure 5.3, the participants of the carbon market in the Guangdong pilot ETS can be categorised into ten groups. The blue images

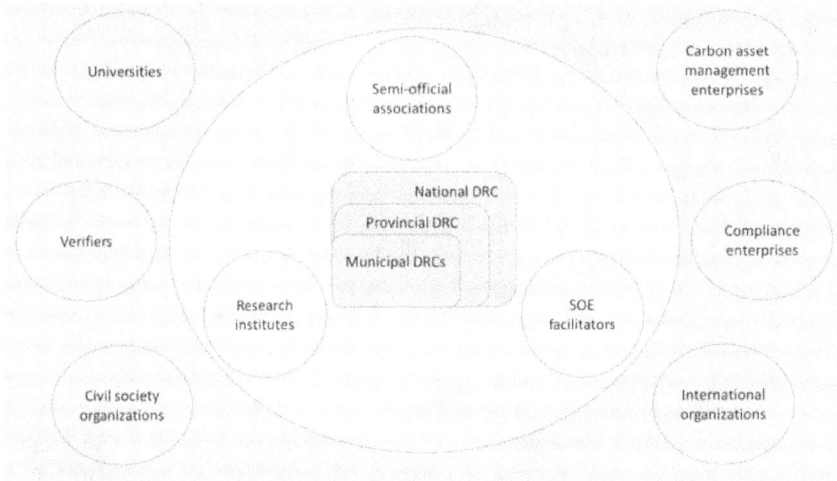

Figure 5.3 Major participants of the Guangdong ETS

represent the competent authority of the Guangdong pilot ETS in the core position. The competent authority is the Development and Reform Commissions (DRCs) from the national, provincial and municipal levels. Moving to the broad central area, the yellow circles represent the supporting organisations of Guangdong pilot ETS governance. The supporting organisations include research institutes, SOE facilitators and semi-official associations. Finally, the outer green circles represent the remainder of the participating institutions, such as universities, verification institutions, civil society organisations, private carbon asset management enterprises, compliance enterprises and international organisations, in the carbon market of the Guangdong pilot ETS. The participating roles of these organisations in Guangdong pilot ETS will be elaborated on in the following paragraphs.

The competent authority of climate governance in the Guangdong ETS encountered the problem of overlapping steering functions between the DRC and the Economic and Information Commission (EIC) until 2014, when the climate change division was established under the provincial DRC (Lo et al., 2018). At the national level, the department of climate change under NDRC took charge of eight ETS pilots and prepared for the national ETS. At a low administration level, a relatively high degree of autonomy was granted to Guangdong DRC from the top-tier authority, and the climate change division took full responsibility for managing the ETS experiment (Zhang, 2015). However, no climate change division was established in municipal DRCs, and the main tasks for local DRCs include policy implementation and updating the latest emission data.

Three types of ETS supporting organisations are defined in this study. These organisations provide theoretical, technical and publicity assistance to the development of carbon trading in Guangdong. The research institutes play a crucial role in mechanism design and system construction. The scope of research institutes adopted herein was limited to the *shiye danwei* (public service agency), which is often established by the state and offers research-based public services. As a relic of the Chinese Communist system, the *danwei* (unit) system currently faces reforms towards a market economy (Schröder, 2012). The *shiye danwei* was forced to become financially and operationally independent of government organisations. The SOE facilitator constitutes the second kind of supporting organisation for the Guangdong pilot ETS. Specifically, the definition of SOE facilitator implies that SOEs and their affiliations facilitate the mechanism construction of the Guangdong pilot ETS and provide technical assistance. The third type of supporting organisations in the Guangdong pilot ETS are the semi-official associations, which take responsibility for connecting government and industrial actors and publicising low carbon concepts. This group concept has two implications. First, the identified group member should be an association, rather than an enterprise or a public service agency. Secondly, the semi-official attribute implies that the association has relatively a close relationship with governmental agencies. To summarise, these three types of ETS supporting organisations present hybrid institutional characteristics. Although they normally operate independently, the government can offer numerous resources for their development and exert considerable influence on them.

In the outer domain, the remaining participants fall into six categories. The universities mentioned in this study refer to educational institutions with specialists in carbon emission trading and low-carbon development. These universities participate in the Guangdong pilot ETS by offering professional consultation, assisting low carbon project review and backing up capacity building activities. The verifiers are mainly third-party organisations responsible for the data verification of carbon emissions in different industrial sectors. The civil society organisation category is used to describe social and civilian groups that aim for environmental protection and climate change mitigation and adaptation. The carbon asset management enterprises are private actors, whose main businesses encompass low-carbon consultation, capacity training, emission trading and voluntary GHG emission reduction project development. The compliance enterprises are the chosen industrial actors that participate in the Guangdong pilot ETS. They are allocated with emission allowances and are required to fulfil emission reduction obligations during each compliance year. The final category is international organisations, the overseas actors play active roles in the Guangdong ETS construction, their engagements are described more fully in the following section.

Participation of the international organisations

The international participants of the Guangdong ETS include government organisations, business actors, universities and NGOs. One of the most influential overseas institutions in the Guangdong carbon emission market is the British Consulate-General Guangzhou (BCG Guangzhou). The climate change and energy division of BCG Guangzhou plays a substantial role in the development of the Guangdong pilot ETS by providing continuous financial and expertise support. In addition, by holding study visits, supporting research projects, organising capacity training and conducting exchange activities, other overseas actors in the pilot ETS such as the German Corporation for International Cooperation (GIZ), the University of Edinburgh, ICIS and BP have established good relationships with the competent authority in Guangdong and introduced international enterprises and research institutions to the Chinese carbon emission trading pilot. Generally speaking, the major role of international organisations in the Guangdong ETS is twofold: capacity building and participating in rule setting indirectly. These two aspects are explicated below.

Engage in capacity building actively

The climate change and energy division of BCG Guangzhou is an important 'transnational actor' and 'sponsor' in the development of Guangdong pilot ETS (Chen, 2017). Supported by the UK Foreign and Commonwealth Office, the China Prosperity Strategic Fund (SPF) is the primary financial source for BCG

Guangzhou to support the carbon trading projects in Guangdong. Clean and low carbon transition is one of the five policy goals of the SPF in China; specifically, it aims to boost green growth and help to prevent dangerous climate change. Apart from the fund support, BCG Guangzhou has advantageous low carbon expertise and bilateral contact resources. Thereby the BCG Guangzhou plays a substantial role in the capacity building activities of the Guangdong ETS.

One example of this is the study tour. The Guangdong provincial government placed high importance on the carbon market establishment after NDRC announced the pilot ETS programme. However, at that time, carbon emission trading was a new concept to most governmental officials as well as experts from supporting organisations. The Guangdong ETS working group was keen to learn from the experience of mature carbon market traders like EU-ETS.

> At the early stage, the preparatory work of Guangdong ETS met many difficulties. The BCG Guangzhou arranged a study tour for us (government officials and supporting organisation experts). We went to Britain and Belgium to learn the experiences of EU-ETS, during the 6 days' visit we attended 14 seminars with the EU officials and experts from government, business and research organisations, we learnt a lot.
>
> (Interview with the research institute, August 2017)

Apart from the study tour, the BCG Guangzhou also organises low carbon research projects to enhance the capability of the Guangdong ETS. By March 2017, a total number of eight research projects in the Guangdong ETS had been initiated by the BCG Guangzhou. These projects are all funded by the SPF and the participants include research institutes, universities and business actors from the UK and Guangdong.

Another important approach to capacity building in the Guangdong ETS is holding training sessions and workshops. Given that international organisations have sufficient funds, advanced technology and look for opportunities to cooperate with Chinese partners, overseas players are keen to become involved in carbon market training sessions in Guangdong. One of the most influential institutions is the Deutsche Gesellschaft für Internationale Zusammenarbeit (GIZ) GmbH, it is a German federally owned enterprise concentrating on international cooperation and professional training for sustainable development. In 2012, the GIZ launched the project to improve capacity building for the establishment of emissions trading schemes in China. As the largest pilot ETS in China, the trial carbon market in Guangdong is one of the main focuses of the GIZ. On behalf of the German Federal Ministry for the Environment, Nature Conservation, Building and Nuclear Safety (BMUB), the GIZ acts as a project practitioner with the support of the German government. The training programme strives to support Chinese stakeholders to learn from the experience of carbon emission trading in German and Europe and to build up a carbon market that is appropriate to China's circumstances.

Our lead executive agency in China is the department of climate change under the NDRC, with its permission, we collaborate with the local stakeholders and organise trainings in the pilot ETS. We invite experts from German and European institutions to introduce carbon trading theory, EU-ETS system building, and the technical skills of carbon emissions monitoring, reporting and verification.

(Interview with the GIZ, September 2017)

With the development of the ETS experiment, international business actors also actively participate in capacity building activities in the Guangdong ETS. ICIS is the world's largest petrochemical market information provider. ICIS' carbon team concentrates on carbon price forecasting and analysis for major ETSs around the world covering EU-ETS, Californian-ETS, Australian-ETS, Reginal Greenhouse Gas Initiative (RGGI) and Chinese-ETS. The ICIS team set up its Chinese office at the end of 2013. It provides price assessment, market analysis and trading consultancy targeting the ETS pilots in China. In Guangdong, ICIS not only provides expertise in the training workshops, but also collaborates with the core supporting organisations to summarise the development of the Guangdong ETS. In 2013, ICIS wrote the *Guangdong Pilot Emissions Trading Scheme Report (2013–2014)* in conjunction with Sun Yat-sen University (Research Center of Low Carbon Technology and Economy); in 2017, ICIS collaborated with the China Emissions Exchange (CEEX) in Guangzhou to publish the *Investment Analysis of Carbon Market in China and the Guangdong ETS Assessment (2013–2015)*.

It is noteworthy that, although ICIS pays close attention to carbon trading activities, it does not engage in market transactions directly. The energy investment divisions of other international enterprises like BP and Shell are eager to participate in capacity training programmes in the Guangdong ETS. However, most of their workshops are for compliance enterprises and investment organisations targeting carbon trading skills to explore new clients and expand the Chinese market.

Participate in rule setting indirectly

The participation of international organisations not only contributes to the capacity building of the Guangdong ETS, but also has an indirect impact on the rule setting of the trial carbon market. The lack of professional capability and resources of the Chinese government generates an opportunity for overseas players to take part in the pilot ETS governance. The international actors are only too keen to maximise their influence and play more significant roles in the emerging Chinese carbon market.

The policy impact of the SPF project is valued highly by the British government. We hope the supported projects can have significant effects on the local climate governance. Therefore, our Chinese partners should have close relationships with local authority and have certain political influence. We even negotiate with the Guangdong DRC while drafting the project proposal to make sure the research outcomes can make contributions to the pilot ETS development.

(Interview with the BCG Guangzhou, July 2017)

Looking at the eight SPF projects in the Guangdong ETS, the research focus closely follows the development of the trial carbon market. At the pilot ETS preparatory stage, the SPF projects were concerned with building the carbon trading system and the carbon emission allowance allocation. When it moved on to the ETS construction stage, capacity building as well as the technical problems of measurement, reporting and verification (MRV) became the prioritised topics. After the Guangdong ETS entered the further development stage, the project emphasis was transferred to the compatibility of different pilots and the innovative green financial tools. Since most of the Chinese partners of the SPF projects are core supporting organisations of the Guangdong ETS and directly participate in the trial carbon market building, the influence of those policy-oriented projects should not be ignored.

The GIZ also attaches great importance to the political impact of its programmes. Compared with the BCG Guangzhou, the German institution puts more focus on addressing the practical problems of China's pilot ETS system by using domestic and European know-how and technologies. In addition to holding capacity training sessions, the GIZ also translated many EU-ETS policy documents as well as technical guidelines to provide reference materials for their Chinese partners. In 2016, the GIZ team translated the book *Emissions Trading in Practice: A Handbook on Design and Implementation*, which rectified the absence of professional textbooks in China. In addition, to enhance its political impact, the GIZ focuses on the key stakeholders of China's ETS pilots such as governmental officials, experts from core supporting organisations and technicians of the compliance industries.

> When the Chinese governments meet practical difficulties, we offer possible solutions and leave them to make decisions, we only function as the service provider. However, since our experts are all quite experienced and many German companies have world-leading technology, our proposed solutions usually receive positive feedbacks from the Chinese governments. We just invited experts from the Deutsche Lufthansa AG to provide suggestions for engaging the airline companies into China's pilot ETS.
>
> (Interview with the GIZ, September 2017)

Through the preliminary collaboration between capacity building activities in the research institutes and the SOE supporters in Guangdong, ICIS established close relationships with some core players in the pilot ETS. In addition, the ICIS carbon team has participated in the EU-ETS for many years and accumulated a great deal of experience in institution design. Therefore, when the

Guangdong DRC held consultation meetings about the trial carbon market, ICIS staff had the opportunity to share their knowledge with the core builders of the Guangdong ETS. Some of their suggestions were accepted by the Guangdong government and affected the pilot ETS rule setting process.

As the only trial carbon market in China with a regular auction of emission allowances, the Guangdong ETS adopts many innovative strategies to stabilise the market operation. The international organisations contribute by sharing experience of the carbon market and strengthening the construction of ETS institutions.

> When introducing the EU-ETS system to the Guangdong DRC, we proposed to set a reserve price for each auction. This approach is quite essential for the nascent market like the Guangdong ETS, because it contributes to the stability of carbon price. The Guangdong DRC finally took our suggestion and some of our colleagues also came to Guangdong to help establish the floor pricing system.
>
> (Interview with ICIS, October 2017)

Conclusions

In this chapter we explore issues about the climate governing regime in China to reduce carbon emissions. We review the development of climate change and governance in China and summarise the changes from four aspects. First, the attitude of the Chinese government towards climate change has transformed from one of skepticism to one of positivity. Secondly, the political approach has shifted from government control to a market-oriented mechanism. Thirdly, an increasing number of participants from home and abroad have begun to engage in climate change adaptation. Fourthly, policy experimentation and pilot projects have been initiated in local areas to find effective methods of addressing climate change problems. In 2011, the Chinese government launched a large-scale ETS experiment programme in seven pilot areas. As the most influential trial carbon market, the Guangdong ETS was selected as our research subject to explore the local approach governing climate change.

We identify ten groups of participating organisations for the Guangdong ETS including the competent authority, research institutes, SOE facilitators, semi-official associations, universities, verifiers, civil society organisations, carbon asset management enterprises, compliance enterprises and international organisations. Realising limited attention has been put on the governance engagement of international entities, we place the research focus on the overseas participants of the Guangdong ETS governance. Our findings reveal that the international organisations not only play an active role in the pilot ETS capacity building, but also have an indirect impact on the rule setting of the trial carbon market. By offering financial assistance and sharing their expertise, the international players make a contribution to the establishment of the carbon emissions trading system in China. Moreover, the international participants also strengthen their influence in China's climate governance and assist their domestic partners to expand new markets in China.

Note

1 The major GHGs are methane, nitrous oxide, the halocarbons and other fluorinated gases, and carbon dioxide (CO_2). The most important anthropogenic GHG is CO_2, which is also the focal point for discussion in this research (UN Human Settlements Programme, 2011).

References

Bäckstrand, K., and Kronsell, A. (2015). *Rethinking the Green State: Environmental Governance Towards Climate and Sustainability Transitions.* Abingdon: Routledge.

Chang, Y. C., and Wang, N. N. (2010). Environmental regulations and emissions trading in China. *Energy Policy, 38*(7), 3356–3364.

Chen, B. et al. (2017). Local climate governance and policy innovation in China: A case study of a piloting emission trading scheme in Guangdong Province. *Asian Journal of Political Science, 25*(3), 1–21.

Chen, J. (2010). Transnational environmental movement: Impacts on the green civil society in China. *Journal of Contemporary China, 19*(65), 503–523.

Chen, L. Y. (2017). How do experts engage in China's local climate governance? A case study of Guangdong Province. *Journal of Chinese Governance, 2*(4), 360–384. Doi: 10.1080/23812346.2017.1379646

Deng, Z. et al. (2018). Effectiveness of pilot carbon emissions trading systems in China. *Climate Policy, 18*(8), 992–1011.

Dickson, B., and Chao, C. M. (2001). Introduction: Remaking the Chinese state. In C. M. Chao and B. Dickson (Eds.), *Remaking the Chinese State: Strategies, Society, and Security.* London; New York: Routledge.

Economy, E. (2007). Environmental governance: The emerging economic dimension. In A. P. J. Mol, and N. T. Carter (Eds), *Environmental Governance in China.* Abingdon and New York: Routledge.

Environmental Defense Fund. (2018). *The Progress of China's Carbon Market 2017.* Retrieved September 26, 2018, from https://www.edf.org/sites/default/files/documents/The_Progress_of_Chinas_Carbon_Market_Development_English_Version.pdf

Environomist. (2017). *China Carbon Market Research Report 2017.* Retrieved July 20, 2017, from www.environomist.com/upload/file/20170216171434_40.pdf

Francesch-Huidobro, M., and Mai, Q. (2012). Climate advocacy coalitions in Guangdong, China. *Administration and Society 44*(6): 43S–64S.

Fu, S., Zou, J., and Liu, L. W. (2015). *An Analysis of China's INDC. International Center for Climate Governance Reflection No.36.* Retrieved September 16, 2018, from www.greengrowthknowledge.org/sites/default/files/downloads/resource/An-analysis-of-Chinas-INDC-ICCG.pdf

Gao, S., Smits, M., Mol, Arthur P.J., and Wang, C. (2016). New market mechanism and its implication for carbon reduction in China. *Energy Policy, 98*, 221–231.

Guangdong Development and Reform Commission (DRC). (2013). *Scheme of Guangdong Province for First Emission Allowance Allocation and Related Work.* Retrieved March 26, 2018, from http://zwgk.gd.gov.cn/006939756/201401/t20140123_463411.html

Guangdong Development and Reform Commission (DRC). (2016). *Notice on the Allocation of Carbon Emission Allowances in Guangdong Province 2016.* Retrieved

August 20, 2017, from http://dtfz.ccchina.gov.cn/archiver/LowCD/UpFile/Files/Default/20160714093558793864.pdf [in Chinese].

Guangdong Development and Reform Commission (DRC). (2017). *Notice on the Allocation of Carbon Emission Allowances of Pulp and Paper, Aviation and White Cement in Guangdong Province 2016.* Retrieved August 23, 2018, from www.cne-mission.com/article/news/ssdt/201701/20170100001201.shtml [in Chinese].

Guangdong Research Center for Climate Change (GDRCCC). (2015). *Guangdong Pilot Emissions Trading Scheme Report (2014–2015).* Guangzhou: Guangdong Provincial Development and Reform Commission.

Guangdong Research Center for Climate Change (GDRCCC). (2017). *Guangdong Pilot Emissions Trading Scheme Report (2016–2017).* Guangzhou: Guangdong Provincial Development and Reform Commission.

Hu, A. G., and Guan, Q. Y. (2017). *China: Tackle the Challenge of Global Climate Change.* New York: Routledge.

Hübler, M., Voigt, S., and Löschel, A. (2014). Designing an emissions trading scheme for China: An up-to-date climate policy assessment. *Energy Policy, 75,* 57–72.

International Energy Agency (IEA). *Key World Energy Statistics 2018.* Retrieved April 6, 2020, from https://webstore.iea.org/key-world-energy-statistics-2018.

Jing, Y. (2015). *The Road to Collaborative Governance in China.* New York: Palgrave Macmillan US.

Keck, M. E. and Sikkink, K. (1998). *Activities beyond Borders: Advocacy Networks in International Politics.* Ithaca, NY: Cornell University Press.

Koehn, P. H. (2016). *China Confronts Climate Change: A Bottom-up Perspective.* New York: Routledge.

Kossoy, A., and Guigon, P. (2012). *State and Trends of the Carbon Market 2012.* Washington, DC: The World Bank. Retrieved from http://documents.worldbank.org/curated/en/749521468179970954/State-and-trends-of-the-carbon-market-2012.

Kostka, G. (2016). Command without control: The case of China's environmental target system. *Regulation and Governance,* 10(1), 58–74.

Kuhn, B. M. (2015). Policies, collaboration, and partnerships for climate protection in China. In Y. Jing (Ed.), *The Road to Collaborative Governance in China.* New York: Palgrave Macmillan US.

Lau, L. J., Qian, Y., and Roland, G. (2000). Reform without losers: An interpretation of China's dual-track approach to transition. *Journal of Political Economy,* 108(1), 120–143.

Liu, L., Wang, P., and Wu, T. (2017). The role of nongovernmental organizations in China's climate change governance. *WIREs Climate Change* (2017), 8: null. Doi:10.1002/wcc.483

Lo, A. Y. (2016). *Carbon Trading in China: Environmental Discourse and Politics.* Basingstoke: Palgrave Macmillan.

Lo, A. Y. et al. (2018). Towards network governance? The case of emission trading in Guangdong, China. *Land Use Policy, 75,* 538–548.

Ma, J. (2017). *The Economics of Air Pollution in China: Achieving Better and Cleaner Growth.* New York: Columbia University Press.

Ma, X., and Ortolano, L. (2000). *Environmental Regulation in China: Institutions, Enforcement, and Compliance.* Lanham, MD: Rowman & Littlefield.

Mol, A. P. J., and Carter, N. T. (2006). China's environmental governance in transition. *Environmental Politics,* 15(2), 149–170.

Morton, K. (2005). *International Aid and China's Environment: Taming the Yellow Dragon.* London and New York: Routledge.

National Development and Reform Commission (NDRC). (2011). *A Circular on Launching Pilot Carbon Emissions Trading*. Beijing: National Development and Reform Commission.

National Development and Reform Commission (NDRC). (2015). *China's Intended Nationally Determined Contribution: Enhanced Actions on Climate Change*. Retrieved September 8, 2018, from www4.unfccc.int/ndcregistry/Published-Documents/China%20First/China%27s%20First%20NDC%20Submission.pdf

Naughton, B. (1995). *Growing Out of the Plan: Chinese Economic Reform, 1978–1993*. New York: Cambridge University Press.

O'Connor, D. (1995). Applying economic instruments in developing countries: From theory to implementation. *Eepsea Special and Technical Paper, 4*(1), 91–110.

People's Government of Guangdong Province. (2014). *Trial Methods of Emission Management of Guangdong (No. 197, Decree of Guangdong Provincial Government)*. Retrieved August 23, 2017, from http://zwgk.gd.gov.cn/006939748/201401/t20140117_462131.html [in Chinese].

Porter, E. (2014, September 24). The benefits of curbing climate change. *The New York Times*, B1, B4.

Qi, Y. (2013). *Annual Review of Low-carbon Development in China*. Beijing: Tsinghua University.

Qi, Y., Ma, L., Zhang, H., and Li, H. (2008). Translating a global issue into local priority: China's local government response to climate change. *Journal of Environment and Development, 17*(4), 379–400. Retrieved from https://doi.org/10.1177/1070496508326123.

Ross, L. (1998). China: Environmental protection, domestic policy trends, patterns of participation in regimes and compliance with international norms. *The China Quarterly, 156*, 809–835.

Saich, T. (2004). *Governance and Politics of China* (2nd Edition). London: Palgrave.

Schröder, M. (2012). *Local Climate Governance in China: Hybrid Actors and Market Mechanisms*. Basingstoke: Palgrave Macmillan.

Schwartz, J. (2004). Environmental NGOs in China: Roles and limits. *Pacific Affairs, 77*(1), 28–49.

Shapiro, J. (2001). *Mao's War Against Nature: Politics and the Environment in Revolutionary China*. Cambridge: Cambridge University Press.

Shuwen, J. L. (2004). Assessing the dragon's choice: The use of market-based instruments in Chinese environmental policy. *Georgetown International Environmental Law Review, 16*, 617–655.

Tang, S. Y., Wing-Hung, L., Wing-Hung., C., Cheung, K.-C., and Man-Keung, J. L. (1997). Institutional constraints on environmental management in urban China: Environmental impact assessment in Guangzhou and Shanghai. *The China Quarterly, 152*, 863–874.

Tao, J., and Mah, N. Y. (2009). Between market and state: Dilemmas of environmental governance in China's sulphur dioxide emission trading system. *Environment and Planning C Government and Policy, 27*(1), 175–188.

UNEP Risoe. (2008). *Increasing Access to the Carbon Market*. Roskilde, Denmark: UNEP.

United Nations Human Settlements Programme (UNHSP). (2011). *Cities and Climate Change: Global Report on Human Settlements 2011*. London: Earthscan.

United Nations Statistics Division. (2018). *2018 Energy Statistics Pocketbook*. Retrieved April 6, 2020, from https://unstats.un.org/unsd/energystats/pubs/documents/2018pb-web.pdf.

Wang, P., Liu, L., and Wu, T. (2018). A review of China's climate governance: State, market and civil society, *Climate Policy, 18*(5), 664–679. Doi: 10.1080/14693062.2017.1331903

West, J., Schandl, H., Heyenga, S., and Chen, S. (2013). *Resource Efficiency: Economics and Outlook for China*. Bangkok, Thailand: UNEP.

World Economic Outlook Database. (2013). *International Monetary Fund*. Retrieved November 22, 2017, from www.imf.org/external/pubs/ft/weo/2013/01/weo-data/weorept.aspx?sy=1980andey=2018andsort=countryandds=.andbr=1andpr1.x=40andpr1.y=0andc=924ands=NGDP_RPCH%2CPPPPCandgrp=0anda=

Wu, F. (2009). Environmental politics in China: An issue area in review. *Journal of Chinese Political Science, 14*(4), 383–406.

Wu, G. G., and Zheng, Y. N. (1995). *Discussion of the Central-local Relationship: An Axe Question of Chinese Institutional Transformation*. Hong Kong: Oxford University Press.

Xinhua. (2007). *Hu Jintao's Report at 17th Party Congress*. Retrieved January 21, 2018, from www.china.org.cn/english/congress/229611.htm

Zang, D. S. (2009). From environmental to energy: Reconceptualization of climate change. *Wisconsin International Law Journal, 27*(3): 548–572.

Zhang, K. M., and Wen, Z. G. (2008). Review and challenges of policies of environmental protection and sustainable development in China. *Journal of Environmental Management, 88*(4), 1249–1261.

Zhang, Q., and Crooks, R. (2015). *Toward an Environmentally Sustainable Future: Country Environmental Analysis of the People's Republic of China*. Philippines: Asian Development Bank.

Zhang, W., and Vertinsky, I. (1999). Can China be a clean tiger? Growth strategies and environmental realities. *Pacific Affairs, 72*(1), 23–37.

Zhang, Z. (2015). Carbon emissions trading in China: The evolution from pilots to a nationwide scheme. *Climate Policy, 15*(sup1), S104–S126.

Zhou, N., Levine, M. D., and Price, L. (2010). Overview of current energy-efficiency policies in China. *Energy Policy, 38*(11), 6439–6452.

Zhou, X., Lian, H., Ortolano, L., and Ye, Y. (2013). A behavioral model of "muddling through" in the Chinese bureaucracy: The case of environmental protection. *China Journal, 70*(1), 120–147.

Zhu, D. M., and Zhou, L. Y. (2017). Crisis and response: The transformation of Chinese environmental governance institutional framework in contemporary China. *Fudan Journal (Social Sciences Edition), 59*(3), 180–188.

6 Governing climate knowledge

What can Thailand Climate Change Master Plan and climate project managers learn from lay northern Thai villagers?

Chaya Vaddhanaphuti

Introduction

Climate change has become one of the global socio-environmental risks in the past two decades. The United Nations Secretary-General António Guterres described the issue as "a direct existential threat", adding that the "urgency of the crisis" requires us to "put the brake on deadly greenhouse gas emissions and drive climate action" (United Nations, 2018). In the context of Southeast Asia, Thailand, one of the fastest growing industrial economies and largest rice exporters (Muthayya et al., 2014), has been assessing its vulnerability to climate change, and its adaptation and mitigation options mostly in the water, agriculture and energy sectors (Snidvongs and Chidthaisong, 2011). As a non-Annex country under the United Nations Framework Convention on Climate Change (UNFCCC), Thailand officially responds to the issue through the Office of Natural Resources and Environmental Policy and Planning (ONEP) under the Ministry of Natural Resources and the Environment. Based on the Second Biennial Update Report (Bhuridej et al., 2018), which Thailand was required to submit to UNFCCC under the Cancun Agreements from COP 16 in 2010, the total greenhouse gas (GHG) emissions of the country rose from 226,086 $MTCO_2eq$ in 2000 to 318,662 $MTCO_2eq$ in 2013, with about 75 percent, 16 percent and 6 percent of emission coming from the energy, agriculture and industry sectors, respectively. Land use, land-use change and forestry were the major carbon sink, contributing about 27 percent to the emission reduction. The net GHG emissions were still positive, increasing from 214,091 $MTCO_2eq$ in 2000 to 232,560 $MTCO_2eq$ in 2013 with an annual increase of 0.64 percent and a total increase by 8.63 percent. Thailand's Intended Nationally Determined Contribution (INDC) was to reduce its GHG emissions by 20 percent from the projected business-as-usual (BAU) level by 2030. This figure was applied to three sectors: energy (20.4 percent), waste (0.3 percent) and industrial processes and product use (IPPU) (0.1 percent). As shown in Figure 6.1, the 20 percent cut, equivalent to about 115.6 $MTCO_2eq$, would potentially reduce the projected GHG emissions from 554.649 $MTCO_2eq$ down to about 439.049 $MTCO_2eq$ by 2030 (ONEP, 2017).

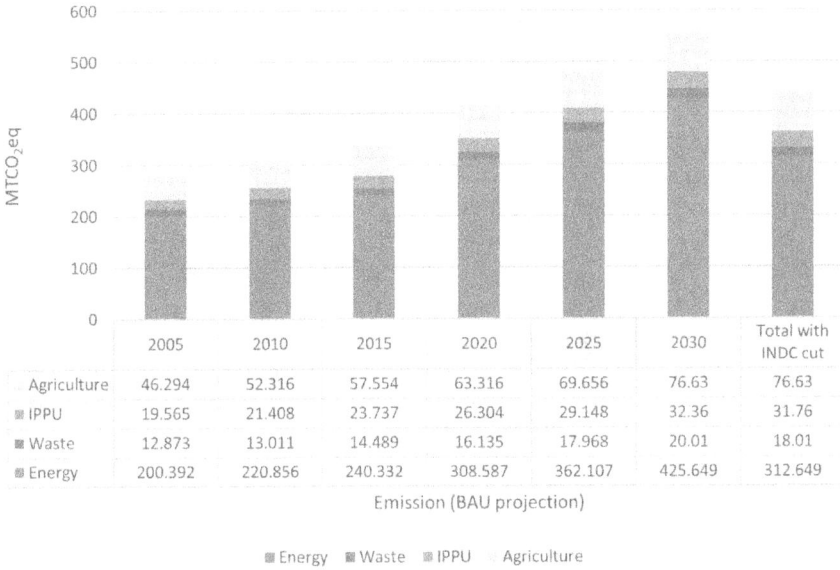

	2005	2010	2015	2020	2025	2030	Total with INDC cut
Agriculture	46.294	52.316	57.554	63.316	69.656	76.63	76.63
IPPU	19.565	21.408	23.737	26.304	29.148	32.36	31.76
Waste	12.873	13.011	14.489	16.135	17.968	20.01	18.01
Energy	200.392	220.856	240.332	308.587	362.107	425.649	312.649

Emission (BAU projection)

■ Energy ■ Waste ■ IPPU ⬚ Agriculture

Figure 6.1 Thailand's emissions projection (under the BAU scenario) and potential emissions target for the energy, waste and industrial processes and product use (IPPU) and agricultural sectors between 2005 and 2030

The emissions target could be achieved through short- and long-term national plans, for example the 12th National Economic and Social Development Plans (2017–2021) and the Thailand Climate Change Master Plan (2012–2050) which ONEP co-drafted with the Deutsche Gesellschaft für Internationale Zusammenarbeit (GIZ) of the German government. The aim in both cases was to pursue a low carbon society while maintaining competitiveness in economic growth and agriculture production and sustainable development (ONEP, 2012). However, for these national plans to remain viable at the ground level requires implementing the national climate policy at sub-provincial levels, coupled with climate change communication and the education of the public.

In a global risk society, according to the late German sociologist Ulrich Beck, climate change demands that Thailand become part of the cosmopolitan climate risk community that is making a concerted effort to find a sustainable future (Beck et al., 2013). As in the case of China's climate governance described by Joy Zhang (2015), global interrelatedness has been instrumental in a number of climate initiatives in Thailand, funded by both international and subnational agencies. More importantly, as Zhang (2015, p. 331) argues, it is the global climate discourses that have transformed the social, political and technological imaginaries of Thailand from within. Using the idea of "cosmopolitan climates",

Hulme (2010) argues that climate change challenges the current global environmental governance. Through the construction of climate knowledge by the coproduction between science and society, the boundaries between nature-culture, present-future and global-local cease to exist (Jasanoff, 2010). Climate change, seen in this light, then, is boundless, boundary-less and everywhere; no longer can climate be governed by a single nation-state (Hulme, 2010). What then can we expect from such an attempt to govern the climate at the international and national level from Thailand's perspective? To be sure, given the diverse approaches towards environmental problems and the complex power relations of different climate actors (scientists, decision makers, NGOs, entrepreneurs, media, lay Thai citizens, etc.), such multifaceted, cross-scale assemblages of climate actors might result in open, contingent and changeable outcomes (Beck et al., 2013). There will not be a smooth transfer of climate ideas, decisions and policy implementation from the West to Asian contexts.

By keeping the concept of cosmopolitan climates in the background, I examine the so-called "climate change" more critically, particularly as Hulme (2010) denotes climate as a plural entity, not a singular one (Rudiak-Gould, 2013). I question whether the communication and circulation of the idea of climate change is easily understood, i.e. is there just one way of understanding climate change? Can we assume that everyone agrees on the meaning(s) of climate change? Does the idea of climate change ever fit with the existing understanding of "weather" and "climate" in the Thai culture and language? How do climate project managers, as they visit their field sites, deal with those who have different ideas about climate change? And what happens when different kinds of climate knowledge meet? Drawing on a series of interview with representatives of different climate organizations operating in Thailand between 2013 and 2014, my aim is to analyze climate change discourses in the context of Thailand, in particular those that were disseminated among the Thai government, national and international non-governmental organizations and the local authorities of Nan province in northern Thailand, one of the provinces that was initially selected as a pilot site for the implementation of the Climate Change Master Plan since 2012. These emerging discourses are then interrogated using my ethnographic data of local Nan villagers. Here I argue that, despite a global risk society where distant evidence of climate impacts is made relatable at Thai villages, and where different climate actors establish themselves to creatively re-define and re-imagine climate mitigation and adaptation, perhaps there remains an ontological obstacle in which, fundamentally, the *scientific* knowledge about climate change does not always make sense to all individuals and cultural groups. This means, the global discourse(s) of climate change can both enhance and disagree with local beliefs, depending on the way it is framed. This explains why increasing climate knowledge and awareness does not always result in a logical translation of climate actions, especially in non-Western contexts (Forsyth &Evans, 2013; Vaddhanaphuti, 2017; Weisser et al., 2013; Zhang, 2015).

Given the fluid nature of climate change itself as well as of the various opinions of climate actors, the aim of this chapter is to examine how different climate project managers perform their climate identities, discourses and actions. This helps to reveal their underlying climate discourses and the necessary resources, as well as interactions between groups ranging from local to international levels (Blok, 2013). In the following sections, I explain the theoretical and analytical framework needed to understand the international knowledge and discourses about climate change (Section 2). This is followed by a short section to introduce a linguistic understanding of the Thai concept of weather and climate, the terminologies that were used by local Thai villagers and experts, respectively (Section 3). In Section 4, I then describe the cultures, goals and practices of three different types of organizations when they conducted their climate projects with the respective participants in different parts of Thailand and Southeast Asia. In Section 5, I use ethnographic data to contrast what and how lay people understood their weather in comparison to the idea of climate change imported and practiced by different organizations. Did they agree? Were they confused? Climate change adaptation has various objectives, but I argue in Section 6 that, fundamentally, it is perhaps adaptation to climate change at the ontological level that is the most essential and also the hardest to achieve. Drawing on the idea of a cosmopolitan world, I end by asking how these organizations can establish a platform on which all participants can creatively reimagine their climate actions.

Deconstructing climate change and its discourses

What is climate change? For anthropologist Tim Ingold, it is definitely *not* a "temperament of our being" (Ingold 2010, p. S112) nor something deeply attached to emotions, identity, livelihood, histories and landscape (Brace &Geoghegan, 2011), like that of weather. From a critical perspective in the fields of geography, anthropology and science and technology studies, weather and climate change are ontologically and epistemologically different. For Mike Hulme (2008) the conventional term "climate change" is a manifestation of seemingly unrelated objects, actors and ideologies, as Lesley Head and Christopher Gibson (2012, p. 701) explain it is an assemblage of weather elements, greenhouse gases, atmospheric processes, discourses, bureaucracies and texts. In other words, climate change is always 'more-than-climate'. Although it emerged as a local concern at first, climate change has transformed into a technical, public and political hybrid entity (Head &Gibson, 2012; Hulme, 2008; Jasanoff, 2010). On the one hand, it transcends the boundaries of nature and culture, local and global and lay and experts, and on the other hand, it emerges from, and is constituted in, networks and practices of language, meanings and politics (Moser, 2016).

The knowledge about climate change is produced and made visible (Rudiak-Gould, 2013) through instrumental quantification, statistical standardization and supranational institution setups like that of the Intergovernmental Panel on Climate Change (IPCC) and the UNFCCC. This form of "eco-managerialism" (Fischer &Hajer, 1999) is based on the assumption that an environmental problem can only be exclusively identified and solved through science and institutional management (Jasanoff, 1987). Clark Miller (2004) likened the IPCC to the 'Empire', the sole governing body that creates order out of climate knowledge through performing hegemonic scientific facts and extending the control of knowledge to its 'clients' across the world.

Nevertheless, the global circulation of such scientific knowledge about climate change has lost its original attachment and contextual relevance to local places and cultures, and at the same time, allows the popularized term "climate change"' to be interpreted in various ways (Hulme, 2008), as observed in recent times where the term has been associated with words like 'polar bear', 'war', 'refugee', 'apocalypse', 'planetary boundary' and so on (Boykoff, 2011). Here are some of the questions that serves as a guide for the rest of this chapter: How might climate change discourse be interpreted, understood and utilized in different political and cultural settings? (Rudiak-Gould, 2012, pp.46–47); In what ways is climate change, as part of development discourse, communicated, and how might it reshape the local conceptualization of weather and climate? (Orlove, 2009; Weisser et al., 2013).

In analyzing the discourse of environmental politics, I follow Hajer and Versteeg's (2005, p. 175) definition of discourse, defined as "ensembles of ideas, concepts and categories through which meaning is given to social and physical phenomena". From the Foucauldian perspective, discourse establishes 'truth' based on knowledge production within a society, and hence forms a reality in which one relates and engages with it (Hajer & Versteeg, 2005). John Dryzek (2013, p. 9) also adds that there are certain preconditions, assumptions and judgments that give rise to the argument and debates for/against a phenomenon. Power, then, is bound up with environmental knowledge and political practice, where an environmental issue is made visible (as 'global' or as a 'risk') through institutions, experts' analytical and methodological frameworks and language (Feindt & Oels, 2005). Hence, the purpose of discourse analysis is to de-naturalize environmental narratives that had been invisibly constructed and made normal in a society. In other words, this method reveals how society makes sense of symbols and experiences of the environmental risks that govern the way people think and act, rather than the problematic physical environment itself (Hajer and Versteeg, 2005). In this chapter which concerns climate change discourse in Thailand, I follow Fischer and Hajer's (1999) analysis of the discourse of sustainable development in how the framing of climate change and its associated actors (NGO, nation states, policy makers, world bank, scientists) interpret, produce and reproduce and transform environmental discourse through sets of institutionalized practice. Below I outline the type of environmental discourses found worldwide.

Dryzek (2013) has identified four broad types (with subtypes) of environmental discourses as follows: Problem Solving (Administrative Rationalism, Democratic Pragmatism, Economic Rationalism); Sustainability (Sustainable Development, Ecological modernization); Limits and Survival; and Green Radicalism (Green Romanticism, Green Radicalism). These are different in terms of the definition and values (re)assigned to the environment, as well as the willingness to alter existing socioeconomic and political structures to solve environmental problems. Further research has applied discourse analysis such as that offered by Dryzek to the issue of climate change, with the additional influence of Foucault's concept of governmentality. Karin Bäckstrand and Eva Lövbrand (2006), for example, found three different discourses on Clean Development Mechanism's tree planting projects: Ecological Modernization, Green Governmentality and Civic Environmentalism. Citing the aforementioned two authors, Angela Oels (2005) also found a shift in climate change discourse from Ecological Modernization to Green Governmentality from the mid-1980s onwards in the context of European climate change policy making.

Some of these discourses are worth mentioning. In Administrative Rationalism (Dryzek, 2013), or what Bäckstrand and Lövbrand (2006), inspired by Foucauldian perspective, call Biopower, the environment is a subject of scientific and managerial control. Scientific knowledge is used to identify and inform decision makers how to fix a particular problem. In Ecological Modernization, there is an assumed compatibility between the economy and the environment that could bring about an environmentally friendly development path through capitalism, private sectors and decentralized government (Dryzek, 2013). In contrast to this capitalist-driven environmental policy making, Green Radicalism discourse (Dryzek, 2013) requires people to challenge and reformulate the existing socioeconomic and political structure to achieve inter- and intra-generational equity and sustainability. In a similar vein, Bäckstrand and Lövbrand's (2006) and Oels' (2005) Civic Environmentalism discourse is about managing the environment through a polycentric, 'glocal' governmental arrangement while allowing the marginalized groups, civic society and NGOs to tackle the nested problems of inequity and unjust development, poverty and the deteriorating environment. The early days of global environmental policy have, as Fischer and Hajer (1999, p. 6) summarized it, "been cut off from the cultural dimension of environmental politics". In other words, cultural diversity and alternatives to technocracy were very much needed (Fischer & Hajer, 1999). Green Radicalism and Civic Environmentalism are, then, a form of "cultural critique" (Fischer & Hajer, 1999) used to challenge the dominant culture that maintained a market-based environmental solution.

With the conceptual and analytical framework provided, using a case study in Nan province, following Fiendt and Oels (2005)'s recommendation, I reveal below how different kinds of discourse make the issue of climate change governable, how a dominant discourse might shift in temporal and spatial dimensions and, finally, how certain discourses might compete and conflict with one another.

Language and climate organizations in Thailand

This section introduces the use and complexity of the Thai terms for 'weather' and 'climate' that are closely related to communication strategies of three different groups of climate organizations. Below, I explain each in turn.

To better understand the difficulties in communicating the term "climate change" in the Thai context, it is vital to understand the structure of the Thai language and five related terms when talking about the weather: air, weather, climate, climate change and global warming. In the vernacular Thai language, there are several interchangeable compound words used to talk about the weather: *aagad* (air), *din fa aagad* (earth-sky-air), *lom fa aagad* (wind-sky-air) and *fa fon* or *fon fa* (air-rain/rain-air). On the other hand, the term *pumi aagad* (geo-air) is rather a technical term, and is used to translate the Western term 'climate'. The Thai term for climate was, in effect, invented to match that of the scientific, English term. Full explanation of these terms is found in my own work (Vaddhanaphuti, 2017, p. 192). As expected, the lengthy translated technical term of climate change, or *karn plien plang sapab pumi aagad* (henceforth KPPA) means something quite different from the common expression *aagad plien plang*, or 'air change'. Interestingly, to maintain and convey an idea of global-scale temperature increase, the Thai media often opt to use the term *lok ron* or global warming instead, because of climate change's wordy Thai translation. Below, I show that in communicating and educating climate change as part of the climate adaptation project, local Southeast Asian people were taught by organizations to think about their weather in scientific terms simply because the travelling idea of climate change, carries with it an assumed universal and dominant status. I argue that climate change is culturally, linguistically and experientially different from the weather in which people continually live in their everyday life.

Using empirical data that I collected in Thailand (Vaddhanaphuti, 2017), i.e. a series of semi-structured interviews with climate organizations that operated at local, national and international levels (see Table 6.1), and ethnographic fieldwork in Nan province, northern Thailand between November 2013 and November 2014 (see Figure 6.2), my goal was to answer the following question: How do climate organizations help local Thai people to make sense of the idea of global climate change? I attempted to uncover the organizational cultures, goals, communication methods and the overall climate change discourse that they tried to deliver to their respective participants. While these organizations initiated their climate change project at different times and for different purposes, I argue that what these organizations brought to their local community members was not only information about climate change, but also the ways in which local people, knowledge and climate were to be governed. Such actions have met with different levels of cooperation and resistance by the local participants as will be explained in the next section.

As seen in Table 6.1, I arrange these climate organizations into three groups based on their traits which I term as follows: "Climate Story Listener", "Climate Service Provider" and "Climate Policy Facilitator" (Vaddhanaphuti, 2017). They were grouped according to the nature of their climate project, and the degree of participation between the organization's members of staff and local villagers. The "Climate Story Listeners" exchanged weather stories and knowledge with local people, and tended not to impose their knowledge upon the locals,

Table 6.1 Characteristics of climate organizations: Climate Service Listener, Climate Story Listener, and Climate Policy Facilitator

Characteristics of climate organizations	Names of climate organizations	Type of organization	Levels of operation	Levels of participation with local villagers in SEA		
				Low	Medium	High
Climate Story Listener	JK	NGO	Provincial			•
Organizational members listen to climate stories from local villagers	RECOFTC	NGO	Provincial			•
	TEI	Non-profit	National			•
	MARCC	United state-funded	International			•
Climate Service Provider	TMD	Thai Government	National/provincial	•		
Experts deliver climate facts	CCKM	Academic	National		•	
Climate Policy Facilitator	GIZ	German–Thai government cooperation	International		•	
Decision makers link climate change to local development issues						

Source: Based on Vaddhanaphuti 2017, p. 189.

effectively achieving two-way, participatory communication. There were four organizations in this category: Joko (JK), The Regional Community Forestry Training Center for Asia and the Pacific (RECOFTC), Thai Environment Institute (TEI), and Mekong Adaptation and Resilience to Climate Change (MARCC). The former two were Nan province-based NGOs specialized in addressing land rights, agriculture and food insecurity. TEI was a non-profit organization based in Bangkok with a particular interest in urban climate resilience, and MARCC was a research collaboration funded by the United States Agency for International Development (USAID) for assessing climate impacts and adaptive capacity. JK, RECOFTC and TEI, whose participants were mostly marginalized communities, saw climate change as one of the drivers and one of the outcomes of unjust socioeconomic environment and political issues. These three organizations constructed climate change discourse as part of local long-term sustainable development. MARCC, on the other hand, needed rigorous scientific studies to reveal the evidence of climate change, before assessing the impacts, educating the local people and drafting adaptation strategies.

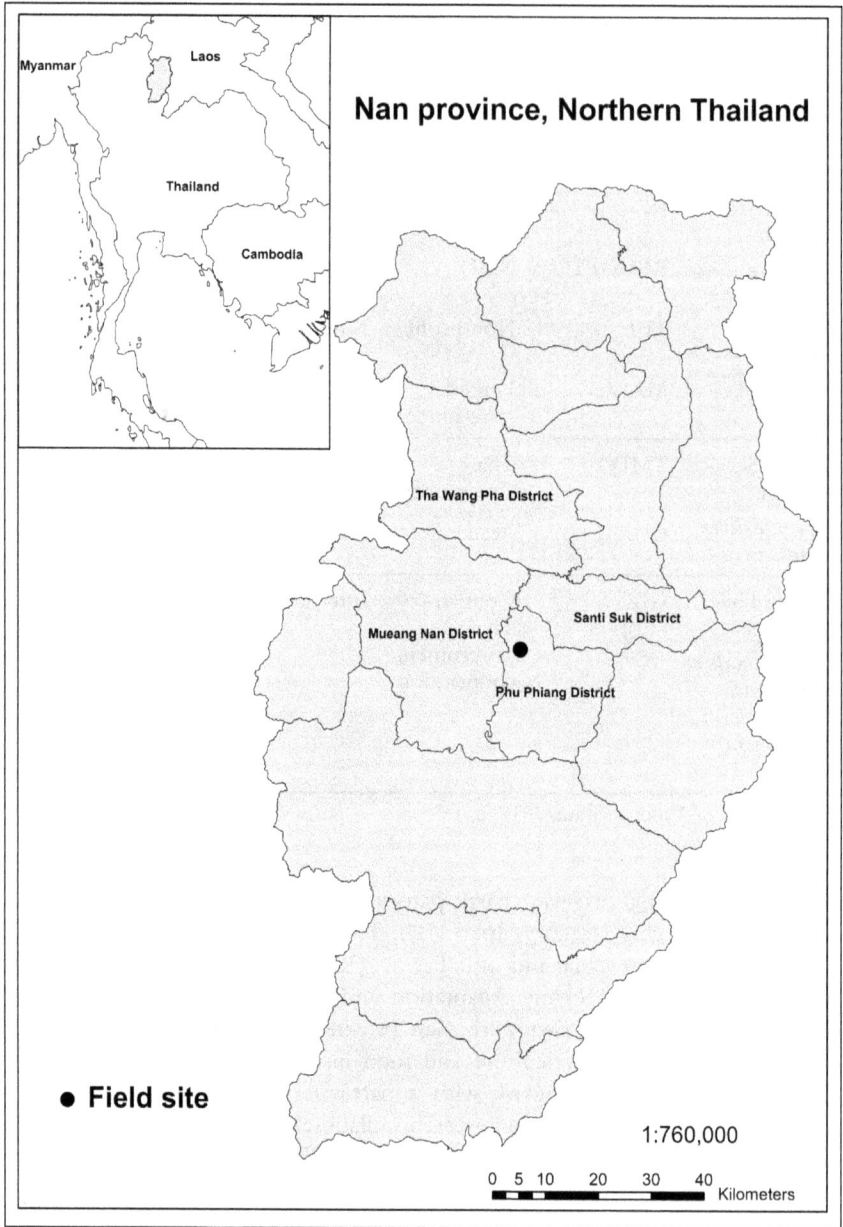

Figure 6.2 Location of ethnographic field site in a village in Muang Chang sub-district, Phu Phiang District, Nan province, northern Thailand

Source: Vaddhanaphuti 2017.

The "Climate Service Providers" were almost the opposite to the "Climate Story Listeners". The former were organizations that gave their recipients information about the physical changes of the climate in the form of pedagogic knowledge transfer usually in a conference or workshop, published text and images. These included Thailand's Meteorological Department (TMD) and Climate Change Knowledge Management (CCKM), two organizations that specialized in weather data management and forecast and planning, both located in Bangkok. The "Climate Policy Facilitators" were those in position to make decisions, connect different stakeholders and bring global climate policy to the provincial level environment-development agenda. Here, I drew examples from a collaboration between Nan City Municipality and the German governmental organization Deutsche Gesellschaft für Internationale Zusammenarbeit (GIZ). The latter was an inter-governmental organization that assisted Thailand's Office of Natural Resource and Environmental Planning in drafting Thailand's Climate Change Master Plan (2015–2050), and chose the former to be a pilot low carbon city (2012–2016). It must be noted that the GIZ team did not want to meet local Nan villagers face to face; the local Nan authorities acted as an intermediary. Essentially what GIZ team did was attempt to integrate top-down and bottom-up climate policy. In the next section, I provide details of the three groups of climate organizations, their communication strategy and climate change discourses.

Construction of climate change discourses

Climate Story Listeners

For these organizations, changing weather, seasons and climate had a crucial impact on f people's livelihood, and so it was vital that people became aware of the problem and took action against these changes. In July 2014, I observed one of JK's Farmer School workshops on the topic of weather observation and climate change held in a village in Muang Chang sub-district, Nan province. About 30 rice farmers from seven northern Thai provinces attended the workshop to gather the local evidence of climate change seen through their eyes and to find possible local solutions, independently of the government. The farmers were able to supply a large amount of evidence of climate change, but the solutions were kept simple, for example by stopping deforestation and the use of chemicals. The JK facilitator attempted to make a connection between the climate and injustice. He said that a few factors, including the incoming of seed corporates into villages made people much worse off. Unpredictable seasonal changes, then, were like rubbing salt into the wound. JK encouraged people to realize the importance of local policy, and to make changes if necessary, to provide some fairness and a safety net for their livelihoods in terms of agriculture, the environment and development.

Later in November 2014, I witnessed RECOFTC's climate perspective at work in a forum entitled 'Holistic Natural Resource Management', held in Nan

Municipality. In a discussion on deforestation, a representative of RECOFTC strongly disagreed with the way the Thai government was blaming ethnic minority villagers for the deforestation and hence global warming. Climate education, therefore, had a political implication, as he put it:

> if the local people had known that [climate change] comes from economic or industrial sectors, and fossil fuel, it would make them feel that they were not the scapegoat, and that they are were not the main agent of the problem.

At the forum, a representative of an upland village in Nan, who worked closely with RECOFTC, told the panel that the villagers were not the cause of climate change, and, in fact, that they could find indicators of climate change in their own forests just as accurately as those in a science lab. On this basis, RECOFTC argued that knowing when and how to use scientific climate knowledge could strategically help the villagers provide a counter-argument to the government's false claims. More importantly, indigenous knowledge could contribute to more advanced climate research, a claim similarly made by JK.

I did not have an opportunity to directly observe TEI and MARCC's work, but through interviews, I established TEI's approach to climate change was similar to that of the previous two organizations, although focused on the urban areas of Chiang Rai (north) and Hat Yai (south) provinces. MARCC, on the other hand, required a team of climate experts and field-based research-ers, and rigorous climate modeling and scenarios for the entire Lower Mekong Basin to locate climate hotspots, and these data were used to formulate adaptation strategies. A spokesperson for MARCC mentioned that she was "proud" of their "strong" scientific impact-vulnerability assessment. Field staff would spend time with villagers to familiarize themselves with local contexts and co-draft community adaptation strategies. Based on their knowl-edge-attitude-practice approach, providing more information about climate change would help their respective villagers to make better adaptation deci-sions. One of their published reports (Oranop-na-Ayuthaya, 2015) indicated that the number of correct and appropriate adaptation strategies had almost doubled compared to the pre-survey period, proving that their climate educa-tion worked.

Despite the differences in the purpose of understanding climate change, all four Climate Story Listeners had somewhat similar (albeit with subtle differences) strategies for communicating and education climate change. First, the organiza-tions would invite their respective villagers to share local weather stories and, second, they would assess impact and vulnerability, and explain abnormal phe-nomena within the frame of climate change. Finally, they would co-draft suitable adaptation measures and local development policies. For the purpose of this chapter, only the first two steps will be discussed. Below, I show that for suc-cessful climate education, the villagers need to be taught these new terms – weather, climate, climate change and global warming – in order to bridge the gap between science and traditional knowledge to help make sense of climate change locally.

For MARCC and RECOFTC, to fully comprehend the complexity of climate change (*karn plien plang sapab pumi aagad* – KPPA), villagers first needed to learn the difference between 'weather' (*aagad*) and 'climate' (*pumi aagad*), hence putting local weather into the global climate system, and appreciate the long-term changes. In their climate projects in Thai, Laos and Vietnamese villagers, members of MARCC showed their participants American cartoons, drawings and games to clarify the terminological differences, despite the apparent absence of such separation in the root of these three languages. This foreign term of *pumi aagad* had reshaped the local understanding of weather ontologically and experientially (Huber & Pedersen, 1997; Miller, 2004) simply because it took away the 'social' part of the weather, and reintroduced the nature-culture divide (Hulme, 2008; Jasanoff, 2010; Rudiak-Gould, 2013). JK and TEI's approach to climate communication was, in contrast, to maintain the authenticity of the local language with a minimum amount of jargon, and therefore preserve local weather stories. Vernacular words like 'changes in the weather' (*aagad plien plang*) was preferred to KPPA. Still there were some further difficulties as seen below.

After listening to the villagers' weather stories, all four organizations would try to explain the links between local and global atmospheric changes, with (MARCC/RECOFTC) or without (JK/TEI's) breaking them down into 'weather' and 'climate' categories. As mentioned above in Section 3, the choice of whether or not to use the terms 'global warming' (*lok ron*) or 'climate change' (KPPA) has merits and drawbacks. From my observation, *lok ron* generally triggered people's notion of heat, in relation to burning, deforestation and waste management. On the other hand, KPPA was more closely linked to extreme seasonal changes (Whitmarsh, 2009). Both JK and TEI conceded that educating people about KPPA was scientifically the more accurate choice, but linguistically a difficult and unfamiliar word, compared to *lok ron*. For practical reasons, they tended to use both terms interchangeably. RECOFTC and MARCC, however, preferred KPPA because using the term gives more scientific credentials, and helps to express the need to appreciate the long-term impact on the community development path.

Clearly, the Climate Story Listeners were adopting a Green Radicalism (Dryzek, 2013) or Civic Environmentalism (Bäckstrand & Lövbrand, 2006) discourse as they aimed to challenge the existing social structure that had partly been the cause of adverse socio-environmental changes. There is a possible exception, however, which is that MARCC might be the only organization whose climate project was driven to a large extent by scientific studies, making its climate discourse a mix of Administrative Rationalism (Dryzek, 2013) and Civic Environmentalism. Nevertheless, all four organizations viewed the impact of climate change as a symptom of broader social injustice. Practitioners of this group of organizations viewed multinational corporates as a destructive force against agrarian livelihoods. Pitted against the government, some of the Nan villagers were also marginalized and inferior. As the name suggests, Climate Story Listeners listened to the voices of local people, and tried to tackle the climate and development aspects simultaneously. Adapting to the impact of climate change by restructuring the agricultural system and farming practice could offer a way out of poverty and inequality, independently of corporate and government assistance. Climate education, in conjunction with other social development projects, was targeted at empowering local people to build capacity and resilience. This would only be possible with the Climate Story Listeners, in local villages, where there is a platform for local people to voice their local development policy and future dreams. Climate change was not an object to be fixed *per se*, but a condition for, and product of, a society that needs to be understood comprehensively.

In terms of communicating the idea of climate change, it should be noted that most but not all organizations (be they Story Listeners, Service Providers or Policy Facilitators) could not avoid imposing scientific authority through the translation and use of foreign meteorological terms. At the most basic level, the local weather had to be dissected and disentangled from culture to make way for a scientific understanding of the terms 'climate' and 'climate change'. Moreover, the terms 'climate change' and 'global warming' in Thai language conveyed different yet related narratives. Practitioners of Climate Story Listener organizations were aware of these complexities, and so tailored their communication very carefully to their audiences, with some successes and failures. As argued further below, these terms were ontologically and experientially foreign to many lay Thai villagers. A successful climate adaptation project may need a linguistic adaptation initially. The other two groups of climate organizations, however, did not pay attention to these linguistic details in climate communication.

Climate Service Providers

As part of the World Meteorological Organization's Global Framework for Climate Services' mission to provide climate information to stakeholders around the world, the Thailand Meteorological Department's (TMD) task was to deliver short-, medium- and long-range weather forecast data by site-visiting selected communities around the country (Climatological Center, 2014; Climatological Center, 2016). A representative of TMD told me that rural Thai farmers were still "lacking the access to forecast data", and that they needed to be "educated" how to make use of such data for the benefit of their crops. In November 2014, I attended a workshop entitled 'Climate Variability and Climate Change: Meteorological Knowledge and Application', which was held at Na Sao village, Nan province. I wanted to see how the idea of climate change was disseminated from the government's perspective. Indeed, it was in stark contrast to the Climate Story Listeners. In a small village meeting room filled with about 40 attendees, the villagers were bombarded with opening videos of the objective of TMD, meteorological jargon like "stratosphere", "greenhouse gases", meteorological definitions of "hot day", graphical representations of the Earth with incoming and outgoing radiation and figures to show the rise in global temperature. Moreover, presented in the official Central Thai dialect, both the terms *lok ron* and KPPA were used indiscriminately and uncritically. The speaker of TDM did not bother patiently teaching these terms in the way Climate Story Listeners did; they assumed the audience would accept and understand them. Of course, these were alien terms which the audience struggled to grasp both conceptually and linguistically.

Perhaps, as far as TMD was concerned, the workshop was a success as they were able to reinforce the authority of their scientific knowledge by putting Nan on the global climate map. Apart from explaining how climate change was caused by greenhouse gas emissions, the speaker of TMD did not mention how climate change might be related to people's agrarian livelihood, nor how to solve the problem locally. Unlike the Climate Story Listeners, it was a one-way communication, and the Na Sao villagers were reduced to a group of tamed, muted students in a lecture room.

While TMD's task was to disseminate information about the climate, a research-oriented organization, Climate Change Knowledge Management (CCKM), was promoting the use of an automatic weather station (AWS) in a remote, vulnerable area in Yasotorn province, north-eastern Thailand (a project between 2010–2016). The purpose of this project was specifically to use the AWS to generate weather data, where there was a gap in spatial coverage, and where commercial rice farmers were highly dependent on the short-term weather forecast. What was interesting about this project was its inclusiveness of local people and their local knowledge. With the AWS, the collected raw weather data was sent to be processed in Bangkok and sent back as weekly mobile phone text messages stating whether it would rain or not in the next seven days. It could be said that the Yasotorn villagers were mere users of weather data. They did not have a role in the weather data process apart from keeping a record of rainfall for data verification the days after they had received the forecast message. The processed weather data between Yasotorn and Bangkok remained a black box.

Despite the villagers showing an interest in learning weather forecasting terminology and interpreting weather maps, it was not easy from the perspective of technicians as, according to the spokesperson of CCKM, "the lay villagers had to be trained to understand the standardized meteorological terms, simply because they were confused about what exactly a 'hot day' actually means". A subsequent interview showed that the CCKM, in its initial phase (2010–2012) was not prepared to include local knowledge, and that local heterogeneous weather interpretations had to be homogenized. Local wisdom and weather lore were never part of their weather calculation.

The confidence in agro-meteorological projects, where experts and machines (AWS) assist decisions about agriculture, and exclude any existing traditional knowledge at the site, has increased for governmental (TMD), private (CCKM) and non-governmental (one of a sub-project of MARCC) organizations. The Climate Service Providers were adopting Administrative Rationalism (Dryzek, 2013) or a Green Governmentality discourse (Bäckstrand & Lövbrand, 2006), a technocratic and totalizing discourse where the problem of climate change was made real through the use of scientific diagrams and terminologies, installation of scientific equipment, monitoring of weather data and a series of workshops and training courses. Coming from a scientific perspective, and seeing climate as an object to be managed, there was very little connection to the community the members of which had always lived in and with their weather. Unlike the work of the Climate Story Listeners, there was a lack of participation by local people in climate education and data processing. Thus far, the Climate Service Providers seem to have reinforced the gap between experts and between science and traditional knowledge; effectively the latter part of the binary was treated as inferior. Lay villagers were reduced to mere users of processed weather forecast data, effectively putting them at the margin of climate knowledge. My concern, however, is whether these organizations could give enough space and appreciation of local weather interpretations and knowledge when it comes to knowledge integration, if an opportunity ever arises.

Climate policy facilitators

As previously mentioned, in 2012, the German governmental organization Deutsche Gesellschaft für Internationale Zusammenarbeit (GIZ) partnered with Thailand's Office of Natural Resource and Environmental Planning to formulate the Climate Change Master Plan. The new objective for between 2018 and 2021 was to "contribute to the decarbonisation of the Thai economy and to strengthen Thailand's resilience to the effects of climate change" (GIZ, n.d.) with a focus climate policy, agriculture, energy, waste and water, and the pursuit of Nationally Determined Contributions (NDCs), effectively reducing GHG emissions and adapting to climate change. Back in 2012, Nan City Municipality was chosen to implement this Master Plan as a pilot low carbon city. Accordingly, in 2014 I interviewed members of GIZ who went to work with Nan Municipality.

To conform to the global climate policy established through the cooperation between the German and Thai governments, Nan stakeholders' ideas and responses to climate change must be enrolled into GIZ's terms, as hinted in a series of workshops in 2012–2013 held in Nan. In Nan, GIZ's staff called themselves "the climate change experts", revealing to the ignorant Nan locals the crisis state of the global environment, backed by well-recognized academic references, and presented themselves as the sole saviour. A representative of GIZ said "[The goal of the series of workshops and of this project] was to make Nan people realise the existence of the climate change problem that they are facing". Accordingly, GIZ's summary reports of the workshops (GIZ, 2013, p. 9) state Nan people were "informed [by GIZ] about the problem of climate change that they had to face". This is Rudiak-Gould's (2013) "constructive visibilism" at work, where various kinds of evidence, found both locally in Nan and internationally, were used to construct a doom and gloom narrative to heighten the awareness of the crisis. But awareness alone was not enough, according to the representative. The emergence and persistence of the problem, from GIZ's perspective, was due to a moral hazard, that local Nan people failed to respond collectively to the problem of climate change they were facing. However, the arrival of GIZ would reveal the problem, find the solution, and positively engage people's attitude and action toward resolving the global problem.

After making the problem visible, GIZ outlined five climate strategies to achieve a low carbon city: urban (re)planning; water resource management; waste management; green industry; biodiversity. The arrival of GIZ gave Nan local authorities an opportunity to link local issues with national and global climate strategies. But, despite sending Nan Municipality officers on training courses in Germany between 2012 and mid-2014, these local officers simply could not follow such a development plan. Disappointingly because of the different "mindsets" of the German and Thai counterparts, the GIZ representative reported, the latter chose to focus on fixing problems of deforestation, food insecurity, haze pollution and waste management, all of which were more immediate pre-existing problems the locals were facing. Such resistance to global (and German) climate hegemony was due partly to GIZ's lack of power to enforce its plans, and an internal conflict of interests between Nan governmental departments. In the later phase of the project (post-2015), GIZ pursued the issue of energy efficiency and alternative energy sources. After sending German technicians from Fraunhofer Institute for Solar Energy Systems to assess potential sources of energy potentials, it was concluded that by 2036 the entire Nan province could become 100 percent reliant on its own (non-renewable and renewable) energy sources, depending on different energy scenarios (Stryi-Hipp, Steingrube & Narmsara, 2015). It was suggested that by building a market for decentralized PV rooftop systems, Nan could become an exemplar of an energy efficient province (Stryi-Hipp, Steingrube & Narmsara, 2015).

Since 2012, GIZ has successfully brought the global climate policy to a very small city in Nan province, Thailand. As a representative of Germany, a country with a highly ambitious climate policy, GIZ was handing on knowledge and know-how, while at the same time enrolling Nan stakeholders into a new climate regime. But ever since the beginning of the German-Thai project, it could be seen that there were a range of, and a shift in, GIZ's climate discourses. Thailand's aim to continue its economic growth in a sustainable way while at the same time turning to a decarbonization policy chimes with a 'strong' version of Ecological Modernization discourse (Bäckstrand & Lövbrand, 2006; Dryzek, 2013). Furthermore, when they visited Nan province, the GIZ team portrayed themselves as an exclusive agent in revealing and solving the climate change problem, making their climate discourse match that of Administrative Rationalism (Dryzek, 2013) or a Green Governmentality discourse (Bäckstrand & Lövbrand, 2006), in a like manner to Climate Service Providers. More recently when GIZ focused on energy efficiency, this was perhaps more in tune with the 'weak' form of Ecological Modernization discourse, as it was based on market-driven regime. Because of the differences between the German and Thai environmental values, knowledge and forms of governance (Jasanoff & Martello, 2004), more stringent rounds of workshops, training sessions and policy negotiations are needed to make sure that the Nan team was in control, and that the project operated in the way the German and Thai governments intended. Unsurprisingly, as shown above, there will always be negotiation, friction and resistance over how climate ideas and policy could and should operate locally.

Perception of the changing weather: lessons learned from northern Thai villagers

I have shown above what and how organizational practitioners disseminated the idea of climate change to lay northern Thai villagers and those living in other Southeast Asian countries. In this section, I use ethnographic data from a village in Muang Chang sub-district, Nan province (as outlined in Section 3) to interrogate the production and reception of climate knowledge and discourse. In the heading of this section, I have intentionally used the phrase "changing weather" rather than "climate change" simply because in vernacular Thai language, local people talked about *aagad plein plang*, not of *karn plien plang sapab pumi aagad*.

After staying in the village for almost a whole year between late 2013 and the end of 2014, I learned that the weather has always played an important role in people's livelihood. This may not be surprising since all of my research participants were farmers whose lives were influenced by small and large changes in the weather pattern and the seasons. I would like to emphasize that it is through multiple sociocultural, non-human things and environmental elements like wearing particular clothes, the frequency of showers, the cost of electricity bills, the behavior of some animals and plants, and so on, that together helped people made sense of the ever-changing sight, might and sound of their weather on this particular agrarian landscape of Muang Chang village (Vaddhanaphuti, 2017). However important this finding may seem, many climate project managers were not prepared to take it into account.

There is also something more to weather than just its sheer physical qualities. Worldwide, there are societies that live in the "moral space" of weather (Huber & Pedersen, 1997), for example, in sacred glacier landscapes (Allison, 2015; Byg & Salick, 2009; Cruikshank, 2005; Paerregaard, 2013), Marshall Island (Rudiak-Gould, 2012) and many African communities (Eguavoen, 2013; Roncoli, Ingram, & Kirshen, 2002; Tschakert, 2007). Similarly for people of Muang Chang in Nan province, northern Thailand (Vaddhanaphuti 2017), the (ir)regularities of their weather were deemed to be "good" or "bad" according to three related sets of religious belief systems. As the weather carried moral significance, there would always be an agent that could be held responsible for the positive and negative changes in the weather. In the Muang-animist belief system, which in itself is a cosmology for traditional northern Thai people (Davis, 1984), the reason for the changing weather was because the deities who reside in the sky are either blessing or punishing people. Cultural rites like the Cat

Parade or spirit houses were needed for people to worship and appease the deities. In the Buddhist belief system, adverse changes in the weather were something to be accepted as natural, rather than to be fixed. In this case people have to learn to be stoic and to cope through livelihood strategies such as crop diversification. In the third belief system, the deeper moral perspective, the explanation for the changing weather was that it was caused by the erosion of cultural traditions and morality. Only by re-living the golden past, caring for nature like the ancestors used to, would the "good" weather return. These three interrelated sets of belief and responses could be very loosely matched with Dryzek's (2013) Administrative Rationalism, Ecological Modernization and Green Radicalism discourses, respectively. The exception is that all three are culturally-driven, showing that positive or negative changes in the weather are reflections of Nan people's intimate relationship with their society and surrounding environment (Vaddhanaphuti, 2017).

What was striking about my ethnographic finding was that nobody in the village mentioned the terms "climate change", "greenhouse gas", "carbon" or any other climate change-related terms offered by external organizations. The data, as it stands, shows that the terms 'weather' and 'climate' are ontologically different and foreign. When a farmer talked about a hot day, he was making a connection between his own physical experience of heat and his emotion, memories, livelihood, the deities of the air and the Earth, the past, present and future consequences. Everything was woven together. More importantly, the weather was not to be governed, except by supernatural beings. Adverse changes in the weather were interpreted through a cultural lens, prompting three related kinds of response that led the people to be in awe of the environment and deities, and to make changes physically, socially, spiritually and morally in tandem with the weather. They ways in which they responded to the changing weather showed that they believed the problem could be solved at a local level, individually and/or collectively within the villages, independently of any supranational organization and unrelated to the low carbon initiatives proposed by the German-Thai governmental cooperation.

So, did the climate change discourse from different organizations heighten people's awareness and encourage them to take climate mitigation and adaptation responses ? Yes, their method raised awareness through education (GIZ) and data provision (TMD and CCKM), following Green Governmentality and Administrative Rationalism discourses respectively. But awareness did not always translate to positive action; it missed one crucial point. By giving climate decision makers and experts a superior role in revealing and fixing the problem, local villagers were limited in the ways in which changes in weather and climate pattern could be known and responded to. More importantly, as argued by Susan Darlington (2014) in the context of the Buddhist environmental movement in Thailand, such a kind of discourse overlooks the religious, ethical and moral dimensions of a socio-environmental solution. GIZ's objective in connecting global climate policy with Thailand and subsequently Nan and other provinces through the Ecological Modernization discourse worked as intended, to a certain extent. At the decision-maker level, the climate discourse reshaped existing local policies in Nan Municipality and focused them towards a more integrated climate-development goal for inter- and intra-generational sustainability. Through a global environmental discourse, GIZ enrolled the Thai government officials in, and provided a platform to take part in, global environmental policy making. At the initial stage of the pilot project, however, the global contribution lost out to the more immediate local environmental-development issues in Nan.

These organizations, Climate Service Providers and Climate Policy Facilitators in particular, effectively did two things to the weather of Nan and Thailand: dissected and re-assembled it, and then rendered it accountable, comparable and governable. As shown above, to talk about the climate is to remove culture from the weather; to make a distinction between experts and lay knowledge; and to make a division between the natural and the cultural, the global and the local and the now and the future (Hulme, 2008). The construction of boundaries and binary oppositions (Jasanoff, 1987) that maintains the authority of certain groups and delegitimizes others emerged from the discursive practices of using scientific terminology, graphs, detection and monitoring instruments, claims to knowledge and discourses through a global institution like the UNFCCC, and in connection with other global policy frameworks like Sustainable Development Goals. In the initial phase when Nan was chosen as a pilot city, governing climate change was very much with the limited participation of and by local people.

While the more recent work of GIZ focuses on pursuing energy efficiency and reducing GHG emissions in Nan, solving the problem then requires looking deeper than carbon emissions or reducing the impact of them, simply because, as argued so far, climate change is essentially more-than-climate. Perhaps the Green Radicalism or Civic Environmentalism discourses offered by the Climate Story Listeners could offer an alternative that forms a bridge between the local interpretation of weather and the global environmental policy?

By using communication and education methods the Climate Story Listeners attempt to exchange scientific knowledge for local weather observation and thus open up space for multiple and hybridized weather knowledge. Essentially JK, RECOFTC, TEI and MARCC to a certain extent were going beyond verbal communication, they were linking climate adaptation with local development to achieve long-term resilience. They were trying to reveal local socioeconomic, environmental, climatic and political factors that have resulted in vulnerability and injustice. Because of the additional climatic dimensions, their approach is basically in line with the concept of community-based adaptation, a local development planning that prioritizes the needs of those who are vulnerable to climate change (Forsyth, 2017). This goes beyond Administrative Rationalism because this approach reverts the agency to local people; it also goes beyond Ecological Modernization in that it tries to challenge existing structural rigidity. The Climate Story Listeners, with their Green Radicalism discourse, have tried to preserve pre-existing local knowledge which is used to maintain current livelihoods, and combine it with long-term climatic information to serve the different kinds of needs and goals of the various groups within a community. As regards the Muang Chang's three related religious frameworks described above, Climate Story Listeners could harness the power of Buddhist-inspired development program to encourage climate adaptation activities that that respect the agency of other non-human beings while tackling broader socio-environmental changes (Darlington, 2014).

Conclusion

In this chapter, I have demonstrated the discursive practices of three types of climate organizations that operated in Thailand, in particular in Nan province. The results show that the Climate Service Providers attempted to control the natural forces of the climate, black-boxing the calculation processes from the lay public; the Climate Policy Facilitators also portrayed themselves as the experts, they needed the fear of global climate change to prove new development policies were necessary; and the Climate Story Listeners tried to make sense locally of the universalizing idea of global climate change. These three groups have constructed environmental discourses of Administrative Rationalism, Ecological Modernization and Green Rationalism, respectively. Climate change, according to these organizations, was: a global problem that the Thai people needed to take seriously and act upon; a result of eroding Thai traditions and identity; and a new opportunity to challenge and transform Thai society, and seek social and political justice. The diverse (yet related) ways in which climate change has been portrayed, seen through Jasanoff (2004)'s concept of the "idiom of co-production", show that organizational identities and worldviews cannot be separated from their institutional arrangements and situated practices that co-produce (and reproduce) the ordering of a society. Hence, the practices of linguistic translation, graphical representation and policy (re)formulation were inevitably needed to secure these discourses of climate change in the Thai context. There were complex arrangements of climate actors that performed their work locally, and consequently there were multiple meanings and narratives of climate on which they did not often agree, and for which they often competed against each other.

Additionally, I have also used my ethnographic data from a village in Nan to challenge the discourses and meanings of the so-called "climate change". The northern Thai people think and talk about their local weather, not about the global climate. Their weather is imbued in their culture and livelihoods, and vice versa. The meanings of, and their response to, the changing climate are consequently cultural- and place-based; there are many narratives. I have shown, too, that the Thai villagers were visited by many types of climate organizations, ranging from local to international, that came to fill in the void into which UNFCCC's power does not extend. It was through attending the climate education programs offered by many of these organizations that the lay people were able to adapt at ontological and linguistic levels the foreign terms for "climate", "climate change" and "global warming" that were the main obstacle to attaining higher awareness and knowledge, which is, from the project managers' perspective, a fundamental step toward putting Nan and Thailand on the global climate agenda.

Given the situatedness, performativity, multiplicity and fluidity of climate change and the ways in which it can be governed, I support Zhang (2015) in her attempt to add theoretical depth to the cosmopolitan theory that no longer can we think of it in terms of power flowing from core to periphery, from Europe to Asia, or from international to sub-national levels. Neither can we assume that all climate organizations and individuals respond to such a global call in the same fashion. Rather, climate change itself should be thought as an assemblage of natural and social entities, and there are networks of climate actors both near and distant, each of which has its own histories, ontologies, social spaces and particular goals, that converge to make climate action possible (Blok, 2013; Head & Gibson, 2012). Climate change, as Hulme (2010) and Beck et al. (2013) put it, transcends all boundaries and categories. What is interesting, then, is how boundaries are drawn, shifted, dissolved and redrawn through discursive practices, given the complex power relations between climate actors. As shown, Climate Service Providers and Climate Policy Facilitators have, through using technical terms and a series of workshops, reinforced the lay-expert, weather-climate and local-global climate binary oppositions in order to assert their authority and worldviews. Although their work conflicted with the 'nature' of climate change observed by local northern Thai people, the Climate Story Listeners tried to blur such boundaries. From the perspective of Nan cosmology, an open and fluid world, this may not be the perfect solution but it is a temporary compromise.

In the context of developing countries like Thailand where lay villagers are often rendered ignorant and helpless under the singular, global climate change discourse, I suggest that community-based adaptation (Forsyth, 2017) operated through a combination of Climate Story Listeners and Climate Policy Facilitators could perhaps empower the subjugated. Its goal is to liberate the multiple ways in which one thinks, feels and acts toward one's weather, and thereby to re-connect the humanities, emotions, histories and geographies of the personal weather stories with the totalizing idea of global climate change. In a cosmopolitan world of global risk, this is to anticipate open futures and embrace cosmopolitan climate knowledge, thereby creating robust and resilient development pathways for those who are, and will be, affected in different ways by the changes in climate (Hulme, 2010). In other words, for Thailand's emissions cut and adaptation efforts to be successfully achieved in the way advocated by the Thailand Climate Change Master Plan where the climate targets are implemented by local communities, the operational staff must start listening more to what the weather means for their people, rather than unidirectionally preaching technical terms. This entails formulating a solid polycentric platform that recognizes and welcomes all types of ontological outlook and discursive practice, of both local people and of different types of climate actors. More importantly, in the pursuit of a local more-than-climate development program, non-climatic local risks, which are more immediate and more relevant to people's daily lives, need be mainstreamed with mitigation and adaptation measures. In this way, through reconnecting people's weather-places with the global climate risk climate actions can be made culturally and socially just.

References

Allison, E. A. (2015). The spiritual significance of glaciers in an age of climate change. *Wiley Interdisciplinary Reviews: Climate Change, 6,* 493–508.

Bäckstrand, K., & Lövbrand, E. (2006). Planting trees to mitigate climate change: Contested discourses of ecological modernization, green governmentality and civic environmentalism. *Global Environmental Politics, 6,* 50–75.

Beck, U., Blok, A., Tyfield, D., & Zhang, J. Y. (2013). Cosmopolitan communities of climate risk: Conceptual and empirical suggestions for a new research agenda. *Global Networks, 13*(1), 1–21.

Bhuridej, R.et al. (2018). *Second Biennial Report Update Report.* Retrieved from https://www4.unfccc.int/sites/SubmissionsStaging/NationalReports/Documents/347251_Thailand-BUR2-1-SBUR%20THAILAND.pdf

Blok, A. (2013). Urban green assemblages: An ANT view on sustainable city building projects. *Science and Technology Studies, 26,* 5–24.

Boykoff, M. (2011). *Who Speaks for the Climate?: Making Sense of Mass Media Reporting on Climate Change.* Cambridge: Cambridge University Press.

Brace, C., & Geoghegan, H. (2011). Human geographies of climate change: Landscape, temporality, and lay knowledges. *Progress in Human Geography, 35,* 284–302.

Byg, A., & Salick, J. (2009). Local perspectives on a global phenomenon–climate change in Eastern Tibetan villages. *Global Environmental Change, 19,* 156–166.

Climatological Center. (2014). Educating programme on climate variability and change for users. Available from Climatological Center, Thai Meteorological Department.

Climatological Center. (2016). History of Climatological Center. Available from Climatological Center, Thai Meteorological Department.

Cruikshank, J. (2005). *Do Glaciers Listen?: Local Knowledge, Colonial Encounters, and Social Imagination.* British Columbia: UBC Press.

Darlington, S. (2014). Environmental justice in Thailand in the age of climate change. In B. Schuler (Ed.), *Environmental and Climate Change in South and Southeast Asia* (pp. 211–230). London: Brill.

Davis, R. B. (1984). *Muang Metaphysics: A Study of Northern Thai Myth and Ritual.* Bangkok: Pandora.

Dryzek, J. (2013). *The Politics of the Earth: Environmental Discourses* (3rd Edition). Oxford: Oxford University Press.

Eguavoen, I. (2013). Climate change and trajectories of blame in northern Ghana. *Anthropological Notebooks, 19,* 5–24.

Eucker, D. (2014). Institutional dynamics of climate change adaptation in Southeast Asia: The role of ASEAN. In B. Schuler (Ed.), *Environmental and Climate Change in South and Southeast Asia* (pp. 281–317). London: Brill.

Feindt, P. H., & Oels, A. (2005). Does discourse matter? Discourse analysis in environmental policy making. *Journal of Environmental Policy and Planning, 7,* 161–173.

Fischer, F., & Hajer, M. (1999). *Living with Nature: Environmental Politics as Cultural Discourse.* Oxford: Oxford University Press.

Forsyth, T. (2017). Community-based adaptation. In H. von Storch (Ed.), *The Oxford Research Encyclopedia of Climate Science.* Oxford: Oxford University Press.

Forsyth, T., & Evans, N. (2013). What is autonomous adaption? Resource scarcity and smallholder agency in Thailand. *World Development, 43,* 56–66.

GIZ. (2013). *Response to Climate Change: Strategies for Responding to Climate Change for Nan Municipality.*Bangkok: Office of Natural Resource and Environmental Policy and Planning.

GIZ. (n.d.). *Thai-German Climate Programme.* Accessed April 5, 2020. https://www.giz.de/en/worldwide/67624.htmlHajer, M., & Versteeg, W. (2005). A decade of discourse analysis of environmental politics: Achievements, challenges, perspectives. *Journal of Environmental Planning & Policy, 7,* 175–184.

Head, L., & Gibson, C. (2012). Becoming differently modern: Geographic contributions to a generative climate politics. *Progress in Human Geography, 36,* 699–714.

Huber, T., & Pedersen, P. (1997). Meteorological knowledge and environmental ideas in traditional and modern societies: The case of Tibet. *The Journal of the Royal Anthropological Institute, 3*(3), 577–597.

Hulme, M. (2008). Geographical work at the boundaries of climate change. *Transactions of the Institute of British Geographers, 33,* 5–11.

Hulme, M. (2010). Cosmopolitan climates: Hybridity, foresight and meaning. *Theory, Culture & Society, 27,* 267–276.

Ingold, T. (2010). Footprints through the weather-world: Walking, breathing, knowing. *Journal of the Royal Anthropological Institute, 16*(S1), S121–S139.

Jasanoff, S. (1987). Contested boundaries in policy-relevant science. *Social Studies of Science, 17,* 195–320.

Jasanoff, S. (2004). The idiom of co-production. In Jasanoff, S. (ed). *States of Knowledge: The Co-production of Science and Social Order.* London: Routledge.

Jasanoff, S. (2010). A new climate for society. *Theory, Culture & Society, 27,* 233–253.

Jasanoff, S., & Martello, M. L. (2004). Knowledge and governance. In S. Jasanoff & M. L. Martello (Eds.), *Earthly Politics: Local and Global in Environmental Governance* (pp. 335–350). Cambridge, MA: MIT Press.Lebel, L., Foran, T., Garden, P., & Manuta, B. (2009). Adaptation to climate change and social justice: Challenges for flood and disaster management in Thailand. In P. Kabat, F. Ludwig, M. van der Valk, & H. van Schaik (Eds.), *Climate Change Adaptation in the Water Sector* (pp. 125–141). Abingdon: Earthscan from Routledge.

Miller, C. A. (2004). Resisting empire: Globalism, relocalization, and the politics of knowledge. In S. Jasanoff & M. Long-Martello (Eds.), *Earthly Politics: Local and Global in Environmental Governance* (pp. 81–102). Cambridge, MA: MIT Press.

Moser, S. C. (2016). Reflections on climate change communication research and practice in the second decade of the 21st century: What more is there to say? *WIREs Climate Change, 7,* 145–369.

Muthayya, S., Sugimoto, J. D., Montgomery, S., & Maberly, G. F. (2014). Trends in global rice trade. *Annals of the New York Academy of Sciences, 1324,* 7–14.

Oels, A. (2005). Rendering climate change governable: From biopower to advanced liberal government? *Journal of Environmental Policy and Planning, 7,* 185–207.

Office of Natural Resources and Environmental Policy and Planning (ONEP). (2012). *Thailand Climate Change Master Plan 2012–2050.* Office of Natural Resources and Environmental Policy and Planning. Retrieved from https://actionforclimate.deqp.go.th/?p=6436&lang=en

Office of Natural Resources and Environmental Policy and Planning (ONEP). (2017). *Thailand's Nationally Determined Contribution Roadmap on Mitigation 2021 – 2030*. Retrieved from www.onep.go.th/wp-content/uploads/Thailand-NDC-Roadmap.pdf

Oranop-na-Ayuthaya, O. (2015). INFOGRAPHIC: Raising climate change awareness of communities in the Lower Mekong Basin. *MARCC*. Retrieved from www.mekongarcc.net/resource/infographic-raising-climate-change-awareness-communities-lower-mekong-basin?page=1&search=1resource_

Orlove, B. S. (2009). The past, present and some possible futures of adaptation. In W. N. Adger, I. Lorenzoni, & K. L. O'Brien (Eds.), *Adapting to Climate Change: Thresholds, Values, Governance* (pp. 131–163). Cambridge: Cambridge University Press.

Paerregaard, K. (2013). Bare rocks and fallen angels: Environmental change, climate perceptions and ritual practice in the Peruvian Andes. *Religions, 4*, 290–305.

Roncoli, C., Ingram, K., & Kirshen, P. (2002). Reading the rains: Local knowledge and rainfall forecasting in Burkina Faso. *Society & Natural Resources, 15*, 409–427.

Rudiak-Gould, P. (2012). Promiscuous corroboration and climate change translation: A case study from the Marshall Islands. *Global Environmental Change, 22*, 46–54.

Rudiak-Gould, P. (2013). "We have seen it with our own eyes": Why we disagree about climate change visibility. *Weather, Climate, and Society, 5*, 120–132.

Snidvongs, A., & Chidthaisong, A. (2011). *Thailand's First Assessment Report on Climate Change 2011: Working Group 2 Impacts, Vulnerability and Adaptation*. THAI-GLOB: Thailand Research Fund. Retrieved from https://www.trf.or.th/div3download/default.html

Stryi-Hipp, G., Steingrube, A., & Narmsara, S. (2015). *Renewable Energy Scenarios for the Thai Provinces Phuket, Rayong, and Nan*. Retrieved from www.thai-german-cooperation.info/.../3155324998f206e29cfe74d9b51721caen.pdf

Tschakert, P. (2007). Views from the vulnerable: Understanding climatic and other stressors in the Sahel. *Global Environmental Change, 17*, 381–396.

United Nations. (2018). 'Direct existential threat' of climate change nears point of no return, warns UN chief. Retrieved from https://news.un.org/en/story/2018/09/1018852

Vaddhanaphuti, C. (2017). *Experiencing and Knowing in the Fields: How Do Northern Thai Farmers Make Sense of Weather and Climate-change?* (Ph.D Thesis), King's College London.

Weisser, F., Bollig, M., Doevenspeck, M., & Müller-Mahn, D. (2013). Translating the 'adaptation to climate change' paradigm: The politics of a travelling idea in Africa. *Geographical Journal, 180*, 111–119.

Whitmarsh, L. (2009). What's in a name? Commonalities and differences in public understanding of "climate change" and "global warming". *Public Understanding of Science, 18*, 401–420.

Zhang, J. Y. (2015). Cosmopolitan risk community and China's climate governance. *European Journal of Social Theory, 18*(3), 327–342.

7 Risk perceptions and attitudes toward national energy choices and climate change in Japan and European countries

Midori Aoyagi

Introduction

In this chapter, I discuss risk perceptions and attitudes toward climate change and energy issues in Japan and the impact of the way the media report on those issues. Then using public opinion survey results I compare the public's attitude toward energy and climate change issues as a result of media reporting with those of four European countries. By comparing public attitude formation in Japan with that in European countries, we are able to discuss whether the Japanese public's awareness of climate change is high or low.

First, I briefly discuss climate change and energy policies in Japan after the government signed the Kyoto Protocol. Much progress was made on these policies during and after the period in which the Kyoto Protocol was in effect, especially those relating to the national energy supply, particularly nuclear power generation and renewable energy. Both the Climate Change Countermeasures Implementation Plan and the Basic Energy Plan use top-down approaches, which is why national campaigns are critical in forming the public attitude toward climate change and energy choices.

After introducing Japan's climate targets and its Basic Energy Plan, I examine mass media coverage of these issues in Japan. These climate change and energy policy areas are closely linked, but they are rarely treated as connected policy targets by the Japanese media, which makes it difficult for the public to see energy and climate change as related policy topics.

Finally, I compare the results of public opinion surveys on energy and climate issues in Japan with those in four European countries. In the four European countries, each country implements their own campaign program under the EU climate change program. The EU has implemented several strong economic and social programs to achieve the Kyoto target, such as EU emission trading. The implementation of these programs has had the effect of strengthening the campaign for the climate change. Climate change policies cover a very long time period compared to other policy issues. For example, there are long-term emission reduction targets of 50–80 percent by 2050 and negative emission targets by 2100. Public support for climate policy is essential, and this comparison should improve our understanding of each country's current differences concerning climate change and energy policies.

Japan's carbon emission reduction target and campaign

Japan's carbon dioxide emissions currently depend on economic growth and technological innovation in energy efficiency. Although the Japanese bubble economy burst in the early 1990s, Japan's emissions continued to increase. In the early 2000s, despite the effort to improve energy efficiency in every sector, it seemed that it would be almost impossible to meet the target set in the Kyoto Protocol for the first commitment period from 2008 to 2012.

However, the 2008 global financial shock hit the Japanese economy hard, economic activity suddenly decreased, then carbon emissions decreased dramatically. With the combined effects of the declining economy and some Kyoto mechanisms before the end of the first commitment period, it seemed that Japan had met its Kyoto target. In the 20 years from 1990 to 2010, carbon emissions steadily decreased in the industrial sector, but they dramatically increased in the commercial and residential sectors (Greenhouse Gas Inventory Office of Japan, 2019).

The Japanese government has undertaken a series of campaigns to promote the reduction of greenhouse gas emissions from the residential sector. The Team Minus 6% campaign began in 2005 and was followed by other related campaign programs, all of which are known officially as the National Movement (Kokumin Undo).

The Team Minus 6% program particularly influenced the public perception of the climate change issue. Many companies joined this campaign and could then use its logo in their own advertisements, including in television commercials, newspaper advertisements, magazine tie-in articles, and event campaigns. Individual participation was also encouraged in the program. Under the Team Minus 6% program, Cool Biz, Warm Biz, and other seasonal campaigns were also promoted.

The government invited ambassadors and celebrities to their Cool Biz fashion shows to help market this concept overseas, especially in Asian countries. This was the first time ever that these types of large campaigns were used to raise public awareness in an environmental policy field in Japan. One reason for this was that the Minister of the Environment at the time had originally worked in mass media as an anchorperson at a popular news program, and she suggested the Cool Biz campaign to the bureaucrats in charge of public relations for greenhouse gas emission reduction in the household sector (author interview, 2013).

After the Kyoto Protocol, several greenhouse gas reduction targets were proposed and submitted to the United Nations Framework Convention on Climate Change (UNFCCC) secretariat prior to the ratification of the Paris Agreement. The National Movement was one of the main tools the Japanese government used to start to raise public awareness of tackling climate change. In 2009 the Hatoyama Cabinet stated at the United Nations Climate Conference that Japan's new national target would be a 25 percent reduction so the Team Minus 6% campaign was replaced by the Challenge 25 campaign. This campaign was succeeded by the Fun to Share campaign under the first Abe Cabinet. After the Paris Agreement, the Japanese government announced the Cool Choice campaign in March 2016 and released its 26 percent reduction target in September 2016.

The basic energy plan

The primary source of carbon dioxide emission in Japan is the combustion of fossil fuels for energy. As such, several Japanese ministries that deal with energy issues had to negotiate the emission reduction target with each other, including the Ministry of the Economy, Trade and Industry, the Agency for Natural Resources and Energy, and the Ministry of the Environment. One of the subjects of these negotiations was the Basic Energy Plan, which is the basis for all Japanese energy policies. This plan also has to be consistent with the Japanese carbon dioxide emission reduction target and any emission reduction implementation plans.

Five Basic Energy Plans have been released so far, in 2003, 2007, 2010, 2014, and 2018. Within a year of the third plan being released in 2010, the East Japan Earthquake occurred, and the Fukushima Daiichi Nuclear Power Plant was lost. Because of this disaster, the government had to recreate its entire energy plan. During the re-planning process, many proposals for increasing renewable energy, such as solar, water, and biomass, were put forward, but the voices promoting these plans were barely heard. One such plan was the Deliberative Poll on the Choice of Energy and the Environment carried out by the Energy and Environment Council under the Cabinet secretariat (the then dominant party was the Democratic Party of Japan) from July to August 2012. The result of this deliberative poll showed clearly that the preference of the Japanese public was the anti-nuclear power generation, pro-renewable. The secretariat issued the report on the "Innovative Strategy for Energy and the Environment", which was mainly concerned with how Japan could build a post-nuclear society in late September. The Energy and Environment Council had meetings until November 2012, but after the Liberal Democratic Party won the General National Election in December, it was closed.

The fourth Basic Energy Plan released in 2014, created after the Fukushima disaster, still emphasized the importance of nuclear power. Although the plan frequently referred to safety issues related to the use of nuclear power, it concluded that nuclear power was the most stable base-load source for the national electricity supply and that Japan needed that stable base in its best mix of energy sources.

After the fourth plan was released, the Japanese government submitted its Intended Nationally Determined Contribution (INDC) to the UNFCCC secretariat. Japan's INDC toward post-2020 greenhouse gas emission reductions was set at a reduction of 26.0 percent by fiscal year (FY) 2030 as compared to those in FY 2013 (https://www.env.go.jp/en/earth/cc/2030indc_mat01.pdf).

The fifth Basic Energy Plan was released in 2018 in response to Japan's participation in the Paris Agreement. The plan emphasizes energy security, the environment, economic efficiency, and safety ("3E plus S"). Here, the "environment" refers to de-carbonization, for example, through the promotion of low-carbon energy sources such as renewables and hydrogen.

Mass media coverage of climate change issues

As noted in the previous section, the Japanese government uses mass media as part of its public awareness campaigns. The mass media plays a crucial role in shaping public opinion in every field of policymaking, not only in those related to the environment. The influence of mass media coverage on public opinion has been researched extensively. The impact of media coverage is not a random phenomenon, scholars in the media and communication studies have analyzed several types of media effect: what the topic is, how the issue is framed, and whether or not mass media reporting amplified (attenuated) the public awareness of the various social issues, including environmental issues. One of the most important effects is the "agenda-setting" effect (McCombs and Shaw 1972, Nelson 2010), that is the media decide which topic is to be discussed by the public. Or we can say, "if the media do not cover an issue at all, many people simply would not know about it. This is especially the case for environmental issues, which can often be invisible .... (Priest, 2015, p. 301). Another role is called the social amplification (or attenuation) of risk framework.

> The media have been identified as one of many institutions that can serve to amplify or attenuate societal perceptions of risk, and it is consistent with our general understanding of media agenda setting that the media have the power to focus our attention on certain risks rather than others.
>
> (Priest, 2015, p. 307)

How people perceive risk depends on how the media communicate about the risk. This role of the mass media is called framing.

The reason why mass media can be so influential is that it can reach such broad audiences in the current internet era. No other means has yet been found that can reach a broader public.

For climate change issues especially, "The mass media, and television, in particular, serve as a primary source of information" (Aoyagi, 2017). Even in this internet era, this tendency is still true in Japan (Aoyagi, 2017). Thus, a useful way to trace public awareness of environmental issues is to follow mass media coverage.

According to Sampei and Aoyagi-Usui (2009), mass media coverage has an impact on raising the Japanese public's awareness of climate change issues. In this section, I discuss the chronological changes in the coverage of climate change in newspaper articles to highlight the changes in the public awareness of Japanese citizens on the assumption that a more significant number of articles indicates more public awareness of climate change (Sampei and Aoyagi-Usui, 2009).

It is important to understand that what the public considers to be "environmental issues" may be phenomena "constructed" by the mass media. The "who, what, when, where, why, and how" of the reporting on this issue matters in terms of public opinion. As, Hansen stated:

> [the] public and political agendas of environmental issues in recent years is a simple reflection of a sudden deterioration of the environment, so too would it be misleading to assume that those environmental issues which figure prominently in media and political agendas are necessarily the most immediately threatening or serious (as defined for example by scientists or environmental activists).
>
> (1991, pp. 443–444)

The climate change issue is particularly complicated, and ordinary people have to trust experts who work in this area, even though those experts may only engage in a tiny part of this huge issue. The role of the mass media is to identify the "important" aspects of this huge issue and convey that to its audience.

Media coverage of climate change in Japan depends on major international and domestic events such as the annual COP (Conference of the Parties) meetings, the UN Climate Change Conference, and seasonal events, including the Cool Biz and Warm Biz promotional campaigns. Figure 7.1 shows the changes

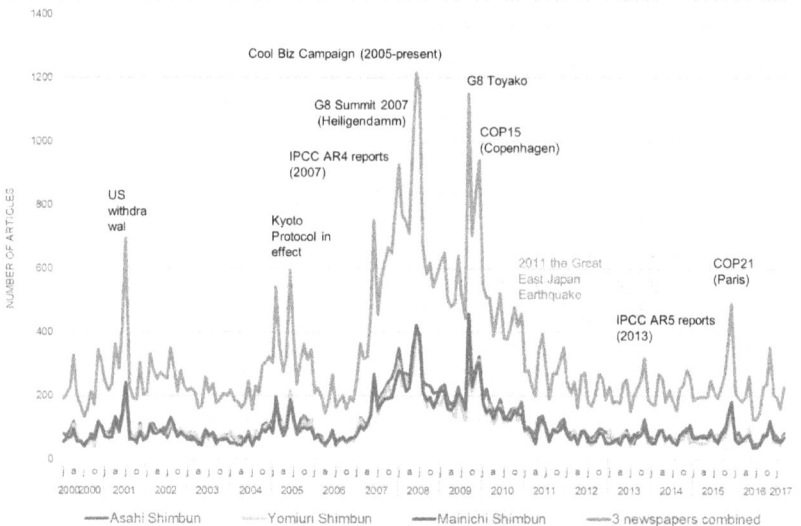

Figure 7.1 Media coverage (number of article) of climate change in three major publications from January 2000 to March 2017

Source: Based on the author's calculation of data using nifty G-search newspaper database.

in the number of articles found by searching a newspaper database using keywords such as "climate change," "global warming," or "greenhouse effect" from 2000 to 2017. According to Sampei and Aoyagi-Usui (2009), not only does the absolute level of media coverage (number of articles published) have an impact on public awareness raising, but so too does the monthly difference.

Several peaks can clearly be seen in the graph. The first occurs in mid-2001 when the United States announced its withdrawal from the Kyoto Protocol. Under the Protocol, the UNFCCC held a meeting in Bonn for the further discussions of the so-called Kyoto mechanisms, including emissions trading, joint implementation, and clean development mechanisms. The Intergovernmental Panel on Climate Change (IPCC) also released its Third Assessment Report in 2001. This series of reports clearly stated that the effects of human-induced warming were already occurring, the influence was widespread, and to stop this progression humans needed to reduce greenhouse gas emissions to below 1990 levels.

The next two peaks are in early and middle 2005. In February 2005, the Kyoto Protocol came into effect after Russia ratified it late in 2004. At that time, the Japanese government was seriously challenged to meet the Kyoto target of a 6 percent emission reduction. The Ministry of the Environment began the National Movement to raise awareness and started the Team Minus 6% and Cool Biz campaigns. The mid-year peak was due to the Cool Biz campaign. The Minister of the Environment repeatedly visited the president of Keidanren (a large Japanese business federation) and the presidents of the large department stores to seek their cooperation in the Cool Biz campaign, especially the proposed new summer business attire. The lower peak of the second half of 2008 coincides with COP11 and related meetings.

There was a series of peaks in 2007 through 2009. In early 2007, the film *An Inconvenient Truth* about former U.S. Vice President Al Gore's campaign to educate people about climate change was released in Japan. In addition, parts of the IPCC's Fourth Assessment Report were released from late January to early May, and the full report was released in November. This fourth IPCC report had a notable influence on people's recognition of the climate change problem (Sampei and Aoyagi-Usui, 2009; Aoyagi, 2017). In June 2007, the G8 held a summit in Germany (Heiligendamm) and it met again in June 2008 at Lake Toyako in Japan, and climate change was a main theme at both summits. A notable peak can be seen at the time of the G8 summit in Japan. The number of articles generally decreased after this, most likely because of the worldwide financial crisis that began in mid-2008. There is a peak again in late 2009, which coincides with the COP15 held in Copenhagen in December. In 2013, the IPCC released its Fifth Assessment Reports, but this did not have a great impact on the number of news articles. The next peak appeared in November and December 2015 during COP21, which was held in Paris.

Japan's Basic Energy Plan and climate change implementation require coordinated and consistent policies. Here I examine how this consistency appeared in media coverage from 2011–2014. Tables 7.1–7.3 summarize how often

Table 7.1 Newspaper coverage: percentage of articles that referred to "climate change" in the "Basic Energy Plan" articles

	2011	2012	2013	2014
Asahi Shimbun	13.4%	6.7%	4.5%	10.6%
Yomiuri Shimbun	21.0%	11.4%	7.5%	10.1%
Mainichi Shimbun	13.1%	17.1%	9.7%	14.1%
Total	15.5%	11.2%	7.1%	12.0%

Source: Based on the author's calculation of data using nifty G-search newspaper database.

Table 7.2 Newspaper coverage: percentage of articles that referred to "Basic Energy Plan" in the "climate change" articles

	2011	2012	2013	2014
Asahi Shimbun	1.7%	0.8%	0.8%	2.8%
Yomiuri Shimbun	2.2%	1.0%	0.9%	1.5%
Mainichi Shimbun	1.4%	1.7%	1.7%	4.0%
Total	1.8%	1.1%	1.1%	2.8%

Source: Based on the author's calculation of data using nifty G-search newspaper database.

Table 7.3 Newspaper coverage: percentage of articles that referred to "nuclear power plant" in the "climate change" articles

	2011	2012	2013	2014
Asahi Shimbun	34.1%	27.3%	15.7%	16.5%
Yomiuri Shimbun	19.6%	13.3%	9.1%	9.6%
Mainichi Shimbun	26.5%	23.3%	16.0%	18.4%
Total	26.9%	21.5%	13.7%	15.1%

Source: Based on the author's calculation of data using nifty G-search newspaper database.

climate change is mentioned in articles that deal with the Basic Energy Plan in three national newspapers. The data in the tables were calculated based on results from a database (the G-search newspaper database offered by nifty) search using keywords such as "climate change," "Basic Energy Plan," or "nuclear power plant."

Table 7.1 shows the percentage of articles that referred to "climate change" in articles where the main subject was the Basic Energy Plan. The average of the three papers combined ranged from 7.1 percent in 2013 to 15.5 percent in 2011. The highest rate in one paper was 21 percent in 2011 (*Yomiuri Shimbun*). Table 7.2 shows the percentage of articles that referred to the

"Basic Energy Plan" in articles where the main subject was climate change. The highest value was 4 percent in 2014 (*Mainichi Shimbun*). It is clear from these results that climate change articles rarely focused on energy issues. This means that mass media framing on either climate change or the Basic Energy Plan was not focusing on the importance of the relationship between climate change policy and energy policy. The Japanese public lost the opportunity to link those two issues. Experts often discussed the risk comparison between nuclear risk and climate change risk, but the general public was oblivious of the comparison.

Table 7.3 shows the percentage of articles that referred to the term "nuclear power plant" in articles where the main subject was climate change. The combined values ranged from about 13.7 percent in 2013 to 26.9 percent in 2011, around the time of the Fukushima Daiichi Nuclear Power Plant accident. These values are relatively high as compared to those for the Basic Energy Plan. This inconsistent media coverage may contribute to the public's difficulty in formulating a consistent view on systems of energy supply.

Risk perception of climate change

In this section, I analyze the differences and similarities between risk perception and attitudes toward national energy choices and climate change among respondents to the surveys from five countries: Japan, the United Kingdom, Norway, Germany, and France. Climate change and the Fukushima nuclear accident are the two major policy factors that impacted energy choices in each country, but the responses vary among these five countries. The surveys, described below, help us to understand public attitudes about energy and climate change. The Fukushima nuclear accident had a significant influence on nuclear power policy throughout the world. Although the accident caused severe damage in Japan, the Japanese 2030 greenhouse gas reduction target still includes nuclear power as the primary source of energy supply, whereas Germany decided to phase out the use of nuclear power in the long run.

The four European countries are very confident about their progress on the Paris Agreement and are encouraging the development and use of renewable energy while phasing out nuclear power. Even France, which depends on nuclear power generation more than any other country, is trying to increase the use of renewables. Germany announced it would stop using nuclear power generation after the Fukushima Daiichi Nuclear Power Plant accident even though previously it had expressed its support for nuclear power (Wittneben, 2012). Hydropower is one of the primary electric power generation sources in Norway, but the country also is rich in oil and exports North Sea gas. The United Kingdom is historically heavily dependent on fossil fuels, and it is now too busy working on the Brexit issue to deal with climate change. The 2010 and 2014 energy plans in Japan were both heavily dependent on nuclear power generation but whether or not they will resume operations at existing nuclear power generation plants is currently unclear.

The survey of the four European countries (Steentjes et al., 2017) was funded in each of the individual countries as part of a larger project (European Perception of Climate Change [EPCC]), and the Japanese survey was done by the National Institute for Environmental Studies. Each survey was fielded in May and June 2016, using comparative questionnaires. The sample size was 1010 in France, 1001 in Germany, 1004 in Norway, 1033 in the United Kingdom, and 1640 in Japan. The survey respondents were adults randomly selected nationwide in each country. The Survey Research Center conducted the Japanese survey, and Ipsos-MORI conducted the European survey.

The EPCC member countries and Japan used the same EPCC survey questionnaires (Steentjes et al., 2017). Each member country was responsible for translating the original version (English) into their local language (Japanese in Japan). The questionnaires covered: (a) perception and awareness, attitudes, and ideas and actions related to climate change risk; (b) energy consumption in everyday life; (c) domestic and international policy on climate change, and support for international negotiations; (d) general support for environmental policies; (e) media exposure; (f) preferences and attitudes toward various types of energy supply; and (g) demographics.

From 83–94 percent of respondents in these five countries think the climate is changing, with Japan having the highest rate (Table 7.4). Germany and the United Kingdom had the highest rate of "not changing" responses, but a large majority of people think the climate is changing even in these countries.

Table 7.5 summarizes the responses to the question, "How worried, if at all, are you about climate change?" The response options of "very" or "extremely" worried were chosen by 63 percent of respondents in Japan, 41 percent in France, 30 percent in Germany, 29 percent Norway, and 20 percent in the United Kingdom. Interestingly, 93 percent of Norwegian respondents thought the world's climate is changing, but only 29 percent selected the "very worried" or "extremely worried" responses (Table 7.5). There was a similar tendency in Germany and the United Kingdom. The reason the Japanese respondents showed a high level of worry might be that they believe climate change is already happening. More than 60 percent of respondents in all five countries chose "We are already feeling the effect (of climate

Table 7.4 "As far as you know, do you think the world's climate is changing or not?"

	Yes, I think that the world's climate is changing	No, I do not think that the world's climate is changing	Don't know
France	92	6	2
Germany	83	16	1
Norway	93	4	3
United Kingdom	86	12	2
Japan	94.1	3.6	2.3

Source: Japan data: Author's survey, European data: Steentjes et al. (2017).

Table 7.5 "How worried, if at all, are you about climate change?"

	Not at all worried	Not very worried	Fairly worried	Very worried	Extremely worried	Don't know	Average
France	5	16	38	29	12	0.2	3.29
Germany	10	21	38	22	8	1	2.96
Norway	9	14	48	24	5	0.2	3.01
United Kingdom	15	23	41	12	8	1	2.73
Japan	1.3	6.6	27.6	41.7	21.6	1.2	3.72

Source: Japan data: Author's survey, European data: Steentjes et al. (2017).

Table 7.6 "When, if at all, do you think [country] will start feeling the effects of climate change?"

	We are already feeling the effects	In the next 10 years	In the next 25 years	In the next 50 years	In the next 100 years	Beyond the next 100 years	Never	Don't know
France	61	16	11	7	1	1	1	2
Germany	60	11	12	5	3	1	5	3
Norway	60	13	9	8	3	2	1	0.3
United Kingdom	61	12	11	6	3	2	3	4
Japan	76.7	11.6	5.4	2.3	0.7	–	–	3.3

Source: Japan data: Author's survey, European data: Steentjes et al. (2017).

change)" (Table 7.6), but more than 77 percent of respondents in Japan chose this option, a much higher rate than those in the other countries.

In all countries, less than 16 percent of respondents think the causes of climate change are either partly or mainly "natural processes" rather than human activity (Table 7.7). Most respondents think that climate change is "partly" or "mainly" caused by human activity. The combined totals for these two categories are 68 percent in Germany, 73 percent in France and the United Kingdom, 78 percent in Japan, and 87 percent in Norway.

More than half of respondents in every country either tend to agree or strongly agree with the statement that they are willing to save energy to help tackle climate change (Table 7.8). About 80 percent of respondents in France and Japan chose these options and about 70 percent did so in Norway. In the United Kingdom and Germany, more respondents chose "neither agree nor disagree" than in the other countries.

These surveys were conducted in May and June 2016, which means the Paris Agreement had already been negotiated, and many countries had said they

Table 7.7 "Thinking about the causes of climate change, which, if any, of the following best describes your opinion?"

Climate change is...	There is no such thing as climate change	...entirely caused by natural processes	...mainly caused by natural processes	...partly caused by natural processes and partly caused by human activity	...mainly caused by human activity	...completely caused by human activity	Don't know
France	1	3	5	36	37	18	1
Germany	7	3	6	34	34	15	1
Norway	0.4	3	6	57	30	4	1
United Kingdom	2	4	8	41	32	11	2
Japan	0.5	3.5	6.3	41.2	36.7	10.1	1.6

Source: Japan data: Author's survey, European data: Steentjes et al. (2017).

Table 7.8 "I am prepared to greatly reduce my energy use to help tackle climate change."

	Strongly disagree	Tend to disagree	Neither agree nor disagree	Tend to agree	Strongly agree	Don't know	Average
France	5	6	9	42	38	0.3	4.02
Germany	7	15	22	39	14	2	3.39
Norway	11	10	7	38	32	2	3.7
United Kingdom	4	14	22	43	16	2	3.55
Japan	3.3	5.4	11.2	46.5	31.7	1.9	3.92

Source: Japan data: Author's survey, European data: Steentjes et al. (2017).

would ratify it or at least expressed positive attitudes toward the agreement. Table 7.9 summarizes the responses to the question, "Do you support or oppose <Respondents' country> being part of this [Paris] agreement?" In all countries except Germany, "strongly support" was the top response: 60 percent in Norway, 48.9 percent in Japan, 43 percent in France, and 35 percent in the United Kingdom. The Japanese government had indicated it would ratify the agreement, but they were slow to take action for ratification. As noted previously, the government submitted its INDC in June 2015 before the COP21 meeting in Paris. This survey was conducted between the Japanese government's submission of its INDC and

Table 7.9 "Do you support or oppose [country] being part of this agreement?"

	Strongly oppose	Tend to oppose	Neither support nor oppose	Tend to support	Strongly support	Don't know	Average
France	3	5	14	32	43	4	4.12
Germany	3	5	17	38	32	5	3.96
Norway	5	3	7	23	60	2	4.32
United Kingdom	2	4	24	32	35	3	3.96
Japan	0.4	1	14.8	29.8	48.9	5.1	4.11

Source: Japan data: Author's survey, European data: Steentjes et al. (2017).

its ratification of the Paris Agreement (November 4, 2016). Mass media coverage of the Paris Agreement was intensive (Figure 7.1). In articles about the agreement, experts frequently discussed the "energy mix" or "energy balance" that would both meet the climate change target and provide an efficient and safe energy supply. But as I showed in Tables 7.1–7.3, when the new Basic Energy Plan was revised in 2014, the mass media coverage on climate change and the Basic Energy Plan were not discussed as a connected issue, most of the Japanese public had idea why those issues were being discussed at the same time. But, in relation to foreign policy, as has been shown by public opinion surveys on diplomacy done by the Cabinet Office of the Japanese government every year since 1975, the Japanese public generally like a positive approach toward the environment in foreign policy. It took the same attitude toward joining the Paris Agreement.

Figure 7.2 shows the distribution of responses in Japan to a question about the preferences of the Japanese public for various domestic energy sources over the next 20 years. Nuclear power received more negative responses than other energy sources. Geothermal, which also had a high level of negative responses, was only offered as a choice in Japan. In the four European countries, hydraulic fracking was shown as one of the options. For reasons of space, I will not show all the graphs of the four European countries, but I will introduce them briefly here. France is well-known for its dependency on nuclear power generation, but the general public does not always support nuclear power generation. In this survey, about 50 percent of respondents chose the "very negative" or "mainly negative" options for nuclear power, while only 20 percent chose the positive options (combination of "very"" and "mainly") for it. Fossil fuels were viewed more negatively in Germany than in France, whereas nuclear power was viewed less negatively. Renewables were viewed as favorable options in all countries including Japan. In Norway, hydropower and solar received the most positive responses (about 90 percent), whereas coal and nuclear had the most negative responses. In the United Kingdom, people generally had a less negative response to nuclear power than respondents in other countries, but coal, oil, and hydraulic fracking received more negative responses. Overall, nuclear power generally had the most negative responses in comparison to the other sources

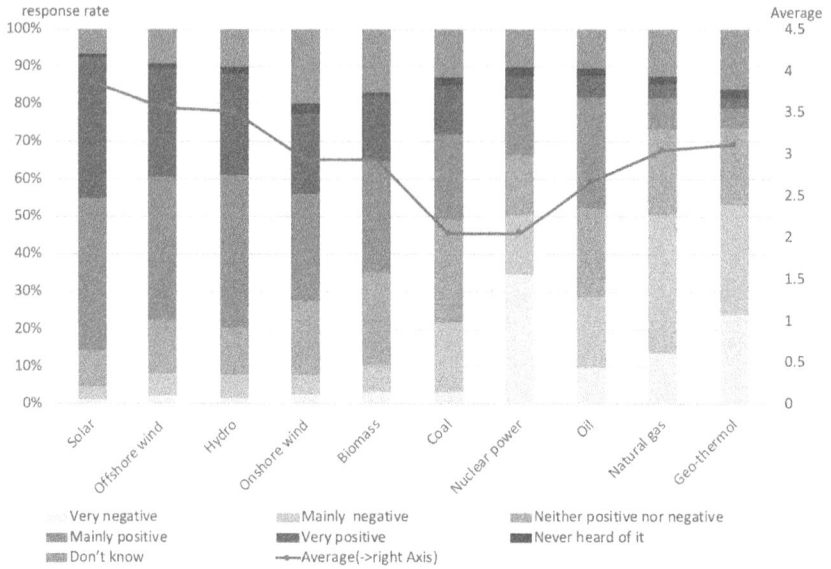

Figure 7.2 Favorability of energy sources (Japan)
Source: Japan data: Author's survey.

in all countries, but responses to the other sources varied from country to country. For example, renewable energy sources had a more positive evaluation in the United Kingdom, Norway, and Germany in comparison to Japan.

The attitudes and risk perception toward climate change and nuclear power are very similar in Germany and Japan, but the political consequences appear to be very different. Although there may be many reasons for this, economic conditions are a possible factor. Germany has had a strong economy for a sustained period, whereas Japan has been suffering from its worst economic conditions in modern history.

Conclusion

In this chapter, I discussed the responses of Japan's mass media, the government, and the public to climate change issues and compared the public's attitude toward climate change and energy to those of people in four European countries. It takes time to go through the complicated negotiations between the various relevant ministries, but the mass media has been paying a great deal of attention to climate change issues and has quickly responded to international and domestic events relating to it. Most likely because of the government campaigns, the public's risk perception of climate change and related issues in Japan is quite high. Governmental campaign have succeeded in raising public awareness toward climate change risk, but as I have shown here, it seems that both the campaign and mass media failed to frame the linkage of climate change issues with energy choice. It takes a long time to effectively and efficiently implement the Paris Agreement, , and it needs careful planning. We have to take into consideration the role of mass media and an effective campaign.

Acknowledgments

This research was supported by the Environment Research and Technology Development Fund (1-1406) of the Ministry of the Environment, Japan. I appreciate the help from the EPCC project (PI: Prof. Nick Pidgeon, Cardiff University, UK), funded by the JPI-Climate Initiative through the separate Research Councils of the United Kingdom, France, Germany, and Norway, which helped make this comparative study possible.

References

Aoyagi M. (2017). Climate change communication in Japan. *The Oxford Encyclopedia of Climate Change Communication.* New York: Oxford University Press.

Boykoff, M. T., McNatt M. B., & Goodman, M. K. (2015). Communicating in the Anthropocene: The cultural politics of climate change news coverage around the world. In Hansen, A., and Cox, R. (ed.), *The Routledge Handbook of Environmental and Communication* (pp.221–231). London and New York: Routledge.

Greenhouse Gas Inventory Office of Japan. (2019). *Japan's GHG Emissions Data (FY1990–2017, Final Figures).* National Institute for Environmental Studies. www-gio.nies.go.jp/aboutghg/nir/nir-e.html

Hansen, A. (1991). The media and the social construction of the environment. *Media, Culture and Society, 13,* 443–458.

McCombs M. E., & Shaw, D. L. (1972). The agenda-setting function of mass media. *Public Opinion Quarterly, 36*(2), 176–187.

Nelson, P. (2010). Reassessing the nuclear renaissance: A historical perspective reveals some unanticipated possibilities for the next 20 years. *Bulletin of the Atomic Scientists,* August 2010. DOI: 10.2968/066004002

Priest, S. (2015). Mapping media's role in environmental thought and action, pp.301–311. In Hansen, A., and Cox., R (ed.), *The Routledge Handbook of Environmental and Communication.* London and New York: Routledge.

Sampei, Y., & Aoyagi-Usui, M. (2009). Mass-media coverage, its influence on public awareness of climate change issues, and implications for Japan's national campaign to reduce greenhouse gas emissions. *Global Environmental Change 19,* 203–212.

Shehata, A., & Falasca, K. (2014). Priming effects during the financial crisis: Accessibility and applicability mechanisms behind government approval, *European Political Science Review, 6*(4), 597–620. DOI: 10.1017/S1755773913000258

Steentjes, K., Pidgeon, N., Poortinga, W., Corner, A., Arnold, A., Böhm, G., Mays, C., Poumadère, M., Ruddat, M., Scheer, D., Sonnberger, M., & Tvinnereim, E. (2017). European perceptions of climate change: topline findings of a survey conducted in four European countries in 2016. Cardiff: Cardiff University. https://orca.cf.ac.uk/98660/7/EPCC.pdf

Wittneben, B. B. F. (2012). The impact of the Fukushima nuclear accident on European energy policy. *Environmental Science and Policy, 15*(1–3). DOI:10.1016/j.envsci.2011.09.002

8 Governing the climate-driven systemic risk in Taiwan – challenges and perspectives

Chia-Wei Chao and Kuei-Tien Chou

Introduction

Extreme weather events and natural disasters have grown in prominence as the main risk that needs to be countered in terms of likelihood and impact based on the Global Risks Perception Survey (World Economic Forum [WEF], 2019). Unlike conventional risks such as chemical pollution or occupational safety, the danger of climate change driven risk is not incremental but an abrupt disturbance to society. Moreover, the ripple effect of climate change will spill over into other political and social challenges. For example, the extreme drought was viewed as a driving force that led to the civil war in Syria, which further creates a refugee crisis in Europe. As the former *U.S. Secretary of State* John Kerry argued that

> I'm not suggesting the crisis in Syria was caused by climate change.... It was caused by a brutal dictator who barrel-bombed, starved, tortured and gassed his own people. But the devastating drought clearly made a bad situation a lot worse.
>
> (VOA News, 2015)

When it comes to Taiwan, the vast majority of citizens are concerned about climate change and it is ranked as seventh in terms of climate risk by Germanwatch (2017). The ripple effect of climate change is not comprehensively analyzed. Hence, this chapter intends to identify the interconnection between climate change and other systemic issues in Taiwan, and to propose a risk governance framework for the nexus.

This chapter is structured as follows, the concept of systemic risk will be introduced, and the interlinkage between climate change and other major systemic risks in modern times. Then the systemic effect of climate change in Taiwan will be analyzed in relation to the carbon bubble and stranded assets issues and the regressive effect of key economic instruments, to provide quantified information about the connection between climate change and financial instability. In the final section, the inability of existing institutions to deal with those systemic risks is explored and the risk governance framework is applied to establish the resilience of the Taiwanese people.

Climate change as a systemic risk

The emergence of systemic risks

As a result of the combination of insurance and safety-enhancing regulation, modern risk management has been a key factor of economic growth and increasing welfare since the beginning of industrial societies. However, the emergence of globally interconnected, non-linear risks such as the global financial crisis, extreme weather events as a result of climate change and growing inequality implies that we are facing non-conventional risks that are not able to be dealt with by existing risk management practices (Renn, 2014, 2017). Those risks are labelled as "systemic risks" which "refers to the risk or probability of breakdowns in an entire system, as opposed to breakdowns in individual parts or components, and is evidenced by co-movements (correlation) among most or all parts" (Organisation for Economic Co-operation and Development [OECD], 2003). This widely cited definition was formulated in the context of the financial system and applied in the process of regulatory reform after the financial crisis of 2008. However, it also attracts the attention of experts in the field of environmental and disaster risk management. At the national level, the Federal Office for the Environment (FOEN) of Switzerland published a report to discuss how to improve the environmental governance to deal with systemic risk in 2015 (FOEN, 2015). At the regional level, the European Environment Agency's (EEA) Scientific Committee organized a seminar to identity the knowledge gap in emerging systemic risks in the context of the green economy and the socio-technological transition in 2016 (EEA, 2016), and elaborate the complexity and uncertainties of systemic risks to health in the European Environment State and Outlook 2020 (EEA, 2019). At the global level, the latest Global Assessment Report on Disaster Risk Reduction published by the United Nations, which is viewed as the comprehensive and authoritative scientific flagship report on disaster risk management, devoted a whole chapter to exploring the approach to integrating systemic risks into the Global Risk Assessment Framework (GRAF) under development (United Nations Office for Disaster Reduction [UNDRR], 2019).

In comparison to conventional risk, systemic risk has four characteristics, they are: (1) global in nature, (2) highly interconnected and intertwined and have complex causal structures, (3) nonlinear in their cause–effect relationships, and (4) stochastic in their effect structure (Renn, 2016, 29–30). Unlike conventional risk, which is familiar not only to experts, but also the general public, and for which the management regimes are therefore relatively stable, the highly interconnected complex causal structure of systemic risk creates a tremendous challenge to cognitive capacity and is difficult to predict, with the result that systemic adverse effects have already happened at the expense of social unrest and ecological collapse.

There are several hazard events that fulfill the above characteristics besides the 2008 global financial crisis, such as the desertification and collapse of the

Aral Sea, fish stocks depletion and pandemics such as Coronavirus (International Risk Governance Center [IRGC], 2018; UNDRR, 2019). However, the authoritative risk governance scholar Ortwin Renn highlights the three main hazardous events that will lead to a systemic effect (Renn, 2017):

1. The growing extent of human intervention in nature, such as the planetary boundaries identified by Steffen et al. (2015), which emphasize the nonlinear and larger scale effect that will happen if the tipping points were crossed.
2. In adequate or ineffective control of the financial system, such as debt crisis happened in 2009.
3. Adverse by-products of globalization and modernization, such as the growing inequality across and within the nations.

The extensive ripple effect caused by systemic risk required a new governance framework to provide guidance for addressing the interlinkage across different domains and drawing the public attention to the non-intuitive phenomenon. Renn (2017) proposed the following supplementary approach to risk governance for systemic risk:

1. Continuous and periodic horizon scanning on the current and possible future contexts which the systemic risk will emerge;
2. Developing narratives and scenarios that provide the causal and functional relationship of the systemic risk;
3. Enhancing the political acceptability of the potential resilience-enhancing strategy to avoid the system breakdown;
4. Including the culture aspect into the management phase owing to the systemic risk will not just be transboundary, but also cross-cultural;
5. Constructing a transparent and two-way communication strategy to involve the stakeholders and public to raise the awareness of non-linear and complex causality of systemic risk, and the benefit of a precautionary approach.

In order to help decision-makers prepare for systemic risks, the International Risk Governance Center published *Guidelines for the Governance of Systemic Risks*, which provides an innovative governance cycle to trigger and facilitate the transition from the previous system to a regime better equipped to cope with systemic risks (IRGC, 2018).

The systemic effect of climate change

As the global risk report identified, the failure of climate-change mitigation and adaptation is listed as the fifth highest risk in terms of likelihood and impact, it will not only intensify the frequency of extreme weather events, but also increase the possibility of food and water crises, which will lead to large-scale involuntary migration (WEF, 2019). These highly interlinked effects highlight the fact that climate change should be prioritized on the global risk landscape.

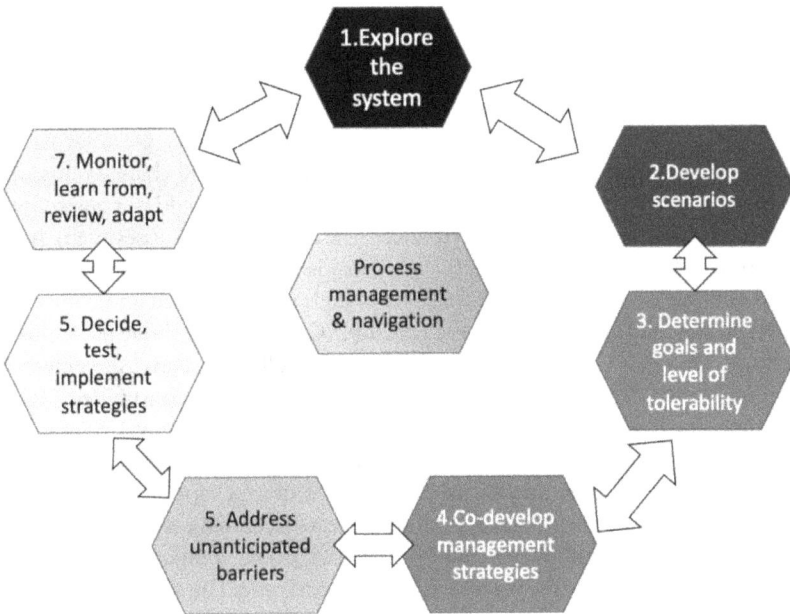

Figure 8.1 Elements of IRGC's systemic risks governance guidelines
Source: Adapted from IRGC, 2018.

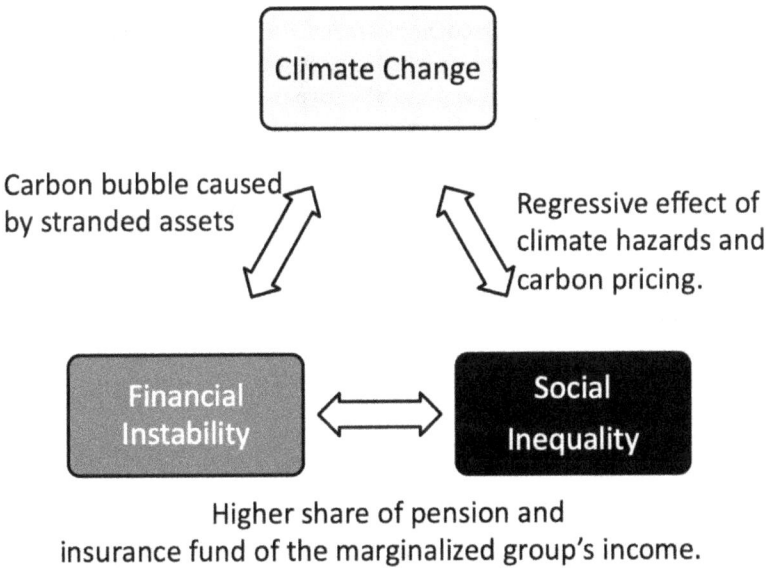

Figure 8.2 The interconnection between climate change and other systemic risks

Furthermore, among the three categories of systemic risks suggested by the experts, climate change has the potential to amplify the other two categories of systemic risks.

Climate change and financial instability

Limiting fossil fuel consumption is an essential task for climate change mitigation, this requires that a significant portion of fossil fuel reserves should not be exploited, and production capacity has to be reduced. According to the assessment by the Carbon Tracker Initiative (2013), the potential listed reserves of coal, gas and oil would emit up to 1541 $GtCO_2$ if those fossil fuels were extracted and combusted. However, the total carbon budget for 1.5 and 2°C of warming above pre-industrial levels are 131–269 GtCO2, which means that to fulfill the Paris Agreement the fossil fuel assets of companies will inevitably have to be constrained. In the meantime, the top 200 fossil fuel companies have a market value of US$ 4 trillion, which will be affected by their capacity to utilize those potential reserves.

Mercure et al. (2018) further investigated the macroeconomic impact of not being able to prepare for stranded fossil fuel assets . According to their study, there would be a global wealth loss of US$1–4 trillion, a loss comparable to the 2008 financial crisis. Moreover, the macroeconomic impact would be distributed unevenly across different sectors and countries. As for the sectors, the extraction industry and utilities are the most vulnerable to the stranded assets. For the existing fossil fuel importers such as the EU, Japan and China, the reduction of fossil fuel imports will rebalance their trade flow and new investment in low-carbon technologies will stimulate their economies. However, the countries that rely on fossil fuel exports, such as Canada, the United States and OPEC countries, will suffer a great loss in terms of economic production and employment. For example, Canada will reduce its economic output by 13 percent in 2035 owing to stranded assets.

The above analysis provides a comprehensive picture of the linkage between climate change and the financial system; however, a more detail assessment is needed to support the risk management strategy. For example, if a large part of fossil fuel reserves are mandated to be kept intact, banks will lose their value by about 1.3 percent, pension funds by 5 percent and insurance companies by 4.4 percent (Weyzig et al. 2014). However, the constraint of the carbon budget not only affects the fossil fuel extraction sector, but also the utilities and other energy intensive industries. If this investment in utilities and other energy intensive industries is taken into consideration, the high carbon exposure will increase to 36–48 percent (Battiston et al., 2017).

2.4   Climate change and social inequality

The increasing income inequality had been a matter of great concern since the 2008 financial crisis. The publication of Capitalism in the Twenty-First Century (Piketty, 2014) and The World Inequality Report 2018 (Alvaredo et al., 2018) drew public attention to this problem. Taxation, education and public expenditure are the main areas of policy connected with this issue. However, any discussion about climate change and inequality is distinctly lacking.

Existing evidence suggests that climate change and social inequality are locked in a vicious cycle. First, climate hazards aggravate the pre-existing socioeconomic inequalities that cause poverty and social exclusion. Second, social inequalities increase the exposure and vulnerability of marginalized communities to climate hazards. Third, the vulnerable and disadvantaged people and communities experience disproportionate losses in terms of their lives and livelihoods. If left unaddressed, the stress induced by climate hazards will make the problem of inequality worse.

At the global level, UNDESA (2016) estimated that 600,000 people lost their lives due to more than 6,400 weather related disasters between 1995–2015, and that low income countries suffered greater losses; the economic losses have accounted for around 5 percent of their economic output.

At the country level, the amplification effect of climate change on inequality can be observed both in developed and developing countries. According to the assessment by Hsiang et al. (2017), the economic damage from climate change in the United States will cost roughly 1.2 percent of GDP per +1°C. Moreover,

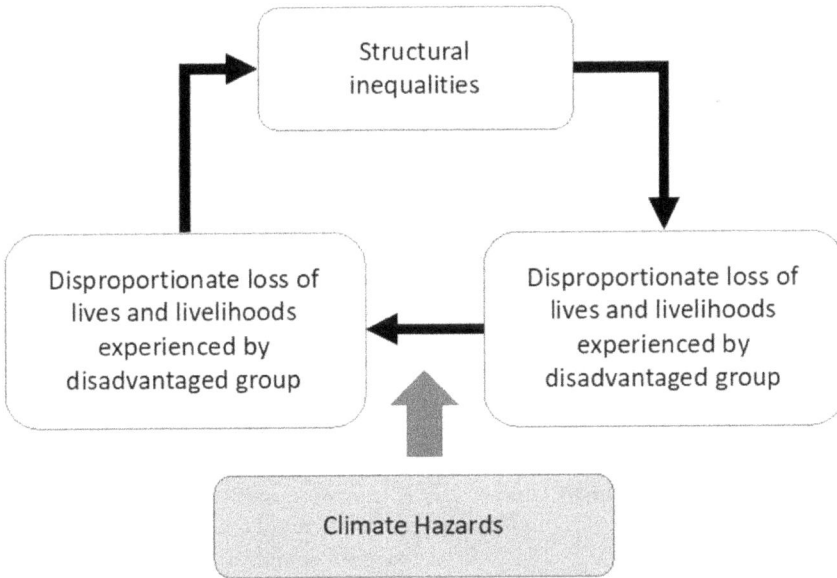

Figure 8.3 The vicious cycle between climate change and social inequality

Source: Adapted from The United Nations Department of Economic and Social Affairs [UN DESA], 2016.

the impact will be distributed unequally, the poorest third of counties are projected to suffer economic damage amounting to between 2 and 20 percent of county income by the end of the twenty-first century. In developing countries, the climate change impact on inequality is prolonged because of inadequate housing. Akter and Mallick's (2013) study shows that 25 percent of poor households in Bangladesh were exposed to the flood caused by Cyclone Aila, however, only 14 percent of non-poor households faced a similar impact.

The nexus between climate change and social inequality also applies to the mitigation measures. Carbon pricing is viewed as a key policy instrument to shape the transition to a low-carbon policy, however, the regressive effect of carbon pricing is a hotly disputed topic. Pricing carbon emissions in developed countries is often believed to harm the poorest part of the population because these households spend a higher share of their income on carbon-intensive goods than the wealthier sections of the population. The analysis of 21 OECD countries revealed that the distribution effect of existing transport fuel and electricity taxes tend to be regressive, as a proportion of net income of the poorest 10 percent these two taxes amount to around 1.7 percent and 0.4 percent respectively, while for the richest 10 percent they amount to around 1.2 percent and 0.1 percent (Flues and Thomas, 2015). Furthermore, the latest study from Rosenberg et al. (2018) shows that a carbon tax will be regressive unless the tax revenues are rebated in the United States. If the tax revenue is used to reduce the budget deficit, the tax burden of the poorest 20 percent will increase by 2 percent in terms of pretax income, while the richest 1 percent will only have to face 0.7 percent increase in their tax burden. Moreover, if the government is forced to use the tax revenues to reduce the corporate tax because of the carbon leakage argument, the progressive effect will be worse. The poorest 20 percent will suffer approximately a 1.6 percent increase in their tax burden, while the richest 1 percent will receive a reduction in tax of about 1.6 percent.

The systemic effect of climate change in Taiwan

Climate change impact in Taiwan

According to the Global Climate Risk Index (Germanwatch, 2017), Taiwan ranks seventh in term of death and property loss caused by extreme weather events, which indicates the its high vulnerability to climate.

Using the highest emission scenario (RCP 8.5) developed by the Intergovernmental Panel on Climate Change (IPCC), Taiwan Climate Change Projection Information and Adaptation Knowledge Platform (TCCIP) assess that the temperature increases at the end of the twenty-first century will range between 3°C and 3.6°C of Taiwan's near-surface temperatures (TCCIP, 2017). For the precipitation, the seasonal mean precipitation will increase by between 10 and 15 percent in the wet season, and decrease by around 20 percent in the dry season in central and southern Taiwan. This shows that increasing precipitation during rainy seasons and decreasing precipitation during dry seasons will pose a great challenge to water resources management. Moreover, as regards typhoons, the most frequent natural disaster in Taiwan, the number of typhoons may decrease owing to global warming, but the intensity and associated extreme rainfall of those that continue will probably increase (TCCIP, 2017).

The carbon bubble of Taiwan's economy

A comprehensive analysis of the linkage between climate change and financial stability in Taiwan should include the financial flow from banks, insurance and pension funds, investment funds and other credit institutions. As the amount of data that can be used is limited, this chapter focuses on the Public Service Pension Fund, the Labor Pension Fund, the Labor Insurance Fund and the National Pension Insurance Fund. Moreover, although there is no big fossil fuel industry in Taiwan, the energy-intensive industries contribute 5.4 percent of total GDP, hence this chapter will assess the exposure of the financial system to climate risk by estimating the proportion of stocks and bonds held by the above pension fund that belong to the main carbon intensive industries.

The first step is to quantify the top carbon emitters in Taiwan. Using the emission data from the greenhouse gas (GHG) inventory registry, the largest ten emitters contribute 108 million-ton CO_2-eq in terms of Scope 2 emission, which accounted for 37 percent of total GHG emissions in 2016.

When the top ten emitters have been identified, the stock and bond holding data of each pension fund provides the basic information to estimate the carbon risk exposure of the financial system. It was only possible to trace the financial flow between the specific company and the fund if the company was listed among the top ten stocks or bonds held by the fund. But in the case of the

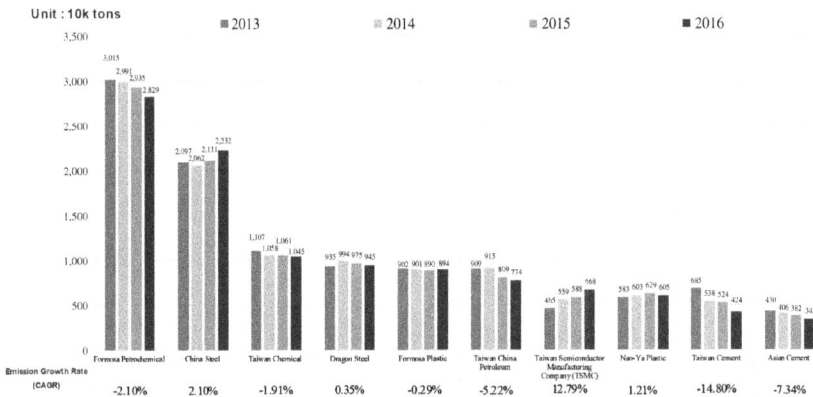

Figure 8.4 The top ten GHG emitters in Taiwan between 2013 and 2016
Source: Risk Society and Policy Research Center (RSPRC), 2018.

labor pension fund and insurance funds, data is also disclosed about the type of domestic stocks held, which provides the information at sector level. Hence, the overall proportion of stock invested in energy-intensive sectors can be quantified to offer a comprehensive picture of the carbon bubble effect of existing pension funds.

Around 25 percent of the stock investment portfolio of the public service pension funds was held in the top ten carbon emitters in 2018, whereas this was less than 14 percent in 2014. The carbon asset risk has increased because of the higher investment flow to TSMC, Formosa Chemical and Taiwan Cement. At the sector level, around half of the overall stock investment portfolio can be attributed to the energy-intensive industries. As for the bond holdings, around 26 percent can be attributed to the top ten carbon emitters and Taipower Company. Compared to 2014, the Formosa plastic bond has vanished and the Taipower bond was slashed by 4 percent, hence the carbon asset risk of public service pension funds is lower in terms of the bond market. To sum up, there is at least 10 percent of the public service pension fund that is exposed to the high carbon asset risk, which implies the performance of fund will be greatly affected by climate policy.

In the case of the labor pension fund, the carbon asset risk reached 23 percent in 2017, in terms of the stock holding in top emitters. The risk was mitigated with the reduction in holding of China Steel stock in 2018, and the holding in Formosa Plastic remained quite stable between 2015 and 2018. In terms of

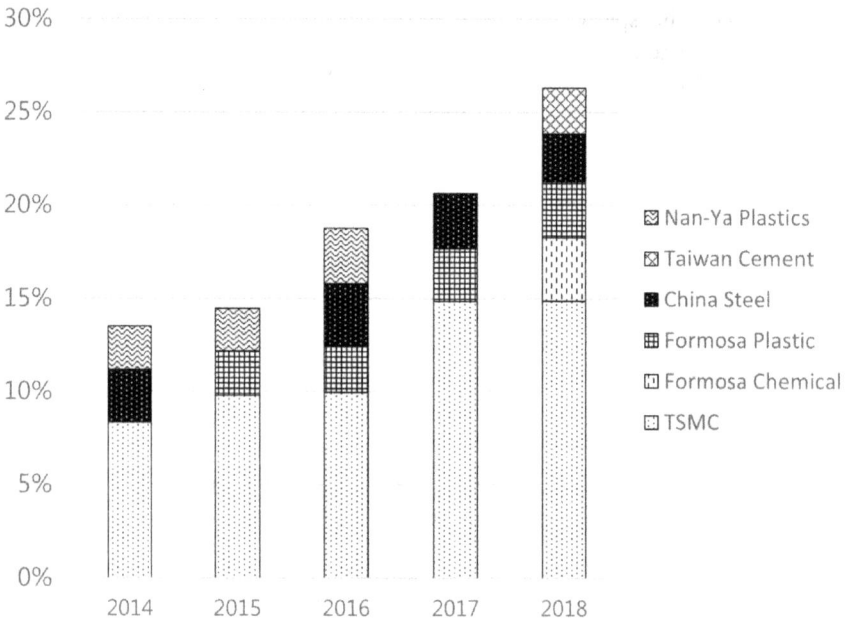

Figure 8.5a The carbon risk assets of the Public Service Pension Fund (stock)

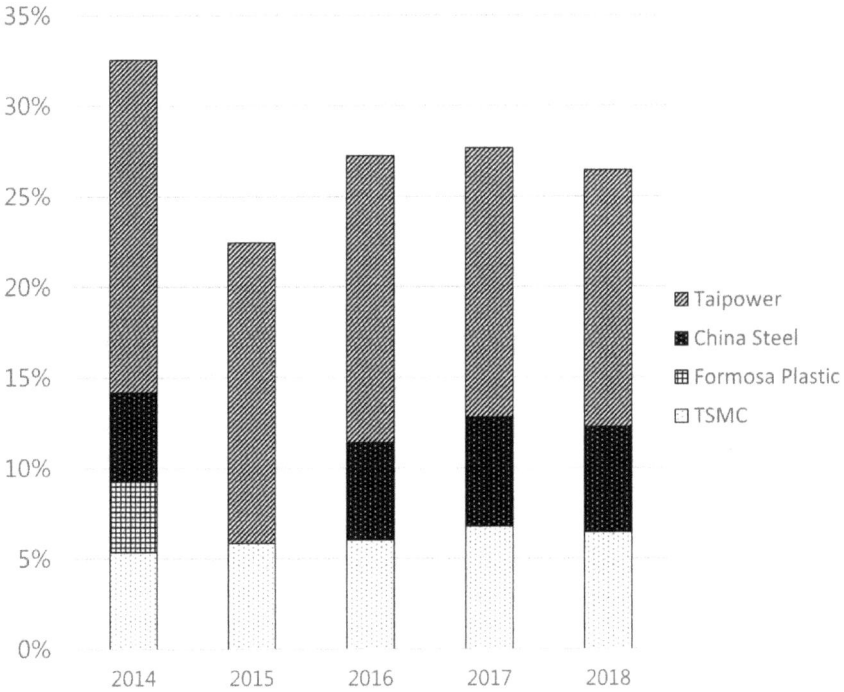

Figure 8.5b The carbon risk assets of the Public Service Pension Fund (bond)

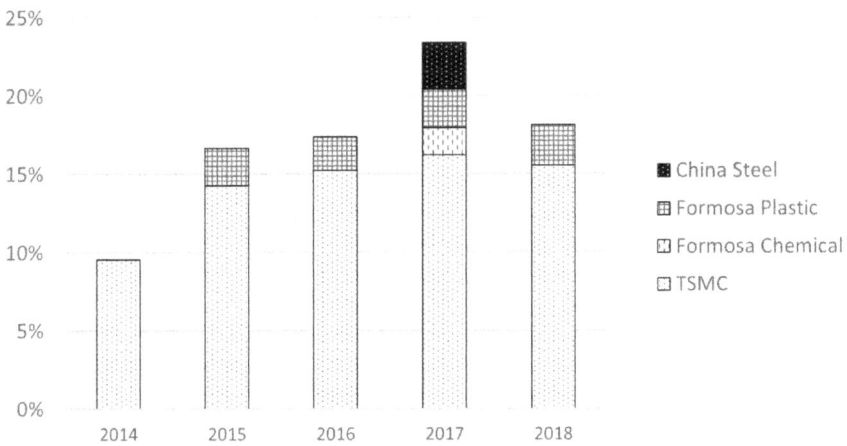

Figure 8.6a The carbon risk assets of the labor pension fund (stock)

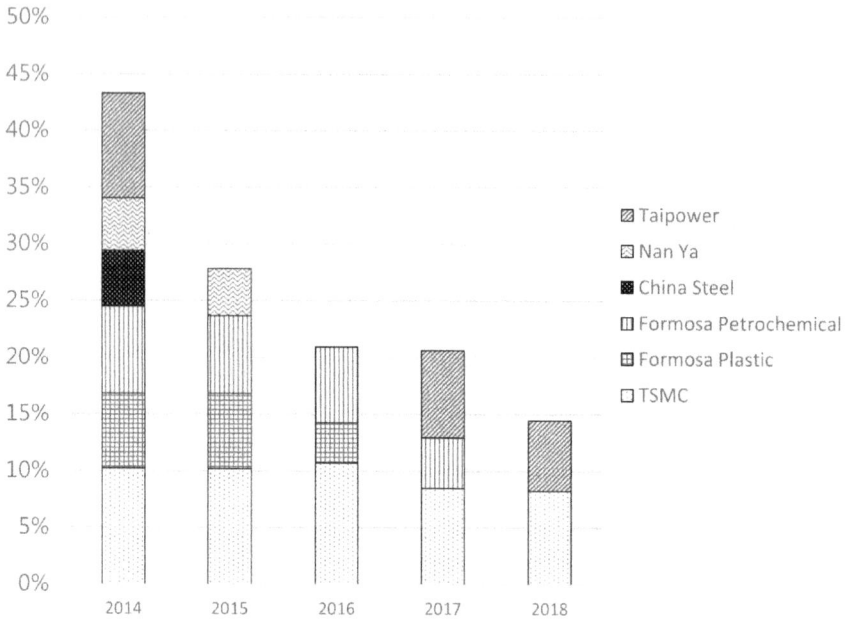

Figure 8.6b The carbon risk assets of the labor pension fund (bond)

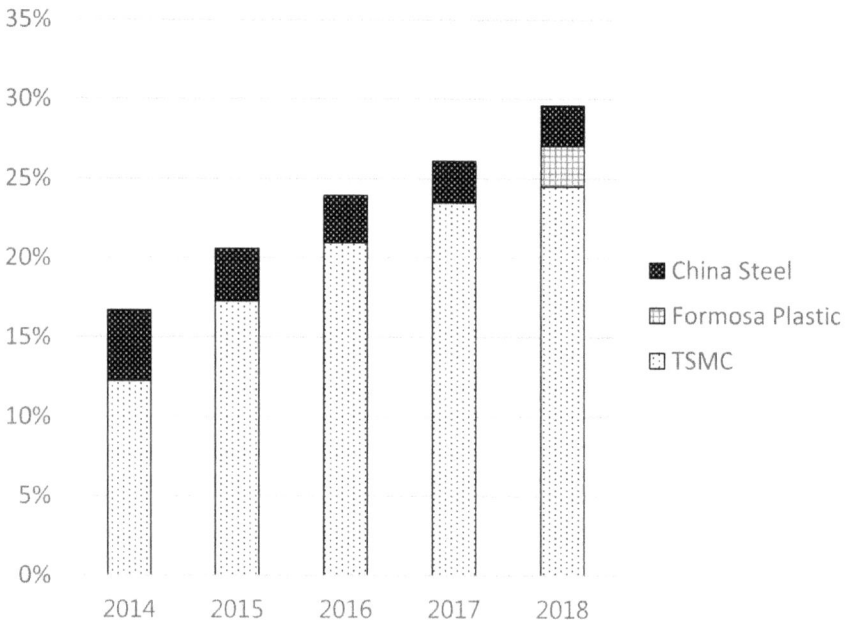

Figure 8.7a The carbon risk assets of the labor insurance fund (stock)

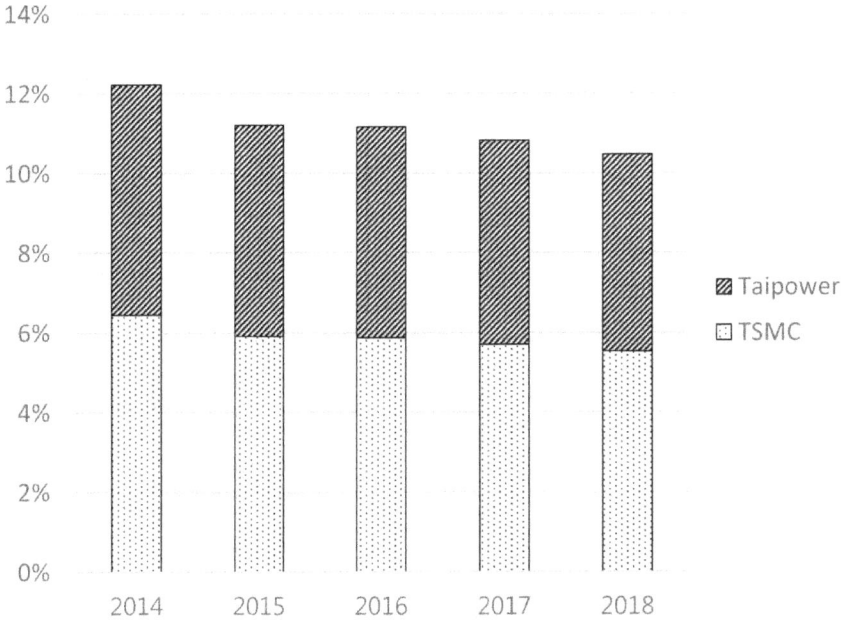

Figure 8.7b The carbon risk assets of the labor insurance fund (bond)

sector level, more than 75 percent of overall stock investment in the portfolio can be attributed to energy-intensive industries, especially the electronics industry. In terms of bond investments, the share of bonds that were invested in the top ten emitters reached 43 percent in 2014, but it was gradually reduced to less than 15 percent, as a result of the bond investment in Formosa Petrochemical being slashed.

In the labor insurance fund, the share of stock held related to top ten emitters grow from 17 percent in 2014 to 30 percent in 2018, mostly due to doubling its share holding in TSMC. However, it also added Formosa Plastic to its portfolio which increased its carbon risk assets. As for the bond market, the trend of carbon risk assets is quite steady and low compared to the portfolio of other funds, range between 10 and 12 percent between 2014 and 2018. However, when it comes to the sector level, around 78 percent of its stock holding was invested in energy-intensive industries, which is the second highest percentage among the four pension funds discussed in this chapter.

In the national pension insurance fund, the financial flow is relatively simple compared to the other funds, it only invested in TSMC, Formosa Plastic, China Steel and Taipower, and most of the investment related to TSMC's stock and Taipower's bond. However, if we compare the portfolio of stock investment and bond market from 2014–2018, the trend is opposite. While the carbon risk assets of stock holding increases in 2018 by adding Formosa Plastic to the portfolio, the carbon risk of bonds is minimized by selling the corporate bond in Taipower.

To sum up, the four pension and insurance funds discussed in this chapter are often viewed as a financial instrument that policy makers can use to stimulate the stock market. From a different perspective, the financial stability of pension fund gets a lot of attention because the existing ruling regime is pushing hard for pension fund reform to avoid the fund going bankrupt. However, there is no discussion about the carbon risk assets of those pension funds, which is contrary to the global narrative and practice. This chapter attempts to scan the carbon risk assets of those four funds through tracking the share of stocks and bonds held in the top ten carbon emitters and energy-intensive industries. The investment portfolio of the public service fund varies, and does that of the other three, the share of non-energy industry is significantly higher. As a result of a large holding in the electronics industry in their portfolios, energy-intensive industry investment accounts for more than 75 percent of the labor pension fund, the labor insurance fund and the national pension fund. Another key piece of energy-intensive industry investment is the petrochemical and plastic industry, which accounts for around 8 percent of those four funds. Hence, it is crucial to include the climate performance of those two industries in the future investment decisions.

The electronic industry contributes 18 percent of the total GDP but simultaneously consuming 18 percent of the electricity consumed nationally. Although the electricity expenditure only represents around 1.9 percent of the total

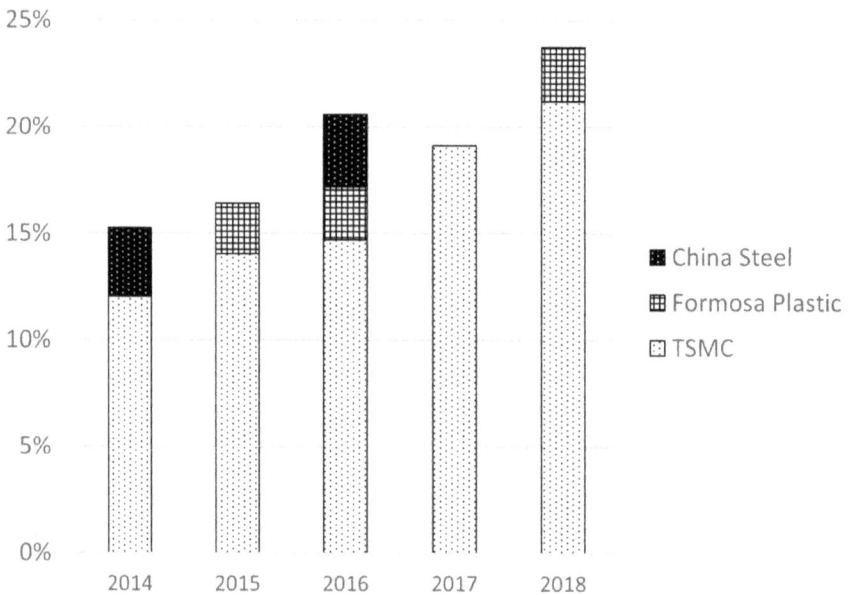

Figure 8.8a The carbon risk assets of the national pension insurance fund (stock)

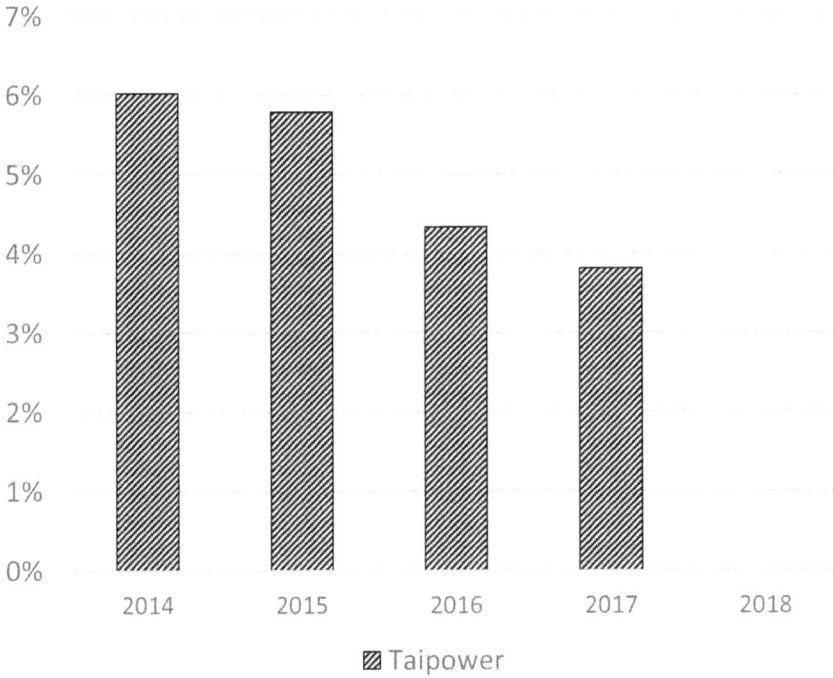

7%

6%

5%

4%

3%

2%

1%

0%

2014   2015   2016   2017   2018

🖾 Taipower

Figure 8.8b The carbon risk assets of the national pension insurance fund (bond)

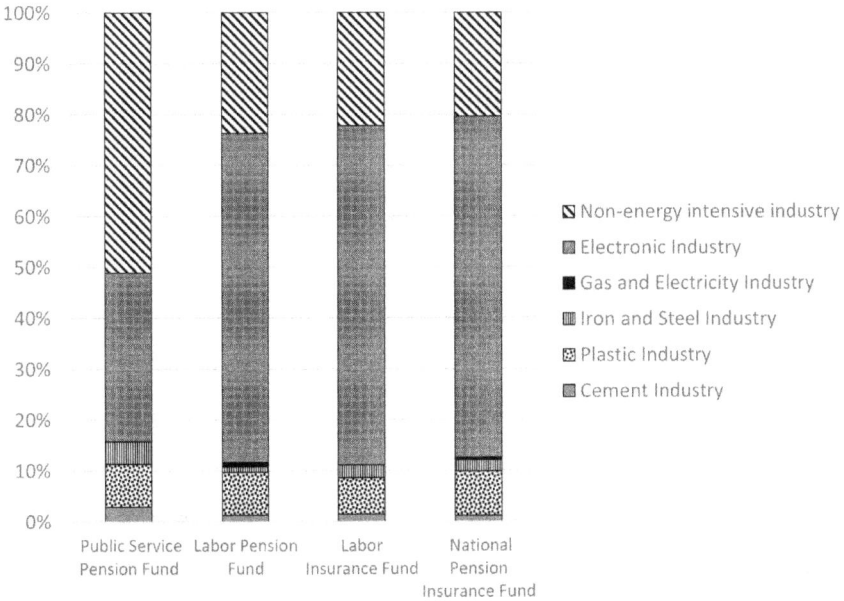

100%
90%
80%
70%
60%
50%
40%
30%
20%
10%
0%

Public Service   Labor Pension   Labor   National
Pension Fund   Fund   Insurance Fund   Pension
Insurance Fund

🆂 Non-energy intensive industry
▦ Electronic Industry
■ Gas and Electricity Industry
▥ Iron and Steel Industry
▨ Plastic Industry
▦ Cement Industry

Figure 8.9 The share of carbon risk assets of the four funds (stock)

production costs, this implies that the incremental effect of climate policy instruments such as carbon pricing is manageable. But energy efficiency regulations such as performance standards or energy saving obligations will still impose a physical constraint on future development. For example, 30 percent of the electronics industry can't fulfill the minimum energy efficiency standard, which is expected to be tightened under the existing policy scheme. If companies cannot comply with that regulation, the expected output growth will be hampered. From a different perspective, TSMC attracts the majority of the investment of the national pension insurance fund, and is graded as the top runner with regard to climate performance. TSMC set up a mid-term climate pledge and committed to purchasing 20 percent of its total electricity demand from renewable power. More importantly, it also adopted internal carbon pricing to prioritize the climate change issue in its business strategy. This approach not only supports the adaptation of TSMC, but also mitigates the carbon risk assets of Taiwan's pension and insurance funds.

The petrochemical and plastic industries, the two biggest petrochemical companies in Taiwan, perform poorly at the Carbon Disclosure Project's (CDP's) climate performance rating: Formosa Plastics and Chinese Petroleum Corporation both get grade C in the CDP ranking in 2017. Moreover, Formosa Plastics ranks bottom out of 22 chemical companies in the climate performance rating (Ferguson et al., 2017). As a result, it is important to adjust the stock holding of Formosa Plastics and its subsidiaries, to lower the carbon risk assets.

The climate change–inequality nexus in Taiwan

With regard to the social inequality issue, the Gini Index of Taiwan has been improved between 2000 and 2013. However, the research team from World Inequality Database suggest that we should use the income share held by the richest 10 percent of the population as the indicator to reflect that the wealth of the economy is distributed unevenly. The national income share of the top 10 percent of the population in Taiwan grew from 31 percent to 36 percent during the same time span, which indicates that the growing inequality should not be neglected.

The linkage between climate change and inequality can be divided into two different impact channels: the disproportionate effect of climate induced disasters and the regressive effect of policy instruments.

In the former case, there is no comprehensive study to assess the distribution of climate impact in specific scenarios in Taiwan. However, the amplification effect of climate change on social inequality can be observed based on a site-specific case study. Lin and Lee (2016) analyzed the effect of Typhoon Marokot, the most deadliest typhoon to impact Taiwan in recorded history, the study estimated the socioeconomic characteristics of affected residents through a regression analysis. According to their research, the farmers, blue-collar labor and the unemployed were more likely to suffer physical and economic loss. For the indigenous people, The risk brought about by the natural disaster is three

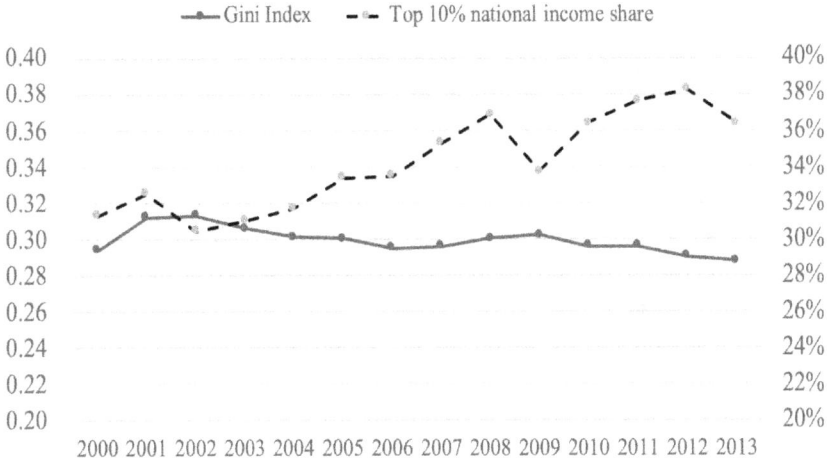

Figure 8.10 The changes in the Gini Index and top 10 percent of national income share

Table 8.1 Share of electricity and fuel spending of consumption expenditures for different income levels

Division of income level	Share of electricity and fuel spending of consumption expenditure
Lowest quintile	3.50%
Second quintile	2.92%
Middle quintile	2.61%
Fourth quintile	2.38%
Top quintile	1.96%
All	2.46%

Source: Estimate based on DGBAS, 2017.

times higher for the indigenous people than the non-indigenous people. Hsu and Hsiao (2017) also examined the linkage between social vulnerability and flood risk. Based on their study on Taipei City and Kaohsiung City, low-income and disabled people are more likely to be affected by serious flooding. The assessment by TCCIP indicated that although the frequency of typhoons might decrease, the intensity will be magnified under the future warming scenario. Based on the above research, it is highly likely climate change will increase the social inequality in Taiwan.

In the latter case, the expenditure on electricity and fuel by the poorest 20 percent is much higher than that of the richest 20 percent of the population in terms of the proportion of overall consumption expenditure. Therefore, the design of any carbon pricing instrument has to take the situation of the poorest into consideration.

According to the Yang and Su (2002), if NTD (New Taiwan Dollar) 572 per ton CO_2 of carbon tax was imposed, it could reduce the overall emission by 6 percent in Taiwan. However, if the tax revenue was used to lower the personal income tax, the Gini Index would increase by 0.38 percent, whereas the income of lowest quintile would decrease by more than 2 percent, the top quintile would decrease around 1.4 percent. Conversely, if the tax revenue was used for lump-sum rebate, the income of lowest quintile would improve by more than 2.7 percent, consequently the Gini Index would be 1.46 percent lower than the baseline scenario.

Bridging the gap of governance

The dysfunctional institution of systemic risk

Based on the above analysis, the interlinkage of climate change, financial stability and inequality, the three thematic systemic risks in Taiwan, cannot be overlooked. More strikingly, those three all rank among the top five risks according to a different risk perception survey carried out on the general public and business leaders (Table 8.2).

Furthermore, the latest survey indicates that more than 86 percent of the Taiwanese agree that the institution investor should take into consideration climate change and low-carbon energy, and a similar proportion of respondents state they don't have any information about carbon risk assets in their pension fund.

To the great concern of the general public, the financial regulator has taken no concrete action to disclose the carbon risk of the investment portfolio. The Financial Supervisory Commission pledged to urge the corporate and public funds to assess and disclose the carbon risk in the Greenhouse Gases Mitigation Action Plan passed in 2018, which is the first time that the carbon bubble was included in the policy agenda.

The National Adaptation Strategy hasn't focused on the link between the climate hazard and socially vulnerable communities, no information and action have been designed specifically to address this issue. However, a very important

Table 8.2 Risk perception of the thematic systemic risks in Taiwan

Survey		RSPRC	Edelman Intelligence	WEF
Year		2015	2017	2018
Risk item	Climate change	68% (1/8)	86% (1/12)	23.3 (6/29)
	Financial crisis	30% (4/8)	–	27.8% (5/29)
	Inequality	44% (2/8)	55% (6/12)	–

report entitled *Climate Change in Taiwan: Scientific Report 2017* (TCCIP 2018), which will be used as the basis for the revision of the National Adaptation Strategy, included a discussion about social and health inequality in its latest version. The report reviewed the local studies of the overlapping of land subsidence and socioeconomic vulnerability, and high-risk of exposure during heatwaves of elderly people who lived alone, but no comprehensive assessment was provided. As a result, the knowledge gap will be the first barrier that needs to be overcome if the inequality issue is to be mainstreamed in the adaptation strategy.

Although a cap-and-trade scheme will be put into practice no later than 2025, and the carbon tax is listed as the priority of energy transition advocates, the regressive effect of carbon pricing receives disproportionate attention in Taiwan. However, it is worth noting that the draft of the energy transition white paper has set up a goal to complete the regulatory impact assessment of an energy tax no later than 2020, in which the distributional impact is specifically listed as a necessary component.

Governance framework for climate change driven systemic risks in Taiwan

There are significant policy gaps that govern the nexus between those three systemic risks—climate change, financial instability and social inequality—in Taiwan, hence this chapter proposes a governance framework based on the suggestions from IRGC (2018).

Step 1.   Explore the nexus between the systems and define the role of the financial regulator and the social welfare department

For this step, the first objective is to frame and define the current and possible future social and political contexts in which the climate-driven systemic risk will emerge. A risk perception survey needs to be carried out by an independent commission to establish public creditability such as the Council for Environment-Related Health Risks in Germany. Moreover, it will enhance the bureaucracy's awareness of its responsibility for risk governance. The risk perception survey should not only present the public concern about systemic risk, but also the interlinkage between megatrends and systemic risk, such as he Global Risks Interconnections Map produced by the World Economic Forum (WEF, 2019).

The second objective is to define the role of each actor in the whole governmental organization, especially the role of the financial regulator and the social welfare department. According to the existing GHG Reduction Action Plan, the financial regulator has already guaranteed to promote the carbon risk asset evaluation and disclosure, but there has been no concrete benchmark for this task. In the social welfare department, the concern about social vulnerability under the forthcoming carbon pricing scheme and the potential impact of the extreme weather events are not well addressed.

Step 2.   Develop scenarios to assess the climate-driven systemic risk

The development of scenarios should be based on an interdisciplinary attempt to provide the causal and functional relationship of the systemic risk. For that reason, the climate change scenario of Taiwan should be developed based on the local narrative on both a mitigation and adaptation strategy. In the meantime, the TCCIP has applied a small scale model to predict the effect on temperature and precipitation of different representative concentration pathways (RCPs). Moreover, an emission scenario that can reflect the shared socio-economic pathways (SSPs) needs to be developed, which can give the public an overview of the transformation of energy and the industrial structure and urban pattern. Based on these scenarios, the carbon risk assets of pension funds can be more predictable, and it would be helpful in the discussion about the carbon bubble effect of the Taiwanese financial system. As requested by the National Adaptation Strategy, the climate relevant authorities have to update the climate change impact assessment every five years. As a result, the impact on social inequality should be targeted as a key research topic during the preparation of the national climate assessment.

Step 3.   Determine goals and level of tolerability for risk and uncertainty

The task of the risk evaluation phase is to diagnose the tipping point and catastrophic effect of the systemic risk to reinforce the political acceptability of the potential resilience-enhancing strategy. The best approach is the mandatory climate change stress test of each pension and insurance fund, which has been widely introduced in the UK, France and the Netherlands (Lehmann, 2020).

Step 4.   Co-develop management strategies and learn from previous programmes

The former steps mainly focus on the role of government authority, but it should be extended to all internal and external stakeholders. There have already been several programmes that focus on carbon risk asset divestment and a community-based adaptation plan which puts great emphasis on social vulnerability. The key actors of those programmes should be invited to share their experience in the multi-stakeholder workshops.

Another important factor is the timing of the strategy formulation. The Greenhouse Gas Reduction and Management Act set up the requirement that the Action Plan should be reviewed every five years (Environmental Protection Administration [EPA], 2018). The climate-driven systemic risk governance strategy is the most suitable example to demonstrate the sector integration and climate-mainstreaming. Hence this chapter suggests that there should be a special chapter to address this issue in the Action Plan.

Step 5.   Address expected and unanticipated barriers of carbon asset risk

The potential barriers to the carbon asset risk evaluation, disclosure and management are the willingness of shareholders of pension funds and the lobby group of energy intensive industries. Therefore, it is important to have a comprehensive disclosure of climate-related risks and opportunities, to rank the climate performance between each asset owner to ease the resistance from peer pressure.

Step 6.   Implement climate-driven systemic risk governance strategy

In order to implement the governance strategy, it is important to break the organizational silo between departments. However, clear accountability about the respective administrative authorities is also essential. The fundamental approach is to set up a time-bound benchmark and require the pension fund operators to complete the carbon asset risk evaluation. The potential impact on socially vulnerable communities can be included as a mandatory element of the impact assessment of a carbon pricing scheme.

Step 7.   Monitor and review the climate-driven systemic risk governance strategy

Governance of systemic risks should be a circular process; hence the monitoring process should be closely connected to the risk perception survey in Step 1. The monitoring framework should not only include a quantitative benchmark but also the relative cognition of different stakeholders. This chapter suggests a risk governance survey of stakeholders should be part of the monitoring framework, which can detect the expert's confidence in the capacity of the governance.

Conclusion

When dealing with transboundary and nonlinear systemic risk such as climate change, policy makers are not able to give certainty to inspire public trust in the regime. Therefore, new mechanisms need to be established to help people to cope with the necessary uncertainty.

As reported by the literature and practical implementations, the nexus between climate change, financial instability and social inequality, the three thematic systemic risks, cannot be overlooked. However, the vulnerability to the climate and the strength of public concern about climate change in Taiwan hasn't led to sufficient and adequate arrangements to deal with climate change driven systemic risk. The high carbon risk assets of the pension and insurance funds is the best illustration of this.

This chapter proposes a governance framework to cope with systemic risks, which includes a risk perception survey carried out by an independent commission that provides the interconnections of each risk; emission and climate scenarios that are constructed based on the local narratives; and a mandatory stress test for publicly owned pension and insurance funds.

References

Akter, S., & Mallick, B. (2013). *The Poverty–Vulnerability–Resilience Nexus: Evidence from Bangladesh. Ecological Economics, 96*, 114–124.

Alvaredo, F., Chancel, L., Piketty, T., Saez, E., & Zucman G. (2018). *The World Inequality Report 2018.* Cambridge, MA: Harvard University Press.

Battiston, S., Mandel, A., Monasterolo, I., Schütze, F., & Visentin G. (2017). A Climate Stress-Test of the Financial System. *Nature Climate Change, 7*(4), 283–288.

Carbon Tracker & Grantham Research Institute on Climate Change and the Environment. (2013). *Unburnable Carbon 2013: Wasted Capital and Stranded Assets.* Retrieved from: www.carbontracker.org/report/wasted-capital-and-stranded-assets

Directorate-General of Budget, Accounting and Statistics (DGBAS). (2017). *Report on the Survey of Family Income and Expenditure.* Taipei: Directorate-General of Budget, Accounting and Statistics.

Edelman Intelligence. (2017). *Green Energy Barometer 2017.* Commissioned by Ørsted. Retrieved from: https://orsted.com/en/Barometer

Environmental Protection Administration (EPA). (2018). *GHG Reduction Action Plan.* R.O.C.(Taiwan): Environmental Protection Administration, Executive (in Chinese).

European Environment Agency (EEA). (2016). *Report of the EEA Scientific Committee Seminar on Emerging Systemic Risks.* Copenhagen: European Environment Agency.

European Environment Agency (EEA). (2019). *The European Environment—State and Outlook 2020: Knowledge for Transition to a Sustainable Europe.* Copenhagen: European Environment Agency.

Federal Office for the Environment (FOEN). (2015). *Systemic Risks and Environmental Governance. Final Report.* Bern: dialog:umwelt GmbH.

Ferguson, C., Crocker, T., & Smyth, J. (2017). *Catalyst for Change: Which Chemical Companies Are Prepared for the Low Carbon Transition?* London: CDP's Sector Research Series.

Flues, F., & Thomas, A. (2015). *The Distributional Effects of Energy Taxes.* OECD Taxation Working Papers 23.

Germanwatch. (2017). *Global Climate Risk Index 2018.* Retrieved from: https://germanwatch.org/de/14638

Hsiang, S. et al. (2017). Estimating Economic Damage from Climate Change in the United States. *Science, 356*, 1362–1369.

Hsu, K. M., & Hsiao, H. H. (2017). Distribution of Flood Control Potential Areas and Socially Vulnerable Groups: Comparison of Northern and High Metropolitan Areas. In Hsiao, H. H., Chou, S. C. and Huang, S. L. (Eds.), *Urban Climate Issues and Governance in Taiwan* (181–225). Taipei: National Taiwan University Press (in Chinese).

International Risk Governance Center (IRGC). (2018). *Guidelines for the Governance of Systemic Risks.* Lausanne: IRGC.

Lehmann, A. (2020). *Climate Risks to European Banks: a New Era of Stress Tests. Bruegel Blog.* Retrieved from: www.bruegel.org/2020/01/climate-stress-test

Lin, T., & Lee, T. Y. (2016). Disaster Risk Cycle: Disaster Potential, Vulnerability and Resilience of Morakot Storm. In: Chou, K. T. (Ed.), *New Theory of*

Sustainability and Ecological Governance (43–86). Taipei: Taiwan Risk Society and Policy Research Center (in Chinese).

Mercure, J. F., Pollitt, H., Viñuales, J. E., Edwards, N. R., Holden, P. B., Chewpreecha, U., Salas, P., Sognnaes, I., Lam, A., & Knobloch, F. (2018). Macroeconomic Impact of Stranded Fossil Fuel Assets. *Nature Climate Change, 8*, 588–593.

Organisation for Economic Co-operation and Development (OECD). (2003). *Emerging Risks in the 21st Century: An Agenda for Action*. Paris: OECD Publishing.

Piketty, T. (2014). *Capital in the Twenty-First Century*. Cambridge, MA: Harvard University Press.

Renn, O. (2016). Systemic Risks: The New Kid on the Block. *Environment: Science and Policy for Sustainable Development, 58*(2), 26–36.

Renn, O. (2017). Risk Governance: Concept and Application to Systemic Risk. In Kasperson, R. E. (Ed.), *Risk Conundrums: Solving Unsolvable Problems (Earthscan Risk in Society)* (243–259). London: Routledge.

Risk Society and Policy Research Center (RSPRC). (2015). *Climate Change Survey by National Taiwan University*. Risk Society and Policy Research Center, College of Social Science, National Taiwan University. Retrieved from: http://rsprc.ntu.edu.tw/fordownload/1061216/2015cop21abstract_%20revise.pdf

Risk Society and Policy Research Center (RSPRC). (2018). *The Overview of the Top Ten GHG Emitters in Taiwan and Their Climate Performance* (in Chinese). Retrieved from: http://rsprc.ntu.edu.tw/zh-tw/m01-3/en-trans/open-energy/925-180411-10ghg.html

Rosenberg, J., Toder, E., & Lu, C. (2018). *Distributional Implications of a Carbon Tax*, Columbia University's SIPA Center on Global Energy Policy. Retrieved from: https://energypolicy.columbia. edu/research/report/distributional-implications-carbon-tax

Steffen, W. et al. (2015). Planetary Boundaries: Guiding Human Development on a Changing Planet. *Science, 347*(6223), 1–10.

Taiwan Climate Change Projection Information and Adaptation Knowledge Platform (TCCIP). (2018). *Climate Change in Taiwan: Scientific Report 2017*. Taiwan Climate Change Projection Information and Adaptation Knowledge Platform. National Science and Technology Center for Disaster Reduction. Retrieved from: https://tccip.ncdr.nat.gov.tw/publish_01_one.aspx?bid=20171220140117

United Nations (UN). (2016). *World Economic and Social Survey 2016. Climate Change Resilience—An Opportunity for Reducing Inequalities*. New York: Department of Economic and Social Affairs of the United Nations

United Nations Office for Disaster Reduction (UNDRR). (2019). *Global Assessment Report on Disaster Risk Reduction (GAR)*. Geneva: UNDRR. Retrieved from: https://www.gar.unisdr.org/sites/ default/files/reports/2019-05/full_gar_report.pdf

Voice of America News. (2015). Kerry: Climate Change Threatens Global Security. Retrieved from: https://www.voanews.com/a/kerry-climate-change-threatens-global-security-worsens-refugee-crisis/3011861.html

Weyzig, F., Kuepper, B., Gelder, J. W. van and Tilburg, R. van. (2014). *The Price of Doing Too Little Too Late; The Impact of the Carbon Bubble on the European Financial System*. Green New Deal Series, Volume 11. Brussels: Green European Foundation.

World Economic Forum (WEF). (2018). *The Global Risks Report 2018* (13th Edition). Davos: World Economic Forum.

World Economic Forum (WEF). (2019). *The Global Risks Report 2019* (14th Edition). Davos: World Economic Forum.

Yang, Z., & Su, H. B. (2002). Tax and Welfare Costs and Structural Effects of Green Tax Reform. *Agriculture and Economy, 29*, 29–54 (in Chinese).

9 Ecological modernization, new technologies and framing in the environmental movement

A climate change mitigation technology (CO_2 capture and storage) and its environmental risk

Hajime Kimura

Introduction

Sustainable development goals (SDGs) set by the United Nations (UN) in 2015 (UN, 2015), are gradually influencing the behavior of various players such as governments, industries and civil societies. For example, the fifth fundamental environment plan in Japan, on which Cabinet decided in April 2018, utilizes the idea of SDGs; the basic idea of the plan is "integrated improvements" in the three spheres: environment, society and economy. Companies too are starting to implement SDGs and to carry out their activities in the context of SDGs as well as corporate social responsibility (CSR). Furthermore, civil societies are expected to be necessary to implement SDGs effectively.

These tendencies (governments, companies and civil societies are starting to be related through SDGs) seem to be preferable in the sense of making it easier to discuss broad topics at the same time in the common framework (using the common language) of SDGs (Kanie and Biermann, 2017). However, there are some problems in deciding how to balance trade-offs intrinsic to our society (Pradhan et al., 2017; Overseas Development Institute, 2017; Intergovernmental Panel on Climate Change [IPCC], 2018). For example, the idea of integrated improvements (of the three spheres mentioned above) in Japan's fifth fundamental environment plan means that the Ministry of the Environment actively considers not only environmental problems (environment sphere) but also aging and depopulation problems in rural area (society sphere), and the development of new environmental technologies and their economic effects (economic sphere). Are trade-offs between these distinct spheres balanced in an appropriate manner? For instance, when companies pursue only a single goal out of 17 SDGs within the range of their business interests, some other aspects of SDGs might fail to be accomplished or even be worse.

In order to balance the trade-offs and to make our society sustainable, in an ideal world, it is preferable to establish 'the universal frame' in a way that reflects various (all the) players' interests appropriately and is also applicable to social decision-making processes. But how do we accomplish this kind of challenge?

In my opinion, an approach that utilizes the perspective of ecological modernization is valid, because it could help us to understand (interpret/explain) the behavior principles both of governments and industries in a realistic and comprehensive manner, while being consistent with the standpoint of civil society. In this context, Hayden (2014) also indicates that ecological modernization "allow[s] for possible common ground among policy-makers, economic elites, moderate environmentalists and others", as described in more detail later.

In this chapter, for the focus is on one of the new industrial technologies, an energy and environmental infrastructure, carbon capture and storage (CCS) technology. This has been developed as one of the climate change mitigation technologies and has received attention both inside and outside Asia as a technology capable of being introduced on a large scale in a relatively short period of time. Various concerns about such things as the use of fossil fuels (coals in particular) and the risk of the CO_2 leakage from storage sites are also noted. How the public and NGOs recognize (or frame) and respond to (or constrain) this CCS technology, and the background of different societies' responses to (or constraints on) CCS technology are analyzed from the perspective of ecological modernization, which could help us to understand (interpret/explain) the behavior principles both of governments and industries in a realistic and comprehensive manner, while being consistent with the standpoint of civil society.

Ecological modernization theory and its modern-day meaning

The concept of ecological modernization

Ecological modernization is a concept that has been given various definitions and used in various ways. Historically, it could be said that ecological modernization has emerged as "the only politically viable response" to the conflict between economic rationality and ecological rationality (Hayden, 2014). Economic rationality is "the secular religion of advancing industrial societies: the source of individual motivation, the basis of political solidarity and the ground for the mobilization of society for a common purpose" (Bell, 1976, pp. 237–238). Consequently, it is one of the driving forces which shape our society. On the other hand, ecological rationality in essence limits economic growth from the viewpoint of the sustainability of our lives. As a result, not surprisingly, it meets resistance to change from business. How can we overcome this never-ending conflict? The only viable solution is the 'win-win' scenarios for both the economy and the environment. In this way, ecological modernization appears to be 'eco-efficiency', which enables us to lower the environmental loads per unit of output, and thus provides us with a way of maintaining our existing lifestyles. It can also be seen as 'green growth', which converts a 'zero-sum' game into a 'positive-sum' game.

According to Hayden, the concept of ecological modernization is used as a social *theory* and a political *discourse* or *program*. As a social theory, it "spotlights positive environmental improvement and seeks to account for the processes behind it" (2014, p. 15). As a political discourse or program, it is used as the expression of political implementation of green growth in the sense described above. Howes argued that

> governments can assist with the transition to a more sustainable low-carbon economy by using ecological modernization to design policies that promote technological innovation, engage with economic imperatives, implement institutional change, improve community engagement and change the public discourse to focus on practical 'win-win' scenarios.
>
> (2018, p. 15)

It also can be said that, in the structure of never-ending conflict, ecological modernization emerged as the social theory that helps us, through its conceptual lens, to discover the opportunity to achieve win-win scenarios and to understand the mechanisms behind it, and then started to work as the political apparatus that put those findings to practical use.

Based on the essence of ecological modernization that tries to reconcile two opposite principles, political attempts always have, to some degree, 'tension' or 'competition' between economic and ecologic rationalities. For example, according to the study by the International Energy Agency (IEA) and the Organisation for Economic Co-operation and Development (OECD) (Unander, 2003), despite the improvement of the efficiency in the carbon emission per unit output in developed countries, "economic growth is the primary driver behind increases in energy-related CO_2 emissions". Thus, gains in eco-efficiency as a result of the policy to promote technological innovation are frequently overwhelmed by economic growth at a faster pace than the improvements. Another criticism of ecological modernization is that the discourse "can obscure the lack of real ecological reform" or "can mask a significant gap between what is said and what is done" (Hayden, 2014, p. 19). Because ecological modernization in essence tries to reconcile two opposite principles, it is always adjacent to the contradiction in the gain and the offset, or the appearance and the actual state.

In order to overcome the limits of ecological modernization above, Hayden suggests the possibility of synthesizing ecological modernization and "sufficiency: questioning the pursuit of ever-growing volumes of production and consumption" in the latter part of the concluding chapter of his book, which roughly means improve ecological modernization by adding 'the limit of growth' to it, while he evaluates "its positive vision promising green jobs, economic opportunity, and technological innovation" (2014, p. 431). However, although he refers to "strong form of ecological modernization" (Christoff, 1996, p. 113) as one of the embryonic versions of such a synthesis (described in detail later for the strong/weak form), it is still necessary to pursue how we can overcome the limit of ecological modernization and construct a modern-day perspective of it that is viable in our practical society.

The modern-day perspective of ecological modernization

According to Buttel (2000), ecological modernization is used as the perspective that capitalism is sufficiently flexible institutionally and capable to be 'sustainable capitalism', under certain social conditions. In this context, the social conditions refer to the pressure from non-governmental organizations (NGOs), the effective policy, etc. In addition, Mol proposed, for an analysis of ecological modernization, the analytical model comprising "distinct areas in society that can be empirically identified": economic, political and socio-ideological spheres. In this context, ecological modernization is explained as "emancipation of an ecological sphere from the economic sphere" (1995, p. 30).

Based on the above understanding, my interpretation of the perspective of ecological modernization is as follows (see Figure 9.1):

(1) Distinct social domains, the economic sphere, the political sphere and the socio-ideological sphere, have their own "behavioral principles (values, goals)", economic principles, political principles and social consideration, which are never explained by each other.
(2) Thus, naturally, the economic sphere will act based on its economic principles, *whether it is ecological or not.*
(3) In this case, it is important to work out how we can constrain the economic sphere to move in an ecological direction. There are two approaches. One is the 'direct' constraint by the social sphere, e.g. environmental movements. The other is the 'indirect' constraint via the political sphere, e.g. regulations.

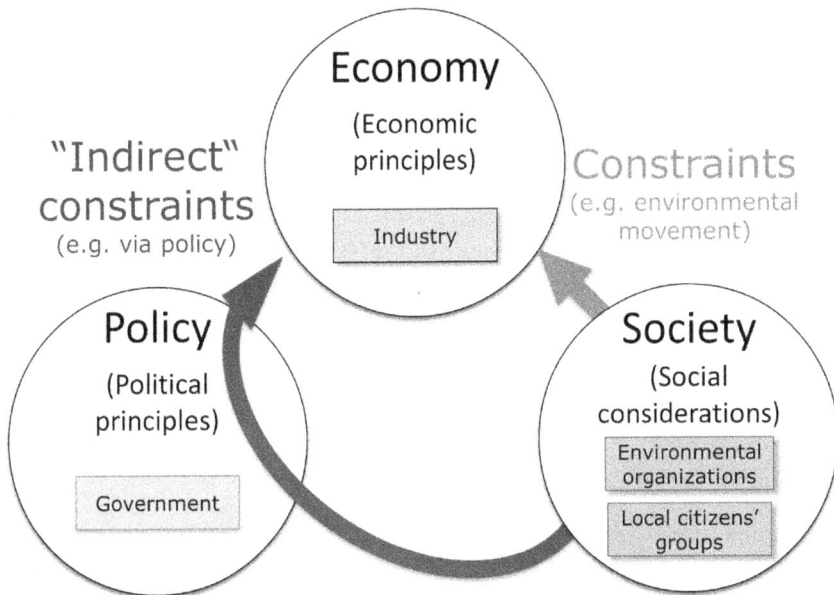

Figure 9.1 The original interpretation of the perspective of ecological modernization

The benefits of taking the above perspective is to be able to understand (interpret/explain) the behavior principles both of governments and industries in a realistic and comprehensive manner, while being consistent with the standpoint of civil society. It could contribute to establishing the universal frame (described above) applicable to social decision-making processes, in order to balance the trade-offs and to make our society sustainable. In this context, Hayden also indicates that ecological modernization allows "for possible common ground among policy-makers, economic elites, moderate environmentalists and others, including much of the public, who are wary of calls to significantly challenge modern consumption patterns" (2014, pp. 16–17).

CCS as a climate change mitigation technology and its environmental risk

Based on the above perspective, the analysis targeted at one of the new industrial technologies, an energy and environmental infrastructure, is shown in the following sections. CCS technology, which has been developed as one of the climate change mitigation technologies, is the energy and environmental infrastructure which reduces the amount of emission of carbon dioxide (CO_2) to the atmosphere, by capturing CO_2 from thermal power plants (using coal in particular), transporting CO_2 to the storage sites located onshore or offshore and putting CO_2 in storage underground (IPCC, 2005). The CCS has attracted attention as the technology capable of being introduced on a large scale and in a relatively short period of time. In fact, IPCC (2018) argues that all of the model pathways for achieving the net CO_2 emissions reductions that would be required to limit global warming to 1.5 °C use carbon dioxide removal (CDR). In these models, CCS makes a significant contribution to CDR, although CDR also includes removals in the agriculture, forestry and other land use sector. One of the reasons why CCS is required to mitigate climate change is that CCS can reduce CO_2 emissions not only from thermal power plants but also from industries such as cement, steel and so on. Furthermore, CCS is the only technology that can realize 'negative' emission beyond 'zero' emission, by being combined with biomass fuels. Thus, CCS is expected to be necessary to limit global warming to 1.5 °C (IPCC, 2018). The report by IEA (2018) also counts CCS among the technologies necessary to meet sustainability goals. It argues that in order to shift from a "central scenario" to a "'sustainable development scenario", CCS must account for a contribution of 9 percent, with energy efficiency improvement accounting for 45 percent, renewable energy 36 percent, fuel-switching 2 percent, nuclear power 6 percent and other 2 percent, respectively. Nevertheless, the progress of CCS deployment both in the power and the industry sector is off track (IEA, 2018).

Although there are great expectations of it, CCS also raises various concerns such as the use of fossil fuels (coals in particular) and the risk of the CO_2 leakage from storage sites (Figure 9.2). For example, the above report by IPCC on

Use fossil fuels

Possible emission of chemical pollutants
(Emission reduction technologies are employed)

Thermal power plant

CO_2 capture plant

Long-time storage of CO_2

CCS :
1. *Capture* CO_2 at power plant
2. *Transport* CO_2 to storage site
3. Put CO_2 in *storage* at underground geological formation

Impermeable layer

Saline aquifer

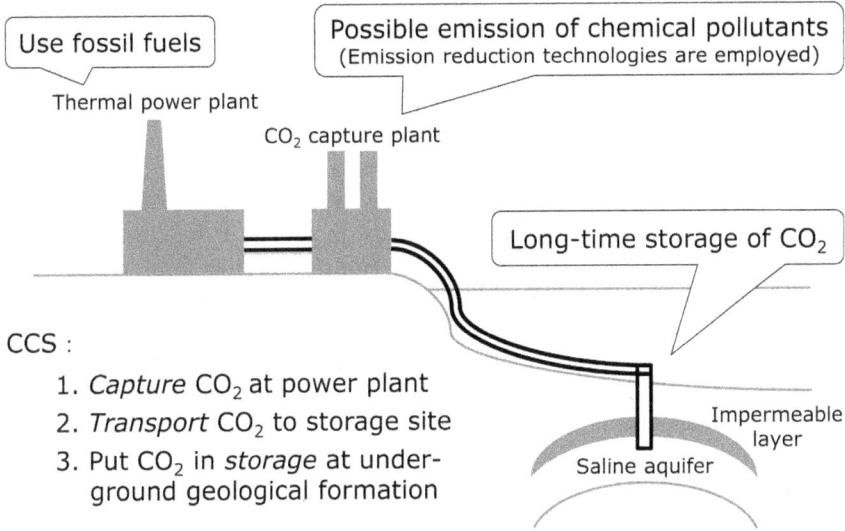

Figure 9.2 CCS technology as a climate change mitigation technology and its environmental risk (simplified schematic)

global warming of 1.5 °C also highlights the linkages between mitigation options and SDGs (IPCC, 2018). There are potential positive effects (synergies) and negative effects (trade-offs) with the SDGs. According to the report, CCS has trade-offs with SDG 3 'Good Health and Well-being' due to "the risk of CO_2 leakage". It suggests also "the use of fossil CCS implies continued adverse impacts of upstream supply-chain activities in the coal sector, and because of lower efficiency of CCS coal power plants, upstream impacts and local air pollution are likely to be exacerbated" (IPCC, 2018, p. 485).

In North America (USA, Canada), CCS has been actively introduced and has been successful commercially, because the existing oil fields are used as CO_2 storage sites where CO_2 is injected underground in order to increase the amount of oil that can be extracted. This technique is called enhanced oil recovery (EOR). EOR using CO_2 (CO_2-EOR) is carried out primarily for economic reasons. However, it also contributes to reducing the amount of emission of CO_2 into the atmosphere. Thus, at least on the face of it, the economic sphere is acting ecologically, although its actions are primarily for economic purposes.

However, CCS has also met with protests in Germany and the Netherlands on environmental grounds, and the research projects were suspended there as a result. In these countries, unlike North America, few oil fields are fitted for EOR. Thus, CCS cannot be promoted purely on economic grounds. Instead, it has to offer a solution in the context of climate change problems. Thermal power plants using fossil fuels (coal in particular) have been swimming against the tide in which CO_2-free energy resources (renewable energy) are preferred to fossil fuels. In this context, from the perspective of economics, CCS technology could be an effective tool for prolonging the life of fossil fuels. This is particularly remarkable in the countries that have abundant natural resources and are highly dependent on them. Norway is one such country but, in fact, also one of the leading players in CCS development.

Some Asian countries, such as China, India, Japan and South Korea, are emitting CO_2 in remarkably large quantities. These countries are also heavily reliant on coal for power generation. Thus, in terms of the climate change problem, these Asian countries in particular could utilize CCS as a climate change mitigation technology. Among these countries, China and India must potentially have economic and/or fiscal reasons to promote CCS, because China and India have vast untapped reserves of coal (IEA, 2017) and China is known to have oil fields fitted for CO_2-EOR. In fact, China's first CO_2-EOR project was initiated in 2009 and China is becoming one of the most active players in CCS development (Asian Development Bank, 2015). In Japan, although CO_2 storage sites offshore are now being surveyed, there are some leading players in CCS technology. For example, Mitsubishi Heavy Industries (MHI), Ltd. delivered the world's largest CO_2 capture plant on a coal-fired power plant to a company in the USA (MHI, 2018). There are also two flagship demonstration projects in Japan run by the Ministry of Economy, Trade and Industry (METI) and the Ministry of the Environment (MOE).

Societies' framing and their reaction to CCS projects

As described in the previous section, CCS technology can be recognized in various ways and thus has proceeded along different paths, depending on the social situations. How do the public and NGOs recognize (or frame) and respond to (or constrain) this CCS technology? Furthermore, what is the background to the different societies' responses to (or constraints on) CCS technology? In the following sections, an attempt has been made to answer these questions based on the above interpretation of the perspective of ecological modernization

Environmental NGOs' framing of CCS technology

Various environmental NGOs have framed CCS technology (Corry and Riesch, 2012). Two axes are used to explain them (Figure 9.3). On one axis there those that are against climate change, from 'direct causes' (on left-hand side) to 'fundamental causes' (on right-hand side). For example, the Bellona Foundation and Green Alliance flame CCS as the approach against the direct cause, namely, as the "effective solution" to climate change (see the lower-left quadrant). On the other hand, Greenpeace focuses on fossil fuel dependency, thus, frames CCS as just as an "end of pipe" solution (see the lower-right quadrant).

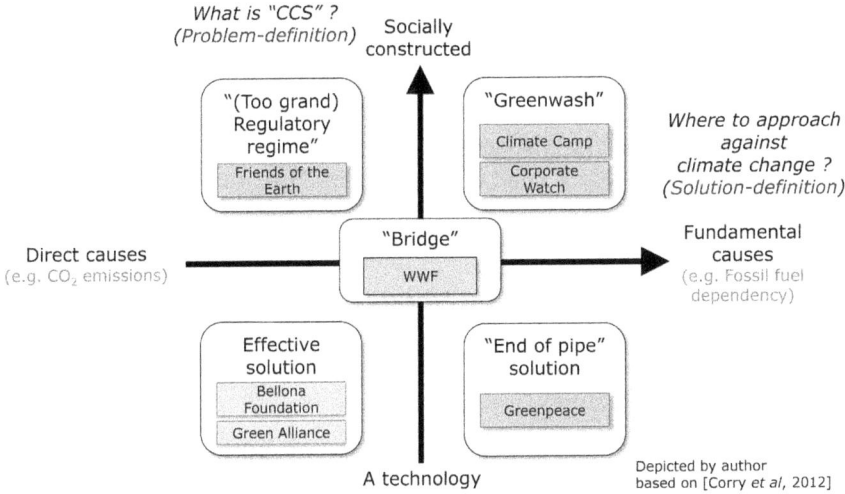

Figure 9.3 The framing of CCS technology by environmental NGOs
Source: Produced by the author based on Corry et al. (2012).

The other axis defines CCS variously as 'just a technology' (at bottom) to 'socially constructed' (at top). For example, Friends of the Earth focuses on the socially constructed aspects of CCS, such as not only CO_2 emission control regulation but also a legal framework for environmental monitoring at CO_2 storage sites and its risk management procedure and other infrastructures needed to implement CCS. In other words, it means that it is too grand to implement all the regulatory regimes needed for CCS implementation (see the upper-left quadrant). CCS is also framed as just a "Greenwash", criticizing vested interests around coal (see the upper-right-hand quadrant).

Supporters of CCS technology often use the frame "effective solution", and another option for supporters is using the frame "Bridge" (see the origin of the axes), which means it is a bridging technology until such time as there is full-scale use of renewable energy. As can be seen, there are various ways of framing CCS technology. Then, how do the public frame and respond to this sort of new technology? Two contrasting cases in Norway, where CCS development is being planned, and Germany, where CCS projects were suspended, and other reference cases in China and Japan are given below.

Case 1: Germany – Framing in local citizen's counter-movements to CCS

In Germany, several local citizens' counter-movements were formed. The frames of four local citizens' counter-movements are shown in Table 9.1. The primary reasons for the movements were concerns about the effect of CCS on the ecosystems and human health, such as using coal, leakage of CO_2 and contamination of drinking water, and therefore NIMBY (Not In My Back Yard). According to Karohs (2013), this local ecological frame also led to a global ecological frame, for example, skepticism about CCS technology, and further, science itself. It was also argued that there was a lack of a democratic decision-making process in CCS and renewable energy was preferable. In other words, to be associated with the frames in Figure 9.3, the frames of "End of pipe" (solution) and "Greenwash" (of using coal and vested interests around it) seem to have been selected in Germany. As a result, CO_2 storage research projects were suspended there.

Case 2: Norway – NGOs have been divided on CCS technology

CCS had a different reception in Norway, where many NGOs have been supporting it for several reasons (Pape, 2015). In Table 9.2, three NGOs (Bellona Foundation, Zero Emission Resource Organization and Friends of the Earth Norway) are supporters of CCS, WWF Norway is in a neutral position, and two NGOs (Greenpeace and Future in Our Hands) are opposed to CCS. A Norwegian environmental foundation, Bellona gets funding from commercial corporations, as well as the Norwegian government. Moreover, historically, a coalition of the environmental NGOs managed to delay the construction of

Table 9.1 Frames in four local citizens' counter-movements in Germany

- Leakage of CO_2 (drinking water contamination, etc.)
- Skepticism about CCS technology, its usefulness as a form of climate change mitigation, and science (credibility) more generally
- Lack of democratic decision-making process
- Against the power company (using fossil fuels)
- Preference for renewable energy

Source: Produced by the author based on Karohs (2013).

Table 9.2 Attitudes of NGOs toward CCS technology in Norway

NGO in Norway	Attitudes toward CCS technology
Bellona Foundation	Leading, permanent and very vocal advocate of CCS
Zero Emission Resource Organization	Actively supporting CCS
Friends of the Earth Norway	Supported and campaigned for CCS
WWF Norway	Largely silent
Greenpeace Future in Our Hands	Openly critical (In recent years, both have been less vocal)

Source: Produced by the author based on Pape (2015)

'conventional (without CCS)' gas fired power plants for about 20 years. In other words, they used CCS as 'rhetoric' for delaying construction of the power plants. Therefore, it could be said that the frame of "Bridge" (Bridging technology) seems to have been selected in Norway, as a whole. Consequently, Norway is still one of the leading players in CCS development.

Case 3: China – public perceptions of CCS technology

In China and Japan, the responses to CCS seem to be relatively silent (compared to Norway and Germany), although CCS projects have proceeded in the both countries. The surveys on the public perceptions of CCS technology in China and the analysis of discourse in Japan are discussed below. In China, the public perception to CCS was lower than most other low carbon technologies. For example, only 38 percent of respondents had ever heard of CCS, while over 90 percent had heard about solar, wind and nuclear energy, and approximately 60 percent had heard about biomass energy and forest carbon sequestration (Li et al., 2014). Another survey suggests that respondents accepted the development of CCS technology in general but were against the implementation of CCS projects near their own houses, in other words NIMBY (Xuan and Wang, 2012). In this context, China's first CO_2-EOR project was initiated in 2009 at Jilin Oilfield, which led China to become one of the leading players in CCS development. Thus, the frame of "Bridge" seems to apply to the Chinese situation, because the project contributes to the reduction of the amount of CO_2 in the air, but the motivation for the CO_2-EOR is not just ecological but also economic.

Case 4: Japan – discourse on CCS technology

In Japan, the framing of CCS in the newspapers and the discourses on CCS governance were analyzed (Asayama & Ishii, 2014). The analysis showed that Japanese newspapers framed CCS as 'Effective solution', neglected the possible frame "End of pipe" solution (it is also called "carbon lock-in" in the original article) and "were very optimistic towards its development/deployment while they discursively neglected the problems such as leakage risk of stored CO_2". Also, the Institute for Sustainable Energy Policies (ISEP), a Japanese NPO, argues that "the realization of Carbon Collection and Storage (CCS) is still a huge technical difficulty and, with the liberalization of energy it's even necessary to reduce the number of coal-fired power plants being considered in Japan and other countries" (ISEP, 2015). Under the circumstances, there are two flagship demonstration projects in Japan, one run by METI and the other by MOE. The project by METI started in 2012, and since 2016, it has "commenced CO_2 injection under the seabed in the offshore area of Tomakomai Port" (JCCS, 2020). The project by MOE started in 2016 and it has designed and constructed the CO_2 capture facility, which could be the world's first power plant to be fueled with bioenergy with CCS on a large scale.

The background of the differences in societies' responses to CCS technology

As described in the previous section, two cases in Germany and Norway had very contrasting receptions (and cases in China and Japan offer other viewpoints). What is the background of the difference in societies' response to (or 'constraints' on) CCS technology in Germany and Norway? It is helpful to use the ideal types of ecological modernization, proposed by Christoff (2009) as follows:

> Different interpretations of what constitutes ecological modernization lie along a continuum from weak (one is tempted to write, false) to strong, according to their likely efficacy in promoting enduring ecologically sustainable transformations and outcomes across a range of issues and institutions.

Table 9.3 represents the ideal types of ecological modernization, which are simplified from Christoff (2009) and the text above. For example, 'weak' is characterized as having a more economic than ecological emphasis. Similarly, technological features are contrasted to institutional ones, and technocratic is also contrasted to deliberative democratic. Christoff (2009) not only suggests that it is possible to emphasize the normative dimensions of ecological modernization, but also indicates that "the political contest between the environmental movement on the one hand and governments and industry on the other is predominantly over which of these types of ecological modernization should predominate".

In the context of the ideal types of ecological modernization, it seems that Norway is in the weak form, regarding CCS technology. In Norway, most NGOs (e.g. Bellona Foundation) tend to be co-opted in exchange for state funding and guaranteed participation in policy making. One of the backgrounds seems to be the energy-intensive industry structure: oil and gas production and its exportation, and heavy power consuming industries (relying on cheap hydroelectricity) like aluminium, silicon and chemical fertilizer. Furthermore, such industries have helped Norway to keep the welfare state. For example, Dryzek et al. (2003) indicated that the pioneering introduction of a carbon tax in Norway (Nordic countries) was for fiscal as much as environmental reasons. In

Table 9.3 The ideal types of ecological modernization

Weak ecological modernization	Strong ecological modernization
Economic	Ecological
Technological (narrow)	Institutional (broad)
Technocratic /neo-corporatist (closed)	Deliberative democratic (open)

Source: Produced by the author based on Christoff (2009).

these circumstances, the promotion of CCS technology, which could be an effective tool to prolong the life of fossil fuels as described above, has its legitimacy to a certain degree. That is because the use of CCS enables the exportation of oil and gas, which, even in the context of climate change problems, leads to (keeping) the welfare state and thus to maintaining citizens well-being.

On the other hand, it seems that Germany is in the strong form of ecological modernization regarding CCS technology. In Germany, even mainstream groups tend to be co-opted less. One of the reasons is that Germany is a heavily industrialized and densely populated country, thus particularly concerned about generating environmental risks. Thus, 'sub-political' spaces of civil society are active (Dryzek et al., 2003). It should also be noted that the CO_2 storage sites were located onshore in the research projects in Germany, whereas CCS development in Norway was planned offshore. In general, onshore CO_2 storage projects tend to be more controversial than offshore, because of the proximity to people's sphere of life (Hammond, 2010). Such a risk perception also probably led to the formation of several local citizens' counter-movements in Germany. In circumstances like this, it is important that there is the trade-off between climate change mitigation and the environmental risks arising from it as described above, and, which should be given priority? There is less motivation to prolong the life of fossil fuels in Germany than in Norway, because the fossil fuels used in Germany are mainly imported, also there is an increasing amount of renewable energy in Germany. Consequently, CO_2 storage research projects were suspended in Germany as priority was given to avoiding environmental risks that might impinge on people's life.

Conclusion

In this chapter, which focused CCS technology as one of the new industrial technologies, the trade-offs intrinsic to SDGs and our society and how they are balanced were discussed, based on the original interpretation of ecological modernization. CCS technology has been developed as one of the climate change mitigation technologies. However, depending on the social background to the technology, it is recognized in various ways and thus has proceeded along different paths. The trade-offs between the environment, society and the economy are dealt with differently in different countries. In North America and China, EOR, which is carried out primarily for economic reasons, also contributes to the mitigation of climate change. In Germany, CCS research projects were suspended on the grounds that priority should be given to avoiding environmental risks that might impinge on people's lives. In Norway, CCS has legitimacy to a certain degree because it enables the exportation of oil and gas, which, even in the context of climate change problems, leads to keeping the welfare state and thus to maintaining citizens' well-being. In Asian countries, what type of paths does CCS proceed? As described above, because China, India, Japan and South Korea are all countries that emit CO_2 in remarkably large quantities and are heavily reliant on coal for power generation, they seem to have the option to utilize CCS as climate change mitigation technology. However, when CCS is developed and deployed both inside and outside Asia, the economic motivation (economic sphere) and fiscal motivation (political sphere) will have to be balanced with social considerations (society sphere).

The benefits of sticking to the original interpretation of ecological modernization proposed in this chapter is to be able to understand (interpret/explain) the behavior principles both of governments and industries in a realistic and comprehensive manner, while being consistent with the standpoint of civil society. As is the case with North America, behavior based solely on economic grounds can lead to environmental benefits, whereas the trade-offs can be remarkable as in the case of Norway and Germany. To establish a universal frame (described above) that can be applied to social decision-making processes, in order to balance trade-offs and make our society sustainable, the ultimate priority should be civil society (social consideration) when facing a choice between the various trade-offs. Furthermore, in order to enhance the ability to reflect civil society's interests in social decision-making processes, it is preferable for civil society to be more active, which is also supported by the fact that active participation of civil societies are expected to be necessary to implement SDGs effectively. Subpolitical spaces in civil society are expected to become increasingly active.

References

Asayama, S., & Ishii, A. (2014). Media Representations and Governance of CCS: Framing and Policy Implications of Japanese Newspapers' Coverage, *Sociotechnica*, *11*, 127–137 (in Japanese).

Asian Development Bank. (2015). *Roadmap for Carbon Capture and Storage Demonstration and Deployment in the People's Republic of China*. Retrieved from https://www.adb.org/publications/roadmap-carbon-capture-and-storage-demonstration-and-deployment-prc

Bell, D. (1976). *The Cultural Contradictions of Capitalism*, New York: Basic Books.

Buttel, F. H. (2000). Ecological Modernization as Social Theory. *Geoforum, 31*, 57–65.

Christoff, P. (2009). *The Ecological Modernisation Reader, Environmental Reform in Theory and Practice*, Routledge.

Corry, O., & Riesch, H. (2012). Beyond 'For or Against' – Environmental NGO-Evaluations of CCS as a Climate Change Solution. In Markusson, N., Shackley, S. and Evar, B. (Eds.), *The Social Dynamics of Carbon Capture and Storage – Understanding CCS Representations, Governance and Innovation* (91–108), New York: Routledge.

Dryzek, J. S., Downes, D., Hunold, C., & Schlosberg, D. (2003). Ecological Modernization, Risk Society, and the Green State. In *Green States and Social Movements Environmentalism in the United States, United Kingdom, Germany, and Norway* (164–191). New York: Oxford University Press.

Hayden, A. (2014). *When Green Growth is not Enough*, Québec: McGill-Queen's University Press.

Hammond, J., & Shackley, S. (2010). Toward a Public Communication and Engagement Strategy for Carbon Dioxide Capture and Storage Projects in Scotland: A Review of Research Findings. *CCS Project Experiences, Tools, Resources and Best Practices.* Retrieved from https://www.google.com/url?sa=t&rct=j&q=&esrc=s&source=web& cd=2&ved=2ahUKEwizy4qhteLoAhWLbN4KHQFKAWAQFjABegQIBRAB&url=h ttps%3A%2F%2Fwww.globalccsinstitute.com%2Farchive%2Fhub%2Fpublications%2F1 79258%2Ftowards-public-communication-engagement-strategy-carbon-dioxide-cap-ture-storage-projects-scotland.pdf&usg=AOvVaw0INhD2f-L_P62o9loJp6pz

Howes, M. (2018). Joining the Dots: Sustainability, Climate Change and Ecological Modernisation. In Hossain, M., Hales, R. and Sarker, T. (Eds.), *Pathways to a Sustainable Economy* (pp. 15–24). Springer.

Institute for Sustainable Energy Policies (ISEP). (2015). [Statement for COP21] Transition to Sustainable Energy by use of 100% Renewable Energy, website: https://www.isep.or.jp/en/304

International Energy Agency (IEA). (2017) *World Energy Balances 2017.* Retrieved from https://euagenda.eu/publications/world-energy-balances-2017

International Energy Agency (IEA). (2018). *Tracking Clean Energy Progress 2018.* Retrieved from https://www.google.com/url?sa=t&rct=j&q=&esrc=s&source=w eb&cd=3&ved=2ahUKEwi7sauNuuLoAhVrJaYKHe65BlQQFjACegQIBBAB&ur l=https%3A%2F%2Fforum.eionet.europa.eu%2Fnrc-energy%2Flibrary%2Feionet-workshops%2F2018-joint-nrc-eionet-meetings-energy-and-climate-change-mitigat ion%2Fpresentations%2F11092018_09_iea_tcep_janoska%2Fdownload%2Fen%2F 1%2F11092018_09_IEA_TCEP_Janoska.pdf&usg=AOvVaw2S21tqvd9GTdF40G HT1MP8

Intergovernmental Panel on Climate Change (IPCC). (2005). *IPCC Special Report on Carbon Dioxide Capture and Storage.* Cambridge and New York: Cambridge University Press.

Intergovernmental Panel on Climate Change (IPCC). (2018). (Special Report) *Global Warming of 1.5 °C.*

(Japan CCS Co., Ltd. (JCCS). (2020). Large-Scale CCS Demonstration Project in Tomakomai Area (Commissioned by METI). Retrieved from https://www. japanccs.com/en/corporate/info.php

Kanie, N., & Biermann, F. (2017). *Governing through Goals – Sustainable Develop-ment Goals as Governance Innovation.* Cambridge, MA: The MIT Press.

Karohs, K. (2013). *A Sustainable Technology? How Citizen Movements in Germany Frame CCS and How This Relates to Sustainability.* Retrieved from www.diva-portal.org/smash/record.jsf?pid=diva2%3A624612&dswid=6351

Li, Q., Liu, L. C., Chen, Z. A., Zhang, X., Jia, L., & Liu, G. (2014). A Survey of Public Perception of CCUs in China, *Energy Procedia, 63*, 7019–7023.

Mitsubishi Heavy Industries (MHI) (2018) MHI's Commercial Experiences with CO_2 Capture and Recent R&D Activities. *Mitsubishi Heavy Industries Technical Review, 55*(1), 32–37.

Mol, A. P. J. (1995). *The Refinement of Production. Ecological Modernization Theory and the Chemical Industry.* Utrecht: International Books.

Overseas Development Institute. (2017). *The Sustainable Development Goals and Their Trade-offs.* Retrieved from https://www.odi.org/publications/10726-sustainable-development-goals-and-their-trade-offs

Pape, R. (2015). Carbon Capture and Storage in Norway – The Moon Landing that Failed. *Air Pollution and Climate series 32.* Retrieved from https://www. airclim.org/publications/carbon-capture-and-storage-norway-%E2%80%93-moon-landing-failed

Pradhan, P., Costa, L., Rybski, D., Lucht, W., & Kropp J. P. (2017). A Systematic Study of Sustainable Development Goal (SDG) Interactions. *Earth's Future, 5,* 1169–1179.

Unander, F. (2003). *From Oil Crisis to Climate Challenge: Understanding CO2 Emission Trends in IEA Countries.* International Energy Agency (Paris) and Organization for Economic Co-operation and Development. Retrieved from https://inis.iaea.org/search/search.aspx?orig_q=RN:35066358

United Nations. (2015). *Transforming Our World: The 2030 Agenda for Sustainable Development.* Retrieved from https://sustainabledevelopment.un.org/post2015/transformingourworld/publication

Xuan, Y., & Wang, Z. (2012). Carbon Capture and Storage Perceptions and Acceptance: A Survey of Chinese University Students, *IPCST, 38.* Retrieved from http://cstm.cnki.net/stmt/TitleBrowse/KnowledgeNet/CDYA201203012022?db=STMI8319

Part III
Local governance on climate change adaptation

10 Tracing sustainability transitions in Seoul governance

Enabling and scaling grassroots innovations*

So-Young Lee, Eric Zusman, and Seejae Lee

Introduction

The South Korean government began to promote Green Growth as an alternative to traditional capital-intensive development models in 2010. Advocates of Green Growth were nonetheless largely silent on the role that citizen participation could play in mobilising support for a sustainable future society. In 2011, the newly elected Seoul Mayor, Won-Soon Park—in his third term in 2018—introduced a series of administrative reforms that opened a two-way dialogue between the citizens and local government. These reforms injected new life into the city's grassroots innovations and either directly or indirectly enabled several progressive social innovations to become part of the city's organisational and institutional architecture. Prominent examples of these innovative additions included: the sharing city project, urban agriculture, one less nuclear plant, a participatory budgeting system, community restoration, a listening open forum, etc. To varying degrees, these initiatives placed an important participatory dimension at their core. By all accounts, this participatory element has helped Seoul transition from a nominally green city into one that preaches and practices sustainability. For many observers, Seoul has begun to take critical steps on the path to a sustainability transition.

How far this participatory Seoul administration has moved down this transitional pathway remains nonetheless an open question. A growing body of sustainability transitions literature can help answer this question. A core insight from the transitions literature is that the critical first step in a transition is the creation of a 'niche' wherein new innovations begin to seed change and take root. The formation of a niche can then pave the way for longer lasting and broader level 'regime' and 'landscape' changes (Frantzeskaki and Loorbach, 2010; Rotmans et al., 2001; Kemp et al., 2001, 2007a, 2007b; Loorbach, 2007; Smith et al., 2005). A complementary body of work on grassroots innovations suggests that 'intermediary' change agents can support and scale not only technological but also social innovations (Hargreaves et al., 2013; Howells, 2006; Seyfang and Smith, 2007; Seyfang, 2009). This chapter adopts insights into sustainability transitions to understand why different innovations enjoyed varying degrees of success in moving along a transitional pathway.

Tracing the processes of change within the Seoul administration underlines the importance of citizen engagement in augmenting the impact of grassroots innovations. It further underscores how enablers in the transition process, i.e. social learning, networking, reflexive governance, and transformative partnerships can help scale changes to and within the regime level. The analysis also illustrates the importance of distinguishing between organisational and institutional change within the regime level. As will be seen, some of the key innovations in the transformation of Seoul's administrative policies, i.e. the listening open forum and citizen communication, have yet to become institutionalised. The chapter thus shines a light on why some innovations induce only organisational change while others accede to a consolidated institutional change. The subtle but critical distinction between organisational and institutional change has implications not only for the transition in Seoul, but also for a broader literature on transition cases outside the public sector (Loorbach and Wijsman, 2013) and beyond Asia's fast-changing cities (Hamann and April, 2013; Magdalena and Eklund, 2015).

The remainder of the chapter is divided into three sections. The next section synthesises literature on sustainability transitions. The following section begins to apply insights from that literature so as to illuminate the drivers and the extent of transformational change across several projects in Seoul governance. The final section discusses the implications of the Seoul participatory administrative governance case for transitions research in and beyond Seoul.

Sustainability transitions and scaling innovations

The imperative to grow economies within ecological limits has yielded myriad sustainability solutions. Work on sustainable transitions has helped better understand the co-evolutionary processes influencing the selection and integration of these solutions into fully-formed development paths. The point of departure for much of this work is 'innovation' (Smith et al., 2005; Frantzeskaki and Loorbach, 2010). Innovation, moreover, does not only involve creating and disseminating new technologies. It can also consist of initiating far-reaching social and lifestyle changes that "may be different from the mainstream but more adapted to [society's] needs" (Akenji, 2014: 21). Moreover, not only technological but economic and social considerations influence the degree and scale of change (Kemp, 1994).

Efforts to clarify how socioeconomic conditions influence the scale of change have led to several useful illustrations of the set of interactions shaping transitions. One of the more illuminating heuristics depicts sustainability transitions as evolving out of path-dependent interactions within and across nested niche, regime, and landscape levels. This multi-level perspective posits that transformative change often involves innovations in micro-level niches. Niche innovations interact with socioeconomic regime level institutions and infrastructures. These mid-level institutions and infrastructures interact with wider landscapes of cultural values and political architecture (Geels and Schot, 2007; Lachman, 2013). In sum, sustainability transitions complete when a constellation of social, economic, and technological factors align at the niche, regime, and landscape levels (Kemp et al., 2001; Geels, 2002).

The sustainability transitions literature has also shown that achieving transformative change is not easy. Established interests may conspire to undermine radical change. The incumbent sociotechnical systems may help vested interests lock in the status quo (Sanden and Azar, 2005; Frantzeskaki and Loorbach, 2010). These barriers to change may however be surmountable. Two branches of action-oriented sustainability transitions literature counsel ways they can be overcome. 'Strategic Niche Management' (Kemp et al., 1998; Raven, 2006; Schot and Geels, 2007, 2008) has looked at "how niches grow, stabilise or decline in interaction with the dynamics of prevailing regime" (Markard et al., 2012: 957; Raven, 2006). Meanwhile 'Transition Management' literature argues that breaking through barriers involves an open-ended *process* that permits stakeholders to determine how to alter that prevailing system (Markard et al., 2012; Loorbach, 2007; Loorbach and Rotmans, 2010). In sum, the extent of such change may rest on the interaction between process and structural dynamics at different levels. For Seoul, the dynamics interplay between the niche and regime levels was particularly important.

To varying degrees, the literature converges on several important sets of factors that could affect these processes and structures moving from the niche to the regime level. 'Social Networks', for instance, can serve as platforms that unite fringe actors with established players to engage in mutually beneficial change processes (Von Malmborg, 2007; Khan, 2013). Networks can help non-governmental organisations (NGOs), businesses, consumers, and academics strengthen niches and adjust the direction of the transition (Hargreaves et al., 2013; Seyfang and Smith, 2007; White and Stirling, 2013). Well-designed networks can help enhance 'Social Learning' processes through "experimentation and pilot projects, the exchange of experiences, [and] training and competence building" (Kemp et al., 2007a: 327; Nevens et al., 2013). 'Reflexive Governance', for another example, involves forging partnerships between varied actors with diverse beliefs. These partnerships can be created through well-managed processes of deliberation on decision-making that help stakeholders arrive at shared visions of a sustainable future (McCormick et al., 2013). Others have noted that 'Transformative Partnerships' can operate on an even wider meta-scale, fostering and orchestrating a broad set of collaborative processes. These structures and processes create synergies between actors and resources that help to seed and spread innovative solutions (Frantzeskaki et al., 2014).

Attention should also be given to how agents support change work at the boundaries of softer organisational and hard institutions. to enable wider scale reform. Particularly at the regime level, mediating agents who work at this boundary between organisations consisting of loose congeries of affected stakeholders and institutions consisting of more consolidated policies, legislation and administrative structures play a potentially critical role in influencing the prospects for wider scale social innovations, as illustrated in Table 10.1. It is further worthwhile to consider how change agents work with the four aforementioned enablers of change – social learning, social networking, reflexive governance, and transformative partnerships to bring to fruition these ever-wider scale changes.

In South Korea (hereafter Korea), a country once lauded for a development miracle that was guided by technocratic bureaucracies, focusing on the interaction between organisational and institutional change may be particularly revealing. The degree to which social innovations alter administrative structures in Korea can arguably give an even greater sense of the implications for broader change than in Europe, where the cases of sustainability urban transitions are concentrated (McCormick et al., 2013). This is particularly true since in many contexts in Europe the bureaucracy is not as autonomous and entrenched as Korea. From this point on, the chapter examines the role of a key agent of change, the Mayor of Seoul, who employed some of the aforementioned enablers to push forward wider scale change for some but not all social innovations.

The narrative that follows begins with a brief introduction to Mayor Park and the life experiences that made him a strong advocate of civic participation. This is followed by examples of the innovations he and the new Seoul administration helped to carry forward. This chapter is limited to five Seoul administrative areas highlighted under Mayor Park, i.e. participatory democracy, administrative service for sharing, social economy, urban environment, and civic communication.[1] These areas are selected because they have been particularly emphasised since Mayor Park was inaugurated and they feature the dynamics between agents and between levels of sustainability transitions.

Table 10.1 A simple analytical framework based on sustainability transitions theory

	Niche	*Regime*		*Landscape*
		Organisational Level	*Institutional Level*	
Project A	√	√	√	√
Project B	√	√	√	
Project C	√	√		

Tracing sustainability transitions in Seoul City government

Mayor Park, agent of change

While the sustainability transitions literature carefully illuminates paths to a more sustainable future, it has looked less closely at how change agents work within and between levels to achieve that future. The importance of a high-profile champion with a strong political will is not always given due attention in the transitions work (Markand et al., 2012). A related set of concerns involves how agents alter structures and processes to initiate and scale innovation and how change agents work at the regime-niche boundaries to align social structures and learning processes from the niche with governance structures and engagement processes within the regime; Mayor Park, in the case of Seoul, was a key change agent precisely because a mayor from the civil society organisations could work within the system to change the system.

Mayor Park was a proponent of grassroots movements to remedy social problems for much of his professional life. His abiding belief in the grassroots was arguably a by-product of his experience as a human rights lawyer and a founder of the NGO People's Solidarity for Participatory Democracy in the 1990s. His work at the Hope Institute—a citizen run think-tank that was established in 2006 to empower communities and drive socioeconomic change—also informed his thinking on these matters. His publications based on field research demonstrate his support for innovating from the ground up (Park, 1999, 2001, 2005, 2009, 2010, 2011). His fieldwork focused on individuals or communities that established niche innovations in various fields, i.e. revitalisation of agricultural village communities for the sustainable future (Park, 2009), communities supporting whole-person education in local areas (Park, 2010), or good practice of social entrepreneurs (Park, 2011). Some of the key messages from those publications exemplify his beliefs; namely, practice in the field can avoid paper administration; restoring communities is critical to fostering creativity in society; and multi-stakeholder communication and participation can supplant a top-down one-way decision-making process.

It is these and other similarly held beliefs that led Mayor Park to take a distinctly different approach to public administration and helped to make several niche-to-regime level innovations. The clearest evidence of these changes is some of the more visible transformations to Seoul Metropolitan Government that have been made since Won-Soon Park was elected in 2011.[2] It is difficult to reform administrative structures in Korea but Mr. Park's embrace of deliberative decision-making, public consultation, and information-sharing have helped forge new channels of communication. The main characteristics of the Seoul administrative policies are: citizen participation, transparency, communication, voluntary resource mobilisation, community restoration, and public-private governance. The changes go beyond simply improving the performance of government agencies; rather, they have fundamentally altered the practice of public administration. There may also be a wider change possible, because other cities tend to follow Seoul. It is therefore important to analyse this new approach affecting several administrative policy areas that support positive economic, social, and environmental interlinkages for citizens.

Participatory democracy

The fundamental goal of Mayor Park's administration is to enhance citizens' participatory planning and management in diverse city policies, i.e. public policy agenda-setting through public-private governance and broadening citizens' participation in decision-making as well as the execution process. One visualised example of participatory governance involves the development of the Honorary Vice Mayor System in 2012—representing the elderly, the disabled, traditional merchants, foreigners, women, youth, small businesses, and arts. The vice mayors actively participate in a diverse range of meetings, forums, ceremonies which totalled 376 events in 2013. This system has offered an effective communication channel that encourages marginalised citizens to voice their opinions.

Another set of sizable changes involved the Participatory Budgeting System. The Public Budgeting System, established in 2013 as the first trial in the country, allowed Seoul citizens to participate in a different stage of the planning process, namely budgeting. As such, the budgeting committee would be comprised of citizens who review community proposals to ensure financial transparency and equitable resource allocation. The system was not warmly welcomed by all the parties involved at first: "It received criticism in the beginning, especially from officials, due to the progressiveness and newness of the system ... but those feelings began to subside ... understood the idea of budget democracy" (Seoul City Councillor, Seoul, 2014).

Arguably there was an acceptance of the participatory system by both groups. According to one citizen participant:

> while the experts group always led the decision-making; but now residents and deliberative participation became more important through town hall meetings and learning to understand the importance of collective intelligence. The experts group develop the first draft for the project plan; then, 100 citizens, who well acknowledge the needs of the local condition than the experts group, review the second draft. Then the experts group finalise the conclusion.
>
> (Citizen Project Planning Committee, Seoul, 2014)

Table 10.2 Features of the participatory democracy administrative: public budgeting system

		Before		After
Public Budgeting System	**Structure**	Budgeting controlled by government officials	Institution created for participatory budgeting	**Institutional level:** System supported by the Local Finance Act and Public Budgeting System Ordinance
	Process	Government develops budget consulting with relevant agencies	Citizens review proposals for financial transparency and equitable resource allocation	

Importantly, the commitment to implementation encouraged additional waves of participation as demonstrated by a sharp 80 percent increase in citizen participation in the budgeting committee. Moreover, the citizen-participatory system selected 223 projects (funded with the equivalent of KRW50.3 billion, approx. US$50 million) in 2013 and the Seoul government underwrote 202 of the proposed projects. The reliability of budgeting under the citizen-participatory system was strengthened further by a review system that consisted of 25 district meetings, sub-committee reviews, and general assembly meetings. Seoul's Public Budgeting System was supported by the Local Finance Act, the Public Budgeting System Ordinance, and a budget of KRW50 billion, and as a result, it is now institutionalised.

The other example that perhaps had the most transient impact on the transition but arguably had the least significance in institutionalisation is the Listening Open Forum. The Listening Open Forum aims for 聽策, *Chung-chec*, which literally means listening policy rather than 政策, *Jung-chec*, which means policy more generally. The Listening Open Forum was a place where Mr. Park's attendance was all but guaranteed; it thereby offered citizens as well as civil servants a chance to voice their concerns directly to the city leadership. It also attracted attention because of its regular scheduling; this enabled citizens from different backgrounds with varying levels of political sophistication to become a source of input; it was held 71 times annually over the period from November 2011 to December 2013. The Listening Open Forum helps advance the participatory decision-making process but it serves more as source of deliberation on daily lifestyle issues rather than weightier matters concerning energy, housing, or social welfare. The Forum also tends to be more of a consultative forum with no direct effect on government decisions; its uncertain legal status also raises questions over whether the Forum will remain in place after Mayor Park's tenure.

Table 10.3 Features of the participatory democracy administrative: Listening Open Forum

		Before	After	
Listening Open Forum	Structure	Autonomous bureaucracies make decisions	Citizens offered opportunity to participate in social experiment	**Institutional level:** None
	Process	Government run webpage for Q&A	Citizens listen, reflect, and adopt opinions	

Administrative service for sharing

A side-effect of Seoul's breakneck growth was that social spaces rapidly faded from view. These shrinking spaces contributed to the growing sense of malaise in Seoul. This sense of disconnect and atomisation led Mayor Park to develop the Sharing City, Seoul Project. The primary aim of this project was to create new economic opportunities, restore relationships, and reduce resource waste. The slogan 'Sharing Ten Million Things, Ten Million Happiness' suggested that happiness could be created between ten million Seoul residents by sharing rather than individual material accumulation. Sharing was not only about promoting the collective use of spaces and buildings but also learning from the experiences and knowledge of others. This learning process adopted from local experimental innovations could be helpful in resolving Seoul citizens' biggest concerns about raising children, locating employment, improving livelihoods, and, ultimately, finding happiness. In terms of environmental protection specifically, sharing could reduce waste and other forms of pollution.

Hence, the Seoul administration began to share its resources with citizens to fulfil their needs rather than to construct new architectural developments for visible achievement which tended to be the priority of the previous administration. Several examples of the practical sharing cases are: the opening up City Hall for public use, i.e. first and second floor for the public library, and opening public buildings for reservation for common use. It provides improved access to inclusive and accessible public spaces.

The Seoul administration proclaimed the sharing promotion rules in 2012 as groundwork to support enterprises and organisations in sharing resources. The Seoul Sharing Promotion Committee was also introduced as the private-public governance mechanism. The committee is composed of representatives from academia, the press, business, non-profit organisations (NPOs), and research institutes, as well as general officials in charge of economic, welfare, transportation, and innovation affairs. The Committee supports the creation of relevant policies for sharing resources, improving existing laws and systems for the

Table 10.4 Features of the administrative service for sharing

		Before		After
Administrative Service for Sharing	**Structure**	Bureaucratic administrative service and physical construction	Institutions reformed to include minorities and other formulations for sharing	**Organisational level:** • Seoul Sharing Promotion Committee established • Seoul Sharing Hub disseminates information while building networks
	Process	Administration formulates plans based on top-down announcement	Private-public partnership and governance mechanism	

promotion of sharing as well as offering direct support to enterprises and organisations on sharing.

Subsequently, several organisational structures were created to help strengthen Sharing City, Seoul. For example, the Seoul Sharing Hub was established to archive, disseminate, and diffuse information while building networks with relevant domestic and overseas organisations, enterprises, media, and other organisations. The Seoul Sharing Hub continued to function as both an intermediary organisation and an incubator affecting the way grassroots multiply and other initiatives scale up.

Social economy

Seoul administration has intensively fostered social enterprises and cooperatives based on its key concept of sharing and established the Comprehensive Support Plan for the Social Economy to promote village enterprises and cooperative vitalisation. Social enterprise policy strengthened especially for the autonomy of the less-privileged groups and it encouraged collecting public procurement, investing in resource expansion, widening networking among relevant stakeholders in order to establish a better environment for social enterprise. Citizens' voluntary participation was welcomed from public policy agenda setting to policy enforcement; various taskforce teams were also established for expert and stakeholder participation. Moreover, when Seoul designated enterprises and organisations, for instance, those qualified for reliability and citizens' participation in activities as of December 2012, it provided a consulting service as well as working space for an incubation programme.

Seoul put priority on policies restoring communities and the administration challenged citizens to build their community with their own initiatives for sustainable local development through social economy implementation, i.e. sharing the common space to raise children, discussing life improvements,

Table 10.5 Features of the social economy

Social Economy		Before		After
	Structure	Institutions organised around achieving economic development	Institutions formulated for less-privileged groups and community revitalisation	**Institutional level:** • Comprehensive Support Plan for the Social Economy • Community Building Division under the Seoul Innovations Bureau • Ordinance for Town Community Support • Basic Plan for Seoul's Town Community
	Processes	Government formulates development plans based on one-way top-down communication	NGOs/CSOs draft plans in two-way consultation with development experts and government officials	

creating work places, and sharing entertainment. This policy is based on existing innovative grassroots experiences, for example, Seongmisan village in Seoul. Seongmisan is the name of a small mountain located in Seoul and about 30 like-minded parents in the village gathered to purchase a place for a cooperative childcare centre and to protect Seongmisan from being developed in the early 1990s. This community-centred village movement became famous as a leading restoration community model in a mega-city and as of 2014 has about 1000 village members working for various cooperatives, Seongmisan School, and other social entrepreneurial projects in the village.

To construct the infrastructure for community restoration, Seoul also established the Community Building Division under the Seoul Innovations Bureau in January 2012; two months later it announced the Ordinance for Town Community Support. In the months that followed, this same Bureau convened seminars and conferences with citizens and experts; research was also conducted by the Seoul Institute. All these efforts helped to announce the Basic Plan for Seoul's Town Community in September 2012. In addition to crafting the legal basis for the project, nine Town Community Support Centres were opened to provide in-kind support and counselling for the local needs when residents drafted and applied for development plans.

Involvement in the drafting of these plans diverged sharply from previous modes of urban planning. According to remarks from a participating Civil Society Organisation (CSO) member (Seoul, 2014):

> Grassroots innovation leaders and activists of NGOs/CSOs based in its local village gathered and developed our village restoration proposal. We

shared our proposal with Mayor Park and developed it further to tailor to the socioeconomic development needs of the town/community. To support this process, we composed a Task Force team and spent two months in meetings and workshops with experts. ... After sharing our final proposal with Mayor Park, he requested that we move to concrete steps to implement the proposal. We accepted support from Town Community Support Centre to guide us in moving to the implementation stage.

The results of this work were clear: the number of small village community support projects registered in 2013 was 2,233, double the number registered in 2012, and 68 percent of them were initiated by residents. As grassroots activists and local residents became more comfortable with building their own communities based on local livelihoods, welfare, education, and cultural needs, the Seoul government began to support the planning of whole-town communities. Locals established Community-Net which introduces Town Projects that will later receive further support from Seoul and expand in number as shown in Figure 10.1.

Urban environment

Seoul has also adopted policies that help to illustrate the benefits of integrating nature into urban communities. The Seoul campaign Blooming Flower, Seoul was created to motivate citizens to become involved in the cultivation of trees and flowers in their daily lives. It launched the Urban Agriculture Festival in 2012, which then generated similar events that featured information exchange and hands-on demonstrations thanks to the full participation of the existing urban agricultural NGOs/CSOs, i.e. direct trade cooperatives, farmers' markets, civic farming community groups, etc., those willing to share their successes and failures and those focused on organic and local agriculture with manual labour rather than heavy dependence on fossil-fuel usage as they believe cultivating the Earth presents a viable solution to the eradication of poverty, degradation of the soil, and the emptying of community while creating a supportive social-cultural environment in cities. To highlight more visible examples, Seoul planted

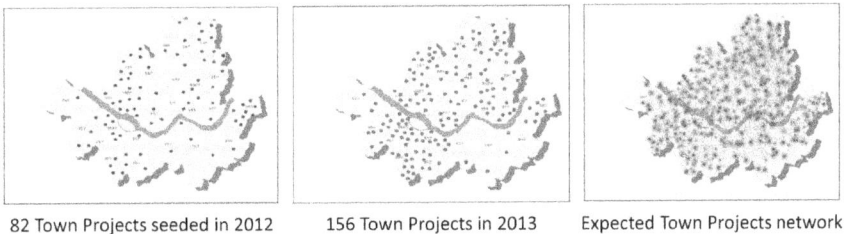

82 Town Projects seeded in 2012   156 Town Projects in 2013   Expected Town Projects network

Figure 10.1 Number of town projects over time
Source: Seoul, 2014.

Table 10.6 Features of the urban environment: urban agriculture

		Before		After
Urban Agriculture	**Structure**	Green space, i.e. park development institutions manage certain parts of Seoul	Institutions organised and implemented programmes/ project of urban agriculture	**Institutional level:** • Proclamation of the first year of Urban Agriculture 2012 • Urban Agriculture Master Basic Plan under the Ordinance of Urban Agriculture Promotion and Support
	Process	Selected or newly established NGOs/CSOs contracted for Seoul nature conservation programme	Existing urban agricultural NGOs/CSOs encouraged to engage in consultation and implementation	

rice paddies in Gwanghwamun Square in the city centre and raised honeybees in the rooftop gardens of City Hall.

These activities then became enshrined in the Urban Agriculture Master Plan under the Urban Agriculture Fostering and Supporting Ordinances. These efforts created an environment favourable to low-impact urban lifestyles. Citizens' voluntary platforms were established to provide a space where diverse discussions and negotiations could take place. These included the Soil and the City Forum to implement practical governance reforms composed of bottom-up multi-stage multi-level arrangements with actors from NGOs, citizens of various backgrounds ranging from academia, civic groups, artists, the press to policymakers, borough municipal officers, firms and business sectors, including national figures and representatives of each field in relation to urban agriculture. By blurring the lines between urban and rural, and ensuring social learning and networking with long-standing grassroots innovations in Seoul, these efforts have since spread across many organisations and institutions nationwide.

Another illustration of a project that brought significant organisational and institutional change was the One Less Nuclear Power Plant. Beginning in May 2012, the programme aimed to save two million tons of oil equivalent (TOE) of energy or a reduction on a par with one nuclear power plant by December 2014. The proposed goal was to be achieved through the development of energy self-sufficient villages, solar-photovoltaic power plants, cooperative-sharing power plants project, car sharing systems, and other small-scale innovations. Most of initiatives originated from existing local niche innovations. Hence, the initiative owed its success to diverse groups of citizens actively participating in the implementation of community-based energy conservation programmes, such as

Table 10.7 Features of the urban environment: One Less Nuclear Power Plant

		Before	After	
One Less Nuclear Power Plant	Structure	Institutions organised around achieving Green Growth	Institutions organised energy efficiency and renewable plant development	**Institutional level:** • One Less Nuclear Power Plant Comprehensive Plan
	Process	Government formulates Green Growth Plan incl. nuclear power plant for low-carbon society	Implemented by citizens' participation	

1.7 million eco-mileage membership, energy clinic programme for individual households, energy guardian angel clubs in schools, good stores serving as leading energy-saving commercial spaces, and so on. It should be emphasised that all these activities were gathered through networking and partnership under the participatory governance system and boosted to scale up their capacity for the transformative result. The end results were noteworthy: the energy savings goal of the project was reached six months ahead of schedule; the average power consumption of Seoul decreased by 1.4 percent in 2013, in contrast to average increases of 1.76 percent across Korea.

The Urban Agriculture and One Less Nuclear Power Plant Projects illustrated the possibility of niches becoming institutionalised through social networking within and among niches to stimulate innovation and firmly nested in a supportive enabling environment. Communities applying alternative energy systems shared their long-term expectations for the future and encouraged each other with their respective visions and these were also widened to include perspectives from outsiders through social networking. Networking aims to create new platforms where diverse discussions and negotiations can take place, allowing both fringe actors and established players to join and promote radical innovation. It is essential to create new platforms where diverse discussion and negotiations can take place, and even established firms promoting radical innovation may join. Hence, to introduce innovation from both fringe actors and established firms, the creation of a social network process and forging partnerships under reflexive governance orchestrating these collaborations is crucial.

Civic communication

Citizens' participation and civic communication for public-private governance is the primary principle of the Seoul administration. Various communication systems have developed including a direct communication channel between the mayor and the citizens via the social network service (SNS), On-site Mayor's Office, Listening Open Forum Local Tour, Open Administration 2.0, etc. Thanks to its emphasis on the process of communication, the implications of the Sharing City, Seoul Project, for instance, extended beyond changes to physical infrastructure. The city government conducted deliberative probes and shared detailed information with citizens on how to communicate, negotiate, and reach conclusions. To quote a government official involved in this process:

> There were over 10 proposals submitted from one village; a group of local proposal developers carried out field-investigation to determine the best proposal. This process itself became networking and learning opportunity among those proposal developers. They eventually selected the best proposal as they reviewed options together during investigation and discussion process.
>
> (Officer of Seoul City Community Support Centre, Seoul 2014)

To a certain extent, changing decision-making structures and processes proved costly and time-consuming (confirming claims in the transition literature, i.e. Kemp, 2007b); however, they cut the rising costs of preserving unsustainable systems and reduced the likelihood of conflict between local government and citizens. That supports the fundamental reason why Mayor Park insisted on communication with citizens of the Open Administration which enables citizens to wholly enjoy their sustainable and sound long-term livelihood.

Mayor Park shared his belief in the importance of citizens' participation through social media and SNS right from the beginning of his involvement in various civic organisations. Even after becoming the Mayor of Seoul, he still replied personally to all followers at least twice per week, and the newly created Social Media Centre offers support for administrative requests. His communication is mutual and interactive resulting in trust and solidarity, which leads to support and legitimacy. Seoul administration also disclosed its public data through the Seoul Open Data Plaza website under the Open Administration 2.0 including white papers, meeting notes, and administrative information in order to share and communicate with citizens. Open communication with citizens has an advantage not only for the individual citizens become to cover the common

Table 10.8 Features of communication with citizens

		Before		After
Communication with Citizens	Structure	Bureaucratic administrative service	Mayor directly communicates with citizens through SNS	**Organisational level:** • Open Administration 2.0
	Process	Administration reply to the Q&A	Disclosure of administrative information	

good beyond individual interests but also for their social learning expenditure among their interactions via communication and participation.

Assessment and discussion

The five areas presented in the previous sections illustrate various degrees of progress along a transition beyond the niche and within the regime level. It is relatively clear that they all have transformed both softer economic and social structures, and have begun to bend the harder institutions and infrastructure that constitute the regime level. They further illustrate the importance of having intermediary agents that help to carry forward the change. Perhaps, most interestingly the case of the Listening Open Forum and citizen communication exemplify the change that is currently in place but may fail to retain its standing after Mayor Park leaves office. This innovation is not pegged to a larger set of harder institutional and administrative reforms. Its prospects for sustainability are therefore arguably less promising. The overall picture that emerges can be succinctly summarised by Table 10.9 below.

This was not solely a function of strategic interventions by Mayor Park even though it is indeed true that he made change possible by fostering existing

Table 10.9 Analysis of transition process

Area	Policy	Niche	Regime Organisational	Institutional	Landscape
Participatory Democracy	Public Budgeting System	√	√	√	
	Listening Open Forum	√	√		
Administrative Service for Sharing	Sharing Seoul	√	√		
Social Economy	Social Enterprise/ co-ops	√	√		√
	Community Revitalisation	√	√		√
Urban Environment	Urban Agriculture	√	√		√
	One Less Nuclear Plant	√	√		√
Communication with Citizens	Communication Policy	√			

Source: Authors' evaluation, based on data published by Seoul, 2014.

emergent niches. Mayor Park and Seoul administration worked with previously mentioned enablers of change.

First of all, in the area of Participatory Democracy, the Public Budgeting System (PBS) is institutionalised as a Municipal Civil Participatory Budget Ordinance as is the Local Finance Act (Article 39). The PBS introduced significant changes to key decision-making structures and processes with wide ranging results. The magnitude of those results is exemplified very clearly in the amount of budget that is now allocated to urban infrastructure and city programmes through a more deliberative process. The implementation of PBS gained momentum from reflexive governance and proved the ideal role of local authority. The PBS left a deep impression on the belief systems of those participating in it. It nonetheless merits stressing that there was some initial resistance to this project which may have limited its downstream effects.

The Listening Opening Forum has been able to enhance the dialogue between the governed and government. This is particularly attractive because it enabled everyday people the chance to interact directly with Mayor Park. The same thing happened in the area of Citizen Communication. Serving as a sounding board for citizens' concerns, the Listening Open Forum did not move beyond its intended purpose and its future prospects appear closely related to the legacy of Mayor Park. It is also still at the consultation level rather than being practical deliberative democracy due to limited opportunity in the decision-making process as well as being on unstable legal grounds. When the consultation process has matured, collaborative decision-making with citizens will be important for the pending issues.

Under the new Administrative Service area, Sharing Seoul introduced significant changes to structures and processes, and yielded positive results that influenced the popular mind-set and stakeholder competencies of those participating in the project. For example, the Seoul Sharing Hub was able to function as an intermediary organisation, as an incubator for multiplying and scaling up grassroots innovations, and as a platform for social learning and networking. To a significant extent, the Sharing Seoul Project offered a platform for cultivating and nurturing other grassroots initiatives although it still needs to secure a more concrete institutional structure for itself.

Social Enterprise and co-ops under the Social Economy area are successful due to the close connection with civil society. It operates under both Seoul governance and citizen initiative innovation, and functions as an innovation cluster. As the Cooperative Act came into effect in 2012, Seoul produced various programmes to become a role model city. The success of Community Revitalisation and enterprise in villages proved the importance of good collaboration between citizens and the Seoul administration—niche incubated, organisations and institutions equipped. However, as there was insufficient human and material capacity, the success of Social Enterprise was minor in comparison to the full support from the Seoul administration. Town Community Support Centres are also still temporary organisations and not yet established in all district offices despite being under the Municipal Ordinance. All these incubated niches institutionalised at the regime level are still in need of continued support from the Seoul administration. They also require more citizen' participation in order to achieve a long-lasting and landscape level transition.

In the case of the Urban Environment area, there has been visible evidence of social learning and networking among and within grassroots innovations and citizens. Both Urban Agriculture and One Less Nuclear Power Plant were projects that were able to produce demonstrable carryover effects. Urban Agriculture has made similar attempts to replicate comparable shows of urban-rural synergies in other cities in Korea and to integrate with the Urban Agriculture Fostering and Supporting Ordinances. In the case of One Less Nuclear Power Plant, this has meant increasing efforts to change lifestyles to conserve energy among Seoul residents. More so than the above examples, however, questions remain over the spread and permanence of these initiatives.

In sum, a transformative social learning and networking process for systemic change, i.e. active citizens' engagement and collective participation, strategically facilitated the creation of communities of practice where an action-reflection approach and user-driven innovation enabled experiential learning as a precursor to transformative social change. As a result, Seoul's sustainability transitions experienced beyond simply recycling or carrying out energy-saving activities to create a culture of sustainable community living. It was only when innovations settled down into an institutional framework and these policies were carried out through stakeholder and citizen participation that the regime transition was completed. Social learning and transformative partnership brought sufficient participation to support the transition, which was closely linked to political opportunity—Mayor Park himself broadened this opportunity. It encouraged transformative partnerships providing a broader context within which synergies across institutions, resources, and governance arrangements could spread new innovations. The cases in five areas reinforce arguments that governance, networking, social learning, and transformative partnerships are critical drivers and enablers of sustainability transitions.

Conclusions

This chapter began with the observation that Seoul, like many other rapidly growing cities in Asia, found itself in a precarious position before 2011. While the national government was championing Green Growth, the city itself was paradoxically losing its connection with the natural environment as well as with its population. Beginning in 2011, the city administration began to strengthen connections with both of these entities, due in part to the astute manoeuvring of Mayor Park. The chapter then raised the question of the extent to which the reforms that Mayor Park helped to introduce maintained momentum. To help analyse the degree and drivers of change, it employed a lens based on sustainability transitions. That analytical lens was used to assess five areas of participatory administrative policies under Mayor Park's watch and found that those insights suggest the scaling of initiatives from niche to regime are partially attributable to the interplay between social and governance processes and structures at the regime and niche levels. Where these synergies were strongest, the innovations managed to create platforms for deeper and wider changes beyond the initial intent of the project. Where they were more limited, the effects remained tightly circumscribed.

However, a bigger question remains: what does the transformation of these particular initiatives say about the future prospects of sustainability in Seoul more generally? At the risk of being speculative, one of the main contentions in this chapter is that the prospects of continuing along the recent transition pathway depend in part on how consolidated the reforms from the recent past are. To a significant degree, crossing the threshold between organisational and institutional change increases the likelihood of maintaining the current trajectory.

A related point involves the prospects for even more permanent and wider changes to the landscape level. On this point, it warrants underlining that the uniqueness of the current Seoul case study is that none of the 'innovations' under the current Mayor were wholly new. The truly new innovation was the mayor's determined effort to elevate existing and scattered grassroots innovations to the point where a critical mass of citizens could connect with their government and environment. In a manner that comes closest to arguments related to transformative partnerships, it was the careful identification and linkage of existing initiatives that helped to pave the way for wider scale change. One might therefore contend that a similar set of transformative partnerships could help set in motion sharing between other cities outside Seoul or even beyond Korea.

The discussion of the landscape level leads to the question of how extending this framework points to other related areas for research. This chapter, while acknowledging the dynamic interplay between different levels, focused chiefly on change moving from the bottom up. It would be interesting to see how changes in more encompassing cultures or political systems make some reforms feasible while preventing bottom up change. It would similarly help to look at for the question of whether or not changes in Seoul can help lead to changes in other cities in Korea or even overseas. A similar set of potentially revealing inquiries would bring in more cases from Seoul with possibly less success in moving from one level to the next.

In addition to expanding the scope of the cases studied, another area for strengthening the analysis would involve incorporating more precise measures of progress from one level to the next. Admittedly the judgement on what is a regime and niche level change is rather subjective. This is partially a function of the transitions approach itself and its reliance on describing an interactive process that is ongoing and does not lend itself to a single snapshot of achievement—hence the need for process tracing. It is also partially attributable to the authors' decision to rely chiefly on qualitative as opposed to quantitative data. Future iterations of this work would potentially benefit from the inclusion of more hard data to support qualitative observations. Time series data, illustrating the rise and fall of measures of sustainability as well as enabling reforms, would be particularly illuminating in this regard.

A third potential need area that relates to transitions work more generally is translating the rather abstract language it uses into prose that is able to communicate with citizens and policymakers. The United Nations has recently approved a new set of global goals, known as Sustainable Development Goals, so it is imperative that both transitions theory and practice can be assimilated by those who do not speak in terms of niches, regimes, and landscapes. In fact, there appears to be ample room for using actual case studies like the ones featured in this chapter to translate transitions theory into a language that is more accessible to policymakers.

With the growing interest in new development paradigms has come a surfeit of slogans for new approaches to development; it is not uncommon to see the development community push for climate resilient, low-carbon, and green growth. In addition to potentially perplexing policymakers, these slogans also run the risk of overlooking the interests of people that growth is supposed to benefit. To make sure that these slogans do not ring hollow, it is necessary to make new approaches to development consistent with often overlooked provisions for participation and engagement.

The sustainable transition approach is increasingly important in an Asian context because of the mounting list of environmental problems confronting the region as well as the significant opportunities to harness the energies of key stakeholders to drive a transition forward. The main factor behind the success will be how these ideas can be combined into practice in an effort to increase awareness and empower citizens to conscientiously contribute to more sustainable equitable lifestyles as well as capacity building tools for policymakers to better consider and integrate approaches for sustainable consumption and lifestyles into wider policy planning for a sustainable society.

Notes

* This chapter draws upon materials (with updates) from the Proceedings of the Urban Transitions Conference, Shanghai, September 2016, *Procedia Engineering* (2017) 198: 293–304. doi:10.1016/j.proeng.2017.07.162.

1 The information used to analyse this process is chiefly qualitative; quantitative data is presented where possible. The cases draw heavily from primary source Korean language municipal ordinances, city policies, white papers, and related government data. Relevant Korean language documents corresponding to case studies are listed in a section in the References. Seoul's information disclosure policy, passed in 2014, made it possible to access and review key documents. The case studies also drew on a 2014 Report on Civic Participant Administration in Seoul (hereafter Seoul, 2014) prepared in Korean by a team of eight scholars, including a co-author of this chapter.

2 Mr. Park was elected in October 2011 and re-elected in July 2014. This research only covers Mr. Park's first term of office. We suggest that subsequent research should update the 'results' of the current policy process with more concrete quantitative data analysis of the 35th Seoul Mayor's later term completed in June 2014. Additional research could investigate how the 'process' during this term moves from more malleable organisational to more concrete institutional change.

References

Books and articles

Akenji, L. (2014). Consumer scapegoatism and limits to green consumerism. *Journal of Cleaner Production, 63,* 13–23.

Collier, D. (2011). Understanding process tracing. *Political Science and Politics, 44*(4), 823–830.

Frantzeskaki, N. and Loorbach, D. (2010). Towards governing infrasystem transitions: reinforcing lock-in or facilitating change? *Technological Forecasting and Social Change, 77,* 1292–1301.

Frantzeskaki, N., Wittmayer, J., and Loorbach, D. (2014). The role of partnershps in 'realising' urban sustainability in Rotterdam's City Ports Area, The Netherlands. *Journal of Cleaner Production, 65,* 406–417.

Geels, F. (2002). Technological transitions as evolutionary reconfiguration processes: a multi-level perspective and a case-study. *Research Policy, 3,*11257–1274.

Geels, F. and Schot, J. (2007). Typology of sociotechnical transition pathways. *Research Policy, 36,* 399–417.

Hamann, R. and April, K. (2013). On the role and capabilities of collaborative intermediary organisations in urban sustainability transitions. *Journal of Cleaner Production, 50,* 12–21.

Hargreaves, T., Hielscher, S., Seyfang, G., and Smith, A. (2013). Grassroots innovations in community energy: the role of intermediaries in niche development. *Global Environmental Change, 23,* 868–880.

Hoogma, R., Kemp, R., Scot, J., and Truffer, B. (2002). *Experimenting for Sustainable Transport: The Approach of Strategic Niche Management.* Spon Press, New York.

Howells, J. (2006). Intermediation and the role of intermediaries in innovations. *Research Policy, 35,* 715–728.

Kemp, R. (1994). Technology and the transition to environmental sustainability: the problem of technological regime shifts. *Futures, 26,* 1023–1046.

Kemp, R., Loorbach, D. and Rotmans, J. (2007a). Transition management as a model for managing processes of co-evolution towards sustainable development. *International Journal of Sustainable Development and World Ecology, 14,* 1–15.

Kemp, R., Rip, A., and Schot, J. (2001). Constructing transition paths through the management of niches, in R. K. Garud (ed.). *Path Dependence and Creation.* Lawrence Erlbaum, London.

Kemp, R., Rotmans, J., and Loorbach, D. (2007b). Assessing the Dutch Energy Transition Policy: how does it deal with dilemmas of managing transitions? *Journal of Environmental Policy and Planning, 9,* 315–331.

Kemp, R., Schot, J., and Hoogma, R. (1998). Regime shifts to sustainability through processes of niche formation: the approach of strategic niche management. *Technical Analysis and Strategic Management, 10,* 175–198.

Kemp, R. and Soete, L. (1992). The greening of technological progress. *Futures, 24,* 437–457.

Khan, J. (2013). What role for network governance in urban low carbon transitions? *Journal of Cleaner Production, 50,* 133–139.

Lachman, D. (2013). A survey and review of approaches to study transitions. *Energy Policy, 58,* 269–276.

Loorbach, D. (2007). *Transition Management.* International Books, Utrecht.

Loorbach, D. and Rotmans, J. (2010). The practice of transition management: examples and lessons from four distinct cases. *Futures, 42,* 237–246.

Loorbach, D. and Wijsman, K. (2013). Business transition management: exploring a new role for business in sustainability transitions. *Journal of Cleaner Production, 45,* 20–28.

Magdalena, F. and Eklund, M. (2015). Towards a sustainable socio-technical system of biogas for transport: the case of the city of Linköping in Sweden. *Journal of Cleaner Production, 98,* 17–28.

Markard, J., Raven, R., and Truffer, B. (2012). Sustainability transitions: an emerging field of research and its prospects. *Research Policy, 41,* 955–967.

McCormick, K., Anderberg, S., Coenen, L., and Neij, L. (2013). Advancing sustainable urban transformation. *Journal of Cleaner Production, 50,* 1–11.

Nevens, F., Frantzeskaki, N., Loorbach, D., and Gorissen, L. (2013). Urban transition labs: co-creating transformative action for sustainable cities. *Journal of Cleaner Production, 50,* 111–122.

Park, W. (1999). *NGO-Citizen's Power Changes the World.* Yedam, Seoul (in Korean).

Park, W. (2001). *Lawyer Park's Japanese Civic Socity Visit.* Arche, Seoul (in Korean).

Park, W. (2005). *Interview German Society.* Nonhyun, Seoul (in Korean).

Park, W. (2009). *Meet the Hope in Communitieis.* Gumdungso, Seoul (in Korean).

Park, W. (2010). *Community is School.* Gumdungso, Seoul (in Korean).

Park, W. (2011). *Shall We Rent Mr. Park?* Imagine, Seoul (in Korean).

Raven, R. P. J. M. (2006). Towards alternative trajectiories? Recongifurations in the Dutch electricity regime. *Research Policy, 35,* 581–595.

Rotmans, J., Kemp, R., and van Asselt, M. (2001). More evolution than revolution: transition management in public policy. *Journal of Future Studies, Strategic Thinking, and Policy, 3,* 15–31.

Sanden, B. and Azar, C. (2005). Near-term technology policies for long-term climate targets: economy-wide vs. technology specific approaches. *Energy Policy, 33,* 1557–1576.

Schot, J. and Geels, F. (2007). Niches in evolutionary theories of technical change: a critical survey of the literature. *Journal of Evolutionary Economics, 17,* 605–622.

Schot, J. and Geels, F. (2008). Strategic niche management and sustainable innovations journeys: theory, findings, research agenda, and policy. *Technology Analysis and Strategic Management, 20,* 537–554.

Seyfang, G. (2009). *The New Economics of Sustainable Consumption: Seeds of Change.* Palgrave Macmillan, Basingstoke.

Seyfang, G. and Smith, A. (2007). Grassroots innovations for sustainable development: towards a new research and policy agenda. *Environmental Politics, 16,* 584–603.

Smith, A., Stirling, A., and Berkhout, F. (2005). The governance of sustainable socio-technical transitions. *Research Policy, 34,* 1491–1510.

Vergragt, P., Akenji, L., and Dewick, P. (2014). Sustainable production, consumption, and livelihoods: global and regional research perspectives. *Journal of Cleaner Production, 63,* 1–12.

Von Malmborg, F. (2007). Stimulating learning and innovation in networks for regional sustainable development: the role of local authorities. *Journal of Cleaner Production, 15,* 1730–1741.

White, R. and Stirling, A. (2013). Sustaining trajectories towards sustainability: dynamics and diversity in UK communal growing activities. *Global Environmental Change, 23,* 838–846.

Korean language primary sources

Seoul City. (2011) *Seoul shall Mentor Reconstruction and Rebuilt (Seoulsi jegebal jegunchok mentoring haedurimnida).* Seoul.

Seoul City. (2012). *Citizens' Forum for Newtown Maintenance Business (Newtown jeonbisaup simintoron).* Seoul.

_____. (no date). *Communication White Paper: Seoul, Communicate with Citizens (Sotongbekseo: siminga sotonghanun seoul).*

Seoul City. (2012). Implementation Plan for Newtown Reconstruction Measure for Control (Newtown jegebal susupbangan sihenggehoek). Seoul.

Seoul City . (2012). *Seoul Citizen's Grand Forum for One Less Nuclear Plant (Seoulsi wonjonhanajulegi simindetoronhoe).* Seoul.

Seoul City . (2012). *Seoul City Newtown Reconstruction Transition toward Social Minority Protection (Seoulsi newtwon jegebal sahoe yakja bohohyung jeonhwan).* Seoul.

Seoul City . (2012). *Seoul City Save One Nuclear Plant through Energy Saving (Seoulsi enoji julyakuro wonjeon hana julinda).* Seoul.

Seoul City . (2012). Seoul City Urban Agriculture Begins (Seoulsi dosinongup bonkyuk sidong). Seoul.

Seoul City . (2012). *Seoul Metropolitan City, Town Community Basic Plan (Seoulsi maulgondonche gibonkehek).* Seoul.

Seoul City . (2012). *Seoul Newtown Reconstruction based on Citizens' Decision (Seoulsi jegebal jumineusae tara gyuljeon).* Seoul.

Seoul City . (2012). *Town Community Citizens' Forum: Setting Direction of Town Community in Seoul City (Maulgondonche simintoronhoe: Seoulsi maulgondonche bijonga banghang suljung).* Seoul.

Seoul City . (2013). Job Creation through Seoul City Urban Agriculture Experts' Training (Seoulsi dosinongup junmonga yangsung iljari chagcul). Seoul.

Seoul City. (2013). *Listen to Citizens: Story of Seoul City Town Community Reconstruction – Citizen Centred (Saramjongsim seoulsi maul jegebal eyagi: juminege dudda).* Seoul.

Seoul City . (2013). *Meet Citizens through Exhibit and Talent Donation for One Less Nuclear Plant (Wonjunhanajuligirulwehan jaenunggibu junsiro siminga mannada).* Seoul.

Seoul City . (2013). *Promote 552+α Job Creation Seoul City Urban Agriculture Vitalization (Seoulsi dosinongup hwalsunghwaro 552+α iljari chagcul chujin).* Seoul.

Seoul City . (2013). *Seoul City Citizens' Participatory Budgeting Review (Seoulsi jumin chamyeo yesan dolabogi).* Seoul.

Seoul City . (2013). *Seoul City Reconstruction Fact-finding Research Practical Guideline (Seoulsi jegebal siltejosa silmu jichimseo).* Seoul.

Seoul City. (2013). *Urban Agriculture Vitalization 2012 Achievements and 2013 Plan Report (Dosinongup walsungwha 2012nun saupsungga 2013nun chojin geheok).* Seoul.

Seoul Municipal Assembly. (2013). *Seoul Municipal Assembly and Seoul City Town Support Centre Forum Report for Sustainable Town Community (Jisokganughan maulgondongcherul wehan Seoulsiuhe mauljiwonsenta toronhoe jaryojib).* Seoul.

Seoul City Town Community Overall Support Centre. (2014). *Have Towns Been Formed? (Mauleun Hyungseongdoego itnunga?).* Seoul.

11 Rethinking adaptation to climate change for the policy landscape of India

Anshu Ogra

Introduction

Adaptation to climate change emerged as a strong policy option post the Marrakech Accords in 2001. The Marrakech Accords are a set of agreements reached at the 7th Conference of the Parties (COP7) to the United Nations Framework Convention on Climate Change (UNFCCC). The Accords called for the development of the National Adaptation Programmes of Action (NAPA) (Dessai & Schipper 2003). Since the Accords there has been a rapid increase in scholarship on adaptation to climate change ranging from peer-reviewed literature (Pelling 2010; Grothmann & Patt 2005; Hunt & Watkiss 2011; Moser & Ekstrom 2010), governmental communiqués (National Adaptation Plans, UNFCCC; European Commission n.d.) to reports from non-governmental organizations (NGOs) (Cimato & Mullan 2010). For the most part, all this literature is univocal in identifying the roots of this concept in evolutionary biology wherein an organism evolves along with the changing environmental conditions to maintain the fit that is required to sustain life (Brookfield 1973; Hoffmann & Sgro 2011). This conceptualization has arguably been found inadequate in academic circles because of: a) lack of common understanding (Ribot 2011), b) not figuring out what adaption entails (Doria et al. 2009) and c) failing to understand the opportunities and limitations offered by the concept for policy making (Smithers & Smit 1997; Adger et al. 2009). There are a few guiding definitions of adaptation provided by various institutions, but they do not seem to help the cause either.[1] The key words used in these definitions for adaptation are 'process', 'outcome' and 'adjustment'. A 2006 study by the Organization for Economic Cooperation and Development (OECD) argues that expectations of a successful adaptation strategy are different when looked at as a process from when they are seen as an outcome (Levina & Tirpak 2006).

Why is adaptation as a concept from evolutionary biology out of tune with anthropogenic climate change? What is it about anthropogenic climate change as an environmental problem that unsettles our understanding of humans as biological beings? The narrative of climate change is informed by global climate models (GCMs). The problem is thus conceived on a global scale. It is argued that the power of this narrative is in its global reach. It is the global temperature and sea-level that are rising, it is the global climate system that is changing, it is GCMs that are telling us so (Hulme 2010). However, adaptation has to be carried out on the spatial and temporal scale of everyday life (Adger et al. 2005). The challenge of conceptualizing adaption to climate change thus lies in the struggle to locate climate change in everyday life (Nightingale 2017). Mike Hulme argues that this continuous struggle to locate local weather in the global climate has rendered climate change not global or universal but cosmopolitan. Global climate and local weather have become mutually embedded categories, a meteorological entanglement which has significant consequences (Hulme 2010). In this chapter I explore one such consequence by showing the implications of having this meteorological entanglement inform an inadequate conceptualization of adaptation policy for the agriculture sector in India.

India's economy is tied crucially to the agriculture sector. In India agriculture contributed 15.4 percent to the total gross domestic product (GDP) of the country in 2018–2019 (Sushma 2018). A 2004 study by National Communications (NATCOM) highlights the agriculture sector as being the most vulnerable to the projected changes in surface temperature, precipitation patterns and rising sea levels. NATCOM assessed and reported the impact of climate change on the agriculture and forestry sectors for the 2050s and 2080s. *Economic Survey* 2018 (Ministry of Finance 2017–2018) argues that the impact of increased temperatures and greater variation in rainfall could reduce annual agricultural incomes by between 15 and 18 percent on average, and between 20 and 25 percent in unirrigated or rain-fed areas (Sushma 2018). The Indian Network for Climate Change Assessment (INCCA) launched in 2010 by the Ministry of Environment, Forests and Climate Change (MoEFCC) published its first study in the same year. The study forecasts the climate change impact on four eco-sensitive zones up to the 2030s, covering the Himalayan region, the North-Eastern region, the Western Ghats and the Coastal region. The sectors covered are agriculture, forests and biodiversity, water resources and coastal zones (INCCA 2010). Key findings from this study indicate: 1) an overall warming for all the four regions being studied, with a net increase in annual temperatures in the 2030s ranging between 1.7°C and 2.2°C, with extreme temperatures increasing by 1–4°C, and 2) a likely increase in extreme precipitation events with the number of rainy days projected to decrease (INCCA 2010). These projections emphasize the need for a timely and effective policy plan for climate change in India.

India's climate change policy landscape: NAPCC and SAPCCs

India's national climate change policy landscape primarily consists of the National Action Plan on Climate Change (NAPCC) and the State Action Plans on Climate Change (SAPCCs). NAPCC was introduced in 2008 by the then Prime Minister of India Dr. Manmohan Singh. It recognized the need to maintain a high economic growth rate to improve the living standards of most people and to reduce people's vulnerability to the impact of climate change (Prime Minister's Council on Climate Change 2008). The need to maintain a high economic growth rate using the arguments around equity has helped India insulate herself from international pressure to mitigate climate change (Dubash & Joseph 2016). At the same time India has offered to proactively take action to reduce the vulnerability of people to the impact of climate change, thus putting adaptation at the centre of her policy structure. The NAPCC was followed by a framework for State Action Plans on Climate Change (SAPCC) by the Ministry of Environ-ment, Forest and Climate Change (MoEFCC) in 2010. This framework empha-sized that States must harmonize their strategies and actions with the missions mentioned in the NAPCC. The goal of harmonization was to achieve a common purpose across SAPCCs.

NAPCC has eight missions: National Solar Mission (NSM); National Mission for Enhanced Energy Efficiency (NMEEE); National Mission on Sustainable Habitat (NMSH); National Water Mission (NWM); National Mission for Sustaining the Himalayan Ecosystem (NMSHE); National Mission for a Green India (NMGI); National Mission for Sustainable Agriculture (NMSA) and National Mission on Strategic Knowledge for Climate Change (NMSKCC). In this chapter I will specifically focus on NMSA and its interpretation in SAPCCs because of the crucial role played by the agriculture sector in the Indian economy and the projected impact of climate change on Indian agriculture.

NMSA acknowledges the risks to the Indian agriculture sector due to increased climatic variability. It mentions that in the absence of appropriate adaptation strategies there would be far-reaching consequences in terms of shortage of food items and rising prices (Ministry of Agriculture and Farmers' Welfare 2010). The mission seeks to transform Indian agriculture into a climate resilient production system through suitable adaptation measures. These adaptation measures focus on ten key dimensions namely:

1) improved crop seeds, livestock and fish cultures;
2) water use efficiency;
3) pest management;
4) improved farm practices;
5) nutrient management;
6) agricultural insurance;
7) credit support;
8) markets;
9) access to information;
10) livelihood diversification.

Alongside NMSA – but independent of it – there are other ongoing centrally funded schemes which carry out work similar to the overall goals of NMSA. Some of these schemes are listed below, followed by Table 11.1 which locates

Table 11.1 Comparing the strategies mentioned in NMSA and SAPCCs with ongoing programmes

Key dimension NMSA	Strategies suggested in SAPCCs	Ongoing programmes
Improved crop seeds, livestock and fish cultures	Seed production and certification Popularization of indigenous varieties	Covered under Sub-Mission for Seed and Planting Material (SMSP) under National Mission on Agricultural Extension and Technology (NMAET) in XII five-year plan
Water Use Efficiency	Rainwater harvesting, stream water conservation	Covered under Command Area Development and Water Management (CADWM), National Watershed Development Project for Rain-fed Areas (NWDPRA)
Pest Management	Promote organic ways to combat weeds, insects, pests and diseases and nutrient management	Organic farming is promoted through the use of bio-pesticides under the NADP and MMA. Practices like green manuring are promoted under programmes like the National Project on Management of Soil Health & Fertility (NPMSF), NADP and MMA
Improved Farm Practices	Farm mechanization	Covered under RKVY, Mission for Integrated Development of Horticulture (MIDH), National Mission for Oil seed and Oil Palm (NMOOP) and National Food Security Mission (NFSM)
Nutrient Management	Reducing the application of synthetic agrochemicals through Integrated Pest Management (IPM) and Integrated Nutrient Management (INM)	Covered under NPMSF and National Project on Organic Farming (NPOF)
Agricultural Insurance	Promote/ strengthen pilot projects for Weather Based Crop Insurance/Index Based Weather Insurance mechanism	Weather Based Crop Insurance Scheme has already been running for 29 crops across India under various pilot programmes
Markets	Creation of policy framework to create viable markets for indigenous species	National Agriculture Market (e-NAM)

Key dimension NMSA	Strategies suggested in SAPCCs	Ongoing programmes
Credit support	Popularizing micro finance for farmers	National Bank for Agriculture and Rural Development (NABARD) under various agriculture credit programmes promotes Kissan credit cards.
Access to Information	Strengthen climate forecast dissemination system for farmers and strengthen Agri-Met services	Under various ongoing programme information is provided about fishery input, irrigation infrastructure, drought relief and management and livestock management
Livelihood Diversification	Provisions for incentives/ subsidies for organic farming, integrated pest management and integrated nutrient management	RKVY, NADP, MMA

Source: Programmes & Schemes | Department of Agriculture Cooperation & Farmers Welfare | Mo A&FW | GoI http://agricoop.nic.in/programmes-schemes-listing (April 2019),

State Action Plan on Climate Change – HOME
http://moef.gov.in/division/environment-divisions/climate-changecc-2/state-action-plan-on-climate-change (16/04/2019).

these ongoing programmes in the adaptation strategies suggested in the SAPCCs[2]:

(1) Macro Management of Agriculture (MMA): It was introduced in 2000–2001. It brought together 27 centrally sponsored schemes relating to cooperatives, crop production programmes, horticulture, fertilizer, mechanization and seeds.

(2) Rashtriya Krishi Vikas Yojna (RKVY): It was launched as a central sector scheme in 2007 to provide support for various kinds of interventions in the agricultural sector to achieve 4 percent annual growth. It focused on strengthening farmers' efforts by creation of an agri-infrastructure.

(3) National Agriculture Development Programme (NADP): It was introduced in the year 2007–2008 to increase the productivity of important crops through focused interventions and by maximizing returns to farmers.

(4) National Horticulture Mission (NHM): It was launched in the year 2005–2006 to promote the holistic growth of the horticulture sector through area based regionally differentiated strategies.

(5) National Project on Management of Soil Health & Fertility (NPMSF): It was introduced in 2000 to focus on soil health management by promoting soil testing and soil nutrient management.

(6) Command Area Development and Water Management (CADWM): This programme was launched in 1974–1975 as Command Area Development Programme. It was restructured and renamed as Command Area Development and Water Management (CADWM) programme in 2004. The main objective of the CADWM Programme is to increase the potential for the utilization of irrigation and to improve agricultural productivity on a sustainable basis.

The overlap between the programmes designed to address ongoing agricultural issues and the identified key dimensions of the NMSA is a deliberate strategy of the NAPCC. Although the NAPCC calls for a qualitative shift in the direction of developmental pathways by mainstreaming or integrating climate policy with developmental goals across the sectors, it also acknowledges that several of the proposed programmes under the eight national missions are part of the current actions (NAPCC 2008). It is important to unpack the different aspects of this approach.

> Aspect 1: Mainstreaming adaptation in development plans: this is the intended and recommended approach (Revi 2008; Sharma & Tomar 2010; Aggarwal 2008; Shah 2009).
> Aspect 2: Taking advantage of the adaptation benefits of ongoing schemes: it is also the intended approach of NAPCC to take notice of the secondary benefits to be had from the ongoing schemes.

However, these aspects when translated into policy intervention on the ground are sometimes misconstrued as tagging existing schemes as adaptation strategies. This interpretation represents what Mike Hulme calls the struggle to locate local weather in the global climate. This struggle has come to define the challenges when thinking through adaptation strategies for climate change. To understand why, I argue that tagging ongoing programmes as climate change adaptation strategies is a misconstrued interpretation of the NAPCC's policy approach. I unpack one such adaptation strategy proposed in SAPCCs which was in place prior to any conversations about climate change. The strategy I refer to is the Index Based Weather Insurance (IBWI) Schemes/Weather Based Crop Insurance Scheme (WBCIS).

Index based weather insurance: the Rainfall Insurance Scheme for Coffee

Out of the 32 SAPCCs available online 23 explicitly mention IBWI as a strategy to address the impact of climate change on agriculture. In IBWI the insurance contract responds to an objective weather parameter (e.g. measurement of rainfall or temperature) at a defined weather station over an agreed time period. The parameters are set to correlate, as accurately as possible, with the loss of a specific crop suffered by the policy holder. All policy holders within a defined area receive payouts based on the same contract and measurement at the same station, eliminating the need for in-field assessment (Dick & Stoppa 2011, 18). The specific insurance scheme selected for this study was the Rainfall Insurance Scheme for Coffee (RISC). It is one of the IBWI schemes introduced in 2007–2008 by the Agriculture Insurance Company of India (AICI or AIC) in consultation with the Coffee Board of India for coffee growers in the three traditional coffee growing states of South India: Karnataka, Kerala and Tamil Nadu. The scheme indexes rainfall by identifying 'normal' rainfall for a particular geographic location in terms of the absolute amount of precipitation over a given period of time.

Before unpacking the rainfall indexing exercise used by RISC it is useful to briefly reflect on coffee plantations in South India where the field work for this study was carried out. In India, coffee is traditionally grown in the Western Ghats, a mountain range that runs parallel to the western coast of the Indian peninsula. This mountain range is one of the world's eight biodiversity hotspots and a world heritage site (UNESCO 2017). Recent studies on the probable impact of climate change on the Western Ghats suggests that precipitation in this region will be more intense with fewer rainy days and that average mean temperatures will see a gradual increase (INCCA 2010). Given that the coffee plant is inherently sensitive to local climatic and weather conditions, coffee growing estates in the Western Ghats will most likely face severe production stress (Vaast 2011; Centre for Social Markets 2012; Chengappa & Devika 2016). Field work for this study was carried out over a period of seven months between 2011 and 2015. The sites selected for the field study were from three districts

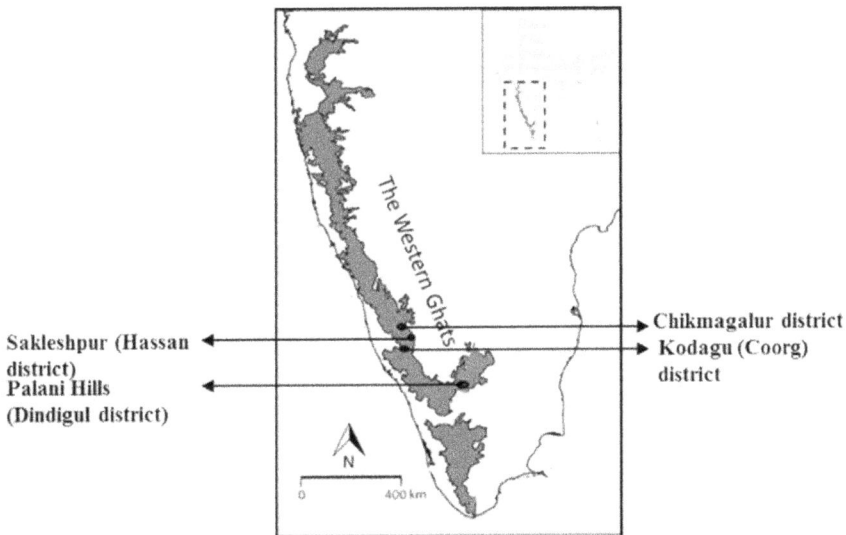

Figure 11.1 Location of the study sites in the Western Ghats belt of South India

in Karnataka (Hassan, Chikmagalur and Kodagu) and one district in Tamil Nadu (Palani Hills in Dindigul district). In-depth interviews were conducted with a total of 72 coffee growers. Additionally 26 other interviews were carried out with Coffee Board officials, meteorologists and insurers.

Please note that from hereon all the field work interviewees cited in this chapter will be referred to as Informants. To differentiate them they will be given a number, for instance Informant 1/2/3, and a brief description such as a coffee grower or a meteorologist or an insurer.

Botanically speaking, the coffee plant is a woody perennial evergreen plant that belongs to the *Rubiaceae* family. Its two main species cultivated today are *Coffea arabica*, known as Arabica coffee, and *Coffea canephora*, known as Robusta coffee. In a coffee plantation each climatic component feeds into another one: temperature, length of dry season, pattern of rainfall, etc. However, rainfall is the trigger that initiates blossoming which decides the entire year's crop cycle; therefore rainfall is very closely monitored by coffee growers. The blossom shower is the first rainfall of the season expected in the month of March or April for Robusta and Arabica respectively. Blossom showers are then followed in quick succession by backing showers; and monsoon and post-monsoon showers are expected between June and October. Figure 11.2 shows the coffee crop cycle in India along with the corresponding rainfall.

RISC provides cover for four types of rainfall: blossom showers, backing showers, monsoon showers and post monsoon showers. Table 11.2 provides the definitions of 'normal' rainfall for the purposes of RISC.

Each definition has three components: a) fixed time frame, b) moving time frame, and c) amount of rainfall. For instance, in case of the blossom showers for Robusta the fixed time frame is March 1 to April 15, the moving time frame is seven consecutive days during the fixed period and the minimum amount of

Coffee Crop Cycle

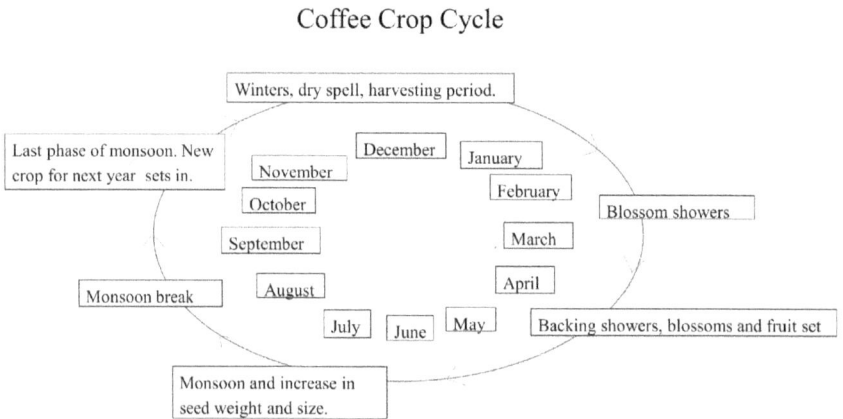

Figure 11.2 Diagrammatic representation of the coffee crop cycle in India

Source: Based on discussions with the late Dr. Vinod Kumar, Central Coffee Research Institute, Balehonnur, Chikmaglur, Karnataka.

Table 11.2 Rainfall definitions as mentioned in RISC document

Rainfall	Definition as per RISC Document
Blossom Showers	shall mean the rainfall received between 1st March to 15th April (Robusta) and 1st March to 30th April (Arabica)for the bud to flower (bud enlargement and anthesis). The normal requirement of rainfall is 25 mm in five consecutive days for Arabica and 20mm in seven consecutive days for Robusta.
Backing Showers	shall mean the rainfall received from 18th day of the starting of blossom shower till 40th day to achieve full fruit development & retention. The normal requirement of rainfall is 12mm in two consecutive days.
Monsoon Showers	shall mean the rainfall received from 1st June to 30th September for the fruit to grow in size. The aggregate rainfall of beyond a specified limit in any seven consecutive days during the period is likely to adversely impact the coffee yield.
Post Monsoon Showers	shall mean the cumulative rainfall of at least 100mm received continuously over a period of 5 days in case of Arabica during 1st November to 31st January. The cumulative rainfall of at least 125mm received continuously over a period of 7 days in case of Robusta coffee during 1st December to 28th February.

Source: Rainfall Insurance Scheme – Coffee (2015–2016), Indiacoffee.org

https://www.indiacoffee.org/sites/coffeeboard.kar.nic.in/files/RISC%20 2016%20Booklet_Karnataka.pdf (September2019).

rainfall expected is 25 mm. The amount of rainfall is further divided into slabs which decide the amount of payoff to be given in the event the grower makes a claim. These slabs are based on the geographic location of the estate. Figure 11.3 shows a payout slab table for the blossom rainfall cover for Arabica in the Karnataka State, Chikmaglur district, Aldur zone for the year 2015.

The rainfall triggers mentioned next to the payouts in Figure 11.3 are arrived at using a network of meteorological organizations. Informant 1, an AIC official, explained:

The AIC uses past rainfall records and yield data to statistically arrive at a proposed set of trigger levels. These proposed triggers are then shared with the Coffee Board.

(Informant 1, January 2014)

RAINFALL INSURANCE SCHEME FOR COFFEE (RISC) 2015		
State: Karnataka	**District**: Chikmaglur Zone: Aldur	Sub Zone: Aldur
Variety: Arabica		
Cover: 1 Blossom Rainfall		
Period: 01-Mar to 30- Apr		
Conditions: The payout will start if the cumulative rainfall is less than 25 mm in 5 consecutive days during the specified period. In case of multiple events and all are less than 25 mm (over 5 consecutive days), the vent with maximum rainfall would be considered.		

Trigger and payout slab:	RF < (In mm)	Payout (In Rs)
	25	2500
	20	3500
	15	5500
	10	7500
	5	10000

Figure 11.3 Blossom cover in RISC for Arabica coffee in Aldur zone Chikmaglur district, 2015

Building on this response Informant 2 a senior Coffee Board official commented:

> Once we at Coffee Board receive the proposed triggers from AIC we in turn consult Karnataka State Natural Disaster Monitoring Centre (KSNDMC) to confirm the rainfall data and check the probability of the occurrence of the proposed trigger levels. Simultaneously, we consult Central Coffee Research Institute (CCRI) to assess different water stress levels for coffee plants under the proposed triggers.
>
> (Informant 2, February 2014)

Speaking about his role in RISC Informant 3, a senior scientist at KSNDMC, added:

> Once we receive proposed triggers to be confirmed for a weather insurance scheme (RISC being one of them) we consult the Centre for Mathematical Modelling and Computer Simulation (C-MAACS).
>
> (Informant 3, February 2014)

Based on the field-work interviews in Figure 11.4 below the institutional network of meteorological and other organizations which work together to identify rainfall triggers for RISC payouts are shown in a diagram.

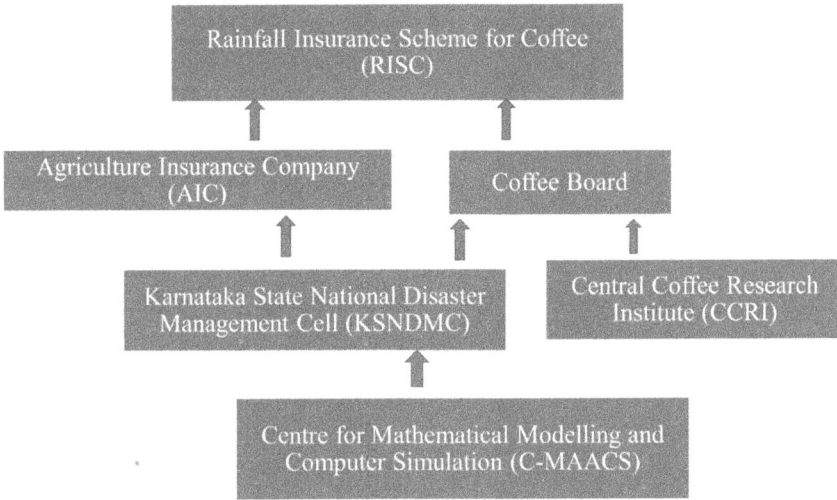

Figure 11.4 Network of meteorological and other organizations informing rainfall triggers for RISC payouts

Tracing the institutional landscape of organizations that supply RISC with information helps one understand that the policy identifies normal/abnormal rainfall based on the correlation between previous years' rainfall and coffee crop yield records. This methodology is appropriate for the scheme as it intends to address normal rainfall variation and provide growers with cover for it.

Coffee growers and rainfall in South India

The concept of a coffee plantation as a monoculture was introduced in India by the British around the mid-nineteenth century. Joseph M. Walsh, a British planter, in a book published in 1894 underlines the "paramount role of climate" in shaping the coffee crop. He writes:

> the principal points that determine the value of location for successful and profitable cultivation of coffee are: a) soil and climate; b) situation and aspect;[3] c) temperature and rainfall; d) proximity to the river; e) shelter from wind and wash.
>
> (Walsh 1894, p. 46)

In their biographies these early age coffee planters often keenly noted their experience of the regional climate and how within the range of natural factors that affected their crop they tried to strike the 'happy mean' by continuously balancing out one natural variable against the other. For example, a planter H. C. P. Hull mentioned how locating the right slope was essential in order to overcome the effects of excessive rainfall (Hull 1877, p. 42); another planter emphasized the importance of shade in counteracting the force and nature of winds beating upon the ground (Elliot 1871, p. 28). Yet another planter by the name of Arnold discussed how locating a plantation at the right height above the sea level could help minimize the effects of heat and drought (Arnold 1881, p. 7).

While exploring the contemporary understanding of what constitutes the 'happy mean' for coffee growers, I asked growers to comment on what they now consider to be 'normal rainfall'. Informant 4, an experienced coffee grower, in response to what he considers 'normal' rainfall commented:

> We (coffee growers) worked towards sustaining a 'band of manageable conditions'.
>
> (Informant 4, November 2011)

The concept of 'a band of manageable conditions' was best explained by Informant 5, a 'progressive' Arabica grower in Ballupet near Sakleshpura, Karnataka. Informant 5 has not been deterred by the fact that his plantation is mostly located at lower elevations, which are not necessarily ideal for growing Arabica coffee. The secret of his success, according to him, lies in how he conceptualizes and sustains his 'band of manageable conditions':

> I have various mutually supporting elements on the plantation: a pond and a well (a natural source of water), a herd of cows, thick canopy of forest cover and a second layer of shade trees. The cows are important not just for milk but, more significantly, for the organic manure (cow dung) which is crucial for fertilizing the coffee crop.
>
> (Informant 5, December 2011)

In other words, for Informant 5 a series of elements need to be sustained as interlinked and interdependent variables for coffee growing. The growers' willingness to use the terms 'a band of manageable weather conditions' or 'happy mean' instead of using the terms 'ideal' or 'normal' weather reflected their situated experience of weather. Central to this approach of 'a band of manageable conditions' is the fact that it recognizes each estate is in several ways topographically and ecologically unique. The Western Ghats receives plentiful rainfall, but the regional topography nonetheless plays a crucial role in shaping the micro-climate within individual estates. The regional topography results in a high degree of variability in the total amount of rainfall received in an area. Therefore, the idea of what constitutes 'normal' rainfall for a grower is variable. Hence, even with their rainfall records, it was very difficult for growers to define normal rainfall for their plantations in terms of absolute precipitation. Informant 6, a coffee grower near Bhadra forest in Chickmaglur district, Karnataka, came closest, in my opinion, to defining a successful blossom shower in terms of the quantity of precipitation. According to him:

> *Revathi* (blossom shower) should not be less than 20 mm but if the downpour was heavy it might result in the dropping of flowers before pollination. However, if the grower followed a misleading forecast about the timing of the *Revathi* (blossom shower) by irrigating his crop as a mitigation strategy, then a late backing shower could also result in the flower dropping.

Though again, he was quick to further qualify, that:

> Even delayed backing showers would not cause any harm if *Revathi* (blossom shower) brings up to 75–85 mm of rain. On the other hand, 40–60 mm of rain during *Revathi* (blossom shower) is not bad but it needs to be immediately followed by backing showers. Also, the coffee flowers do not open properly in case there is an average rainfall (40–60 mm) during *Revathi* (blossom shower) which is then followed by delayed backing showers.
>
> (Informant 6, December 2011)

The most favoured pattern, however, for all of the growers seemed to be a 'running' blossom shower: instead of receiving 65–75 mm in one go they preferred to receive 30–40 mm every day for a week. As is evident, for a grower, there are different ways in which a blossom shower can afford success. A very heavy or very weak blossom shower (less than 20 mm) is not useful. 40–60 mm of blossom shower is successful if it is followed immediately by backing showers. Whereas 75–85 mm of rainfall gives growers time in case the backing showers are delayed. From this discussion it is evident that for a coffee grower the success of the blossom shower is not decided by the absolute quantity of precipitation, rather it depends on a range of natural elements critical for growing coffee that precede or succeed the rainfall event.

Informant 7, a coffee grower in the Palani Hills of Tamil Nadu, told me that he identifies the 'failure of rainfall' in his estate by describing what he calls the 'bubble effect' in the coffee cherry.

> Upon squeezing the fruit if there is a gap or air between the pulp and the outer skin of the fruit that means either the fruit has not developed properly or has shrivelled due to insufficient water availability. While underperformance of rainfall can lead to this situation, it can also result from rains which come as a downpour and give little time for the soil to absorb the moisture.
>
> (Informant 7, February 2014)

For him it was not the amount of rainfall received but the bubble effect in the coffee fruit, which indicated drought conditions in his estate. It is important to note that rain gauges which measure the actual amount of rainfall are not able to indicate the impact of the precipitation on the soil. This is a particular cause of concern in the Palani Hills region. In this region, especially, because these estates are located on a steep slope, a heavy precipitation event might amplify soil runoff just as much as moisture retention. Emphasizing the importance of rainfall for coffee plantations, Informant 8, a former president of the Karnataka Growers Federation (KGF), said:

> Like all fruit crops, coffee is a gamble on the weather. The rubber planter gets his crop in any case, tea may be held up by drought but has chance of making up later in the season, but anything happens to coffee growers' blossom he is done for until another year comes around.
>
> (Informant 8, January 2015)

Growers regularly monitor rainfall through a calibrated rain gauge usually placed in the drying yard in the estate. A drying yard refers to a flat cemented surface that is usually used for drying washed coffee. The rainfall records thus maintained could go back several years or decades. This was evident through the records which the growers happily made available to me. While records were systematically maintained they were not at the centre of growers' perception of the rainfall. For growers, records came into the conversation only when they were specifically asked about them. This strongly suggested that the growers' idea of normal rainfall and their perception of it was situated in the unique topographical, ecological and climatic contexts in which the estate was located. Their perceptions about rainfall, hence, appeared to emerge from their unique ecological contexts. Comprehending a rainfall event by locating it in a context is markedly different from RISC's comprehension of rainfall driven by calibrated rain gauges and instruments and statistical assessments of rainfall and yield data.

Experience vs index: unprecedented weather variations

This section explores the difference in the two perceptions of rainfall identified above. One perception is of the lived experience of coffee growers situated in the context of their plantations and the second is that of RISC which is informed by instruments and statistical assessments. This section identifies temperature trends, monsoons, dry spells and pest and disease cycle as the points of comparison.

Temperature trends

The ideal temperature for growing Arabica is between 15°C and 26°C. The plant overall, however, prefers a relatively cool climate. Robusta, on the other hand, does reasonably well even at relatively higher temperatures. The ideal temperature for Robusta is 20°C–30°C and it also prefers humid conditions. Informant 9, a grower from Sakleshpur region, known for his extensive investments in irrigation systems, felt that there had been a visible rise in overall temperature over the years. According to him:

> If 20 years ago the temperature touched 32°C it meant that there would be a downpour but in 2010, even when the temperature went as high as 33°C the rains did not follow.
>
> (Informant 9, November 2011)

Commenting on these temperature trends, Informant 10, a senior coffee grower in Anemahal, Sakleshpur, shared an anecdotal observation:

> In summers, during my childhood, if we were to sweat while sitting inside our house it was sure sign of approaching rains. Now there is no correlation. Summers are definitely much hotter now.
>
> (Informant 10, November 2011)

Informant 11, another seasoned coffee grower from the region, further supported this observation by adding that:

> Growers regret not maintaining temperature records. While we can feel the change, it is difficult to support our experience with data.
>
> (Informant 11, December 2011)

To support his observations, Informant 11 has started maintaining temperature records on his plantation. Speaking about the importance of temperature for total coffee output Informant 12, a scientist at the Central Coffee Research Institute (CCRI) commented:

> Temperature trends are directly related to vegetative growth and more specifically with the growth of leaves. Leaf growth shows periodicity with maximum number of leaves initiated in August/September. The leaves that grow during this period tend to be larger and are found to be associated with a maximum temperature range of 23° C–27° C and minimum temperature range of 11° C–12°C. However, high temperature inhibits leaf expansion and causes the formation of smaller leaves. Smaller leaves limits the ability of the plant to grow and manufacture food for itself. Irrespective of the performance of rainfall, higher temperatures in fact limits the net crop output.
>
> (Informant 12, December 2011)

Although such observations are reflected in the growers' situated experience of weather on the plantations they are not captured in the indexed understanding of rainfall which informs RISC.

Monsoons

Informant 13, a coffee grower and a researcher from the Hassan district, shared that the beauty of the monsoon pattern in the shade growing coffee regions is its distribution. He said that:

> In coffee, the quantum of rainfall is not important rather the distribution pattern over the period of five months is crucial. There has been a lateral shift in the Monsoon. Monsoon rainfall instead of being distributed over five months is now concentrated within two months. This kind of downpour,

he shared, leads to soil erosion, making coffee plants more fragile and thus introducing new kinds of pests and diseases.

(Informant 13, December 2011)

This observation is in line with the climate change projections made for this region by the INCCA report (INCCA 2010). RISC, however, is not able to capture this variation because, cumulatively speaking, the rains have not failed but their uneven distribution leaves the grower unsatisfied both during the early phase of the monsoon and during the later one. One of the most discussed impacts of the accumulation of monsoonal rainfall in a two-month period was the absence of a monsoon break. Speaking about the monsoon break Informant 14, a coffee grower, said:

> Traditionally after onset of monsoon there has always been one or two breaks that can last from a couple of days to over a week. These breaks are important because a) they give plants a respite from the continuous showers; b) soil gets time to absorb water and c) plant prepares for next round of monsoonal rainfalls. But with the concentration of monsoons in two-month period monsoonal breaks are arguably not observed.

(Informant 14, November 2011)

In monsoonal cover RISC provides cap for cumulative rainfall received over any seven consecutive days but is not designed to take into account the role of monsoonal breaks or the lateral shift in the distribution of rainfall patterns. These essential components of coffee growers' evolving risk narrative are not considered in RISC.

Dry spells

A dry spell is a period of 90 days observed from the end of the monsoons until the commencement of the blossom shower the next year. Informant 15, a coffee grower in Hassan district, Karnataka, spoke about the importance of the dry period. According to him:

> Dry period causes water stress in plants which results in uniform blossoming upon being exposed to first rainfall (blossom showers) in the months of March and April. It is, thus, important in order to generate maximum impact of blossom showers. However, sporadic rainfalls are being increasingly experienced during this period.

(Informant 15, February 2014)

Informant 16, another grower from the same region, while mentioning the importance of the dry period in determining the success of blossom showers, talked about untimely blossoms at her plantation. She explained:

> Early December is the period when the soil had been recently manured, and this is halfway into the period of the dry spell. Thus, even a sprinkle would

initiate blossoming. But this untimely blossoming will wither away. This untimely blossom in turn dilutes the impact of the *Revathi* (blossom shower). By the time the full force of the *Revathi* comes these buds would already be lost and hence there would be a net loss even if *Revathi* delivers as per expectation.

(Informant 16, December 2011)

Due to this unprecedented variation even if blossom showers do perform as expected the grower would already have lost part of the crop. RISC is not designed to address the factors preceding the rainfall.

Pest and disease cycle

Plantations located in Hassan district in Karnataka State were found to be struggling with the infiltration of the White Stem Borer (WSB) in Arabica plants. Coffee White Stem Borer, *Xylotrechus quadripes*, is a serious pest of Arabica coffee causing a yield loss up to 40 percent in all the coffee growing areas in India (Jayaraj & Muthukrishnan 2013).

According to Informant 17, a grower from the region:

In coffee plantations the impact of rainfall is dependent on the temperatures being experienced. The most intricate relation of the two is observed in fighting pests like White Stem Borer (WSB). Exposure to high temperatures for a prolonged period results in high pest incidences like WSB. The worst infestation of White Stem Borer in South-West Monsoon belt was experienced during the drought of 2002–2005. Since then Arabica has not been able to recover in this region, this is because temperatures are relatively higher. WSB stays dormant during the rainy season but appear again once the rainfall stops. To break this cycle of infestation growers are increasingly switching to Robusta. Although Robusta is not entirely resistant to the pest, but it prevents pest cycle from kicking in which if instigated pushes growers in debt for five years straight.

(Informant 17, December 2011)

This period of changing climate drought is, thus, associated not just with the failure of one or two crop cycles but with the loss of the entire plant as a result of infestation. More importantly, as is evident now, the risk assessment of drought for a grower also includes the investment in redesigning the plantation from Arabica to Robusta. RISC identifies the rainfall failure, but it is not designed to capture the extent of the damage incurred by growers because of increased temperatures. However, Robusta, which is planted because of its higher resistance to WSB, faces another set of problems. Honeybees are crucial for pollination in Robusta plants. Informant 18, a Robusta grower fondly remembered:

Till few years back, during flowering season one could hear the hum of honeybees for miles. However, honeybees are becoming scarce due to oddly high temperature. Reduction in number of honeybees result in reduced degree of pollination affecting fruit set and net crop production irrespective of how rainfall performs.

(Informant 18, February 2014)

These observations show that through situated experience of local weather, especially rainfall, growers have been able to register unprecedented changes that are affecting their crop output. However, RISC which is informed by past rainfall patterns and corresponding coffee yield outputs is not able to register these changes. This disconnect is a significant shortcoming of a strategy which is tagged as an adaptation measure for climate change.

Analysis and discussion

The implications of these observations become more evident when growers' increasing concern about the unprecedented variation in weather (especially rainfall) is compared with their decreasing interest in insuring with RISC. Figure 11.5 shows that unprecedented variation in weather is one of the top three concerns of coffee growers. Following this, Figure 11.6 shows growers' decreasing interest in buying RISC.

What is indicated by the growers' decreasing interest in RISC in light of their increasing concern about the unprecedented variation in weather, especially rainfall? One immediate interpretation is that RISC as an IBWI scheme is not designed to address unprecedented changes in weather (rainfall) pattern. It is structured around the idea of 'normal' rainfall derived from previous rainfall

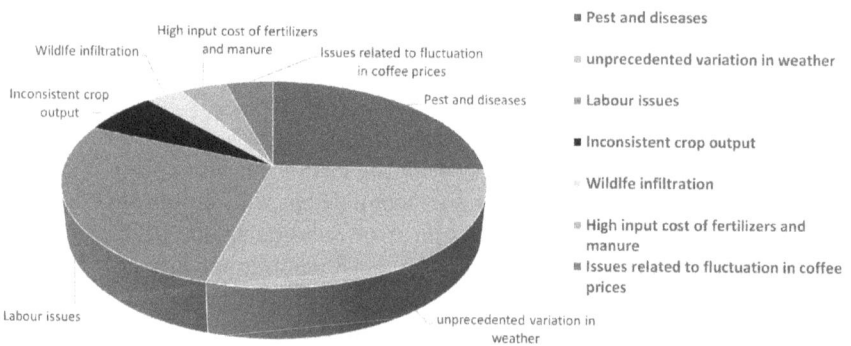

Figure 11.5 Pie chart showing key problems faced by coffee growers on their plantations

Source: Based on primary data. The survey was conducted on January 18, 2015 at the University of Agriculture and Horticulture Sciences, Shimoga University, Mudigere, Chikmaglur, Karnataka.

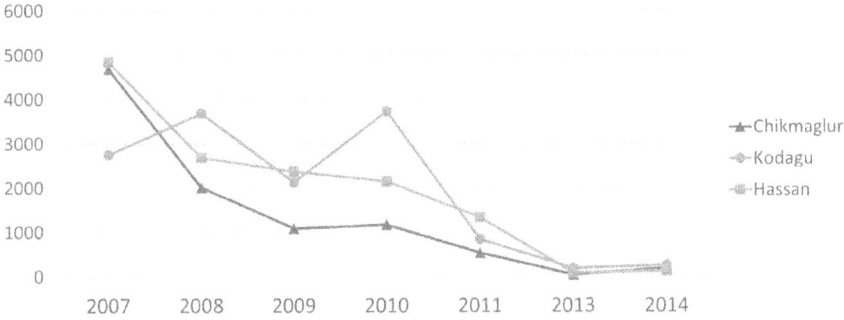

Figure 11.6 Number of growers insured with the RISC in Hassan, Chikmaglur and Kodagu districts of Karnataka for the period of 2007–2014

Source: Based on primary data collected from by the Agriculture Insurance Company (AIC) office, Bangalore.

records. That being the case, what qualifies it as an adaptation strategy? An IBWI scheme has three defining characteristics:

1) it is based on indexing the weather by using scientifically calibrated instruments and previously held weather records;
2) it co-relates the indexed weather with the crop output which is an everyday concern of farmers;
3) using this correlation, it statistically attributes a monetary value to the loss a farmer is likely to incur.

In summary an IBWI scheme indexes the weather, corelates the indexed weather with an everyday concern (crop output) and then statistically attributes a monetary value to this correlation (payoff).

These characteristics draw a direct correlation between weather variation and its impact on everyday life (crop output) and through monetary intervention it suggests reducing the adverse effects of this impact. Thus, for a strategy to qualify as an adaptation strategy it has to relate weather variation to an everyday concern. However, most significantly, the scheme is not designed to differentiate between precedented or unprecedented rainfall variation. What does overlooking this significant shortcoming suggest about the conceptualization of adaptation in policy domain? It suggests that this conceptualization of adaptation equates weather with climate and engages with the known narrative of weather compiled from past rainfall records in order to respond to the unknown climate.

This conceptualization comes from what Mike Hulme identifies as the struggle to locate the 'local' in the 'global' narrative of climate change. Hulme argues that responding to the global narrative of climate change instinctively calls for the need to find identity and meaning in the global narrative. He further adds that in our struggle to do so we are reacting to the sterile idea of a disembodied and unsituated global climate with its universalizing demands on our imaginations and behaviours (Hulme 2008) to tow the lines of internationally agreed targets and timelines. The regional interpretation of the global climate is thus overpowered by known experiences of weather. The anonymity and remoteness of narratives of global climate change is subverted by re-introducing localized narratives of weather variation to which we have greater psychological attachment (Macnaghten 2006). Thus, in the policy domain, 'adaptation' has been conceptualized as a means of re-introducing the known localized narratives of weather variation in the global narrative of climate change.

How do we inform/modify/improve this conceptualization of adaptation? One way is to identify the challenges offered by this conceptualization followed by a reflection on the factors which create the need for this conceptualization. As a challenge, this conceptualization does not have the scope to differentiate between a 'normal' variation and an 'abnormal' variation of weather. It also does not dwell on the relationship between climate and weather. There is an assumption that identifying the concerns about known weather variation is the same as locally situating the global climate change concerns. Going further and reflecting on the factors which create the need for this conceptualization requires rethinking the existing narrative of climate change.

The current narrative of climate change conceives climate on a global scale using scientific instrumentation like Global Climate Change Models (GCMs). GCMs conceive 'global' which is given to us all immediately. The consequence of using this narrative is to acknowledge a global and a local binary. While 'global' represents nature as the physical reality which is a given and exists independent of human beings the 'local 'represents the world in which we live and which we create through our actions and which in turn affects us. It is people, their engagements and interactions. Tim Ingold argues that the distinction between the 'global' and the 'local' is not one of hierarchical degree – in scale or comprehensiveness – but one of kind. Local is not a more limited or narrowly focused apprehension than the global; it is one that rests on an altogether different mode of apprehension, based on practical perceptual engagement with components of a world that is inhabited or dwelt-in, rather than on the detached disinterested observation of a world that is merely occupied (Ingold 2011, p. 100).

The issue of local knowledge in relation to climate change was covered by the IPCC's Working Group II in the Fifth Assessment Report (2014). The term 'local knowledge' was used simultaneously with 'indigenous knowledge' and 'traditional knowledge' in order to indicate their interchangeability. Traditional, indigenous or local knowledge, as argued in the IPCC reports, represents observations and interpretations of meteorological phenomenon which have guided seasonal or inter-annual activities of local communities for millennia. This is reflected in the studies carried out to gather traditional knowledge on the ground. More recently there have been calls for community-based adaptation which requires going beyond the scientific representation of climate change and exploring these local/ traditional/indigenous knowledges.

The attempt to bring together science and traditional/ local/indigenous knowledge is problematic because any such attempt will first need to identify traditional/local/indigenous knowledges; which will be derived from a secondary set of non-factual expertise whose defining characteristic will be unexposed/ untainted by the scientific knowledge in the field. The governing scientific system then goes on to tap into this knowledge system to cherry pick the information snippets which best fit the requirements of science.

In order to engage with the global and local binaries with regard to climate change adaptation I argue for the need to draw insights from the perspective of Science Technology Studies (STS). The concept of situated knowledge best captures the complexity at hand. It argues that knowledge and understanding are contextually generated and simultaneously embody understandings of both the natural and social worlds (Haraway 1988). This thinking is closely in line with Mike Hulme's idea of a cosmopolitan climate which focuses on understanding the ways in which science and society are being mutually constructed through the phenomenon of climate change.

Reconsidering climate change adaptation from an STS perspective, I argue, means it is necessary for us to engage with different kinds of knowledge that can span the range from individual experience to one based on strict scientific and technical protocols. The knowledge or the site of entanglement between the 'natural' and the 'social' is thus where the meanings about a weather event can be generated. Different knowledge about the weather event in particular becomes one of the active and critical sites for these entanglements between the 'natural' and the 'social'.

Concluding remarks

This chapter unpacks the conceptualization of adaption as a policy response to climate change in India. While there is a visible lack of clarity about this concept in academic circles, it has come to closely inform the policy landscape of developing countries like India. Here I use Mike Hulme's (2010) concept of cosmopolitan climates which provides an interesting space to differentiate between the problem of climate change from its scientifically informed global narrative. This chapter has explored the following questions: 1) How is adaption to climate change being conceptualized in India? 2) Which factors inform this conceptualization? 3) What are the challenges offered by this conceptualization? and 4) How can we inform it in order to improve it?

The chapter begins by identifying the major structures in the Indian climate policy landscape: the (NAPCC and the State Action Plan on Climate change (SAPCCs. Focusing on agriculture, a specific sector in these policy documents, the chapter recognizes the trend of identifying ongoing schemes as adaptation strategies. The chapter argues that such a conceptualization is an attempt to localize climate change in everyday life. This conceptualization, as Hulme argues, represents the instinctive struggle to find identity and meaning in the global narrative of climate change. The primary goal of an adaptation strategy, thus, is to find the individual identity of everyday problems in the framework of the global narrative of climate change.

Building on a specific example of the Weather Based Crop Insurance Scheme (one of the ongoing schemes identified as an adaptation strategy) the chapter argues that this conceptualization (borrowing ongoing schemes to identify as adaptation measures) has inherent challenges. It does not have the scope to differentiate between 'normal' variation and 'abnormal' variation of weather which can potentially result in a lower intake of such policies as the increase in the incidences of unprecedented weather variation.

The chapter argues that in order to go beyond the 'global' and 'local' binary of the existing climate change narrative we need to rethink adaptation as an issue of a growing knowledge divide. I argue that it is necessary for us to engage with different kinds of knowledge that can span the range from individual experience to one based on strict scientific and technical protocols. I suggest using Haraway's (1988) concept of situated knowledge to go beyond the scientific and non-scientific divides (local/indigenous/traditional knowledge) and rethink adaptation as a knowledge space where science and society are being mutually constructed through the phenomenon of climate change.

Notes

1 Please see definitions provided by Intergovernmental Panel on Climate Change (IPCC) (2014), United Nations Framework Convention on Climate Change (UNFCCC), United Nations Development Programme (UNDP) and United Kingdom Climate Impact Programme (UKCIP).
2 For more details on these schemes please see: Department of Agriculture & Cooperation, Ministry of Agriculture, Government of India http://agricoop.nic. in/programmes-schemes-listing (16 April 2019).
3 Situation is defined as a nominal location of one place in relation to others. Aspect is defined as the compass direction a slope faces.

References

Adger, W. N., Arnell, N. W., & Tompkins, E. L. (2005). Successful adaptation to climate change across scales. *Global Environmental Change, 15*(2), 77–86.
Adger, W. N. et al. (2009). Are there social limits to adaptation to climate change? *Climatic Change, 93*(3–4), 335–354.
Aggarwal, P. K. (2008). Global climate change and Indian agriculture: impacts, adaptation and mitigation. *Indian Journal of Agricultural Sciences, 78*(11), 911.
Arnold, Edwin Lester (1881). *On the Indian Hills: Coffee Planting in Southern India.* London: Sampson Low, Marston, Searle and Rivington. Retrieved April 6, 2020 from: https://archive.org/stream/onindianhillsorc01arnoiala#page/n3/mode/2up.
Barnett, B. J., & Mahul, O. (2007). Weather index insurance for agriculture and rural areas in lower-income countries. *American Journal of Agricultural Economics, 89*(5), 1241–1247.
Brookfield, H. (1973). Introduction: explaining or understating? The study of adaptation and change, in H. Brookfield (ed.), *The Pacific in Transition: Geographical Perspectives on Adaptation and Change,* 3–24. . Canberra: ANU Press.

Centre for Social Markets. (October 2012). *Coffee to Go?: The Vital Role of Indian Coffee Towards Ecosystem Services and Livelihoods.* CSM and KGF publication. Retrieved April 5, 2020 from: http://csmworld.org/templates/CSM/images/pdf/coffee_to_go_uploads.pdf.

Chengappa, P. G., & Devika, C. M. (2016). Climate variability concerns for the future of coffee in India: an exploratory study. *International Journal of Environment, Agriculture and Biotechnology, 1*(4): 819–826.

Cimato, F., & Mullan, M. (2010). Adapting to climate change: analysing the role of government. *Defra Evidence and Analysis Series, Paper, 1.* Retrieved from: https://documents.wfp.org/stellent/groups/public/documents/communications/wfp242409.pdf.Dandekar, V. M. (1976). Crop insurance in India. *Economic and Political Weekly, 11*(26), A61–A80.

Department of Agriculture & Cooperation, Ministry of Agriculture, Government of India. (2019). *Operational Guidelines for XII Five Year Plan.* Retrieved April 16, 2019 from http://agricoop.nic.in/sites/default/files/RKVY_Guildlines_%28XII_Plan%29-2014.pdf

Dessai, S., & Schipper, E. L. (2003). The Marrakech Accords to the Kyoto Protocol: analysis and future prospects. *Global Environmental Change, 13*(2), 149–153.

Dick, W., & Stoppa, A. (2011). Weather index-based insurance in agricultural development: a technical guide. *International Fund for Agricultural Development (IFAD).*

Doria, M. D. F., Boyd, E., Tompkins, E. L., & Adger, W. N. (2009). Using expert elicitation to define successful adaptation to climate change. *Environmental Science & Policy, 12*, 810–819.

Dubash, N. K. (2013). The politics of climate change in India: narratives of equity and cobenefits. *Wiley Interdisciplinary Reviews: Climate Change, 4*(3), 191–201.

Dubash, N. K., & Joseph, N. B. (2016). Evolution of institutions for climate policy in India. *Economic & Political Weekly, 51*(3), 45.

Elliot, Robert H. (1871), *The Experiences of a Planter in the Jungles of Mysore.* London: Chapman & Hall. Retrieved April 6, 2020 from: https://archive.org/stream/experiencesapla00elligoog#page/n6/mode/2up.

European Commission. (n.d.). *Adaptation to Climate Change – Climate Action.* Retrieved April 6, 2020 from: https://ec.europa.eu/clima/policies/adaptation_en

Grothmann, T., & Patt, A. (2005). Adaptive capacity and human cognition: the process of individual adaptation to climate change. *Global Environmental Change, 15*(3), 199–213.

Haraway, D. (1988). Situated knowledges: the science question in feminism and the privilege of partial perspective. *Feminist Studies, 14*(3): 575–599.

Hoffmann, A. A., & Sgro, C. M. (2011). Climate change and evolutionary adaptation. *Nature, 470*(7335), 479. Retrieved April 6, 2020 from: https://www.nature.com/articles/nature09670.Hull, H.C.P (1877), *Coffee Planting in Southern India and Ceylon.* London: E & F.N. Spon.

Hulme, M. (2008). Geographical work at the boundaries of climate change. *Transactions of the Institute of British Geographers, 33*(1), 5–11.

Hulme, M. (2010). Cosmopolitan climates. *Theory, Culture & Society, 27*(2–3), 267–276.

Hunt, A., & Watkiss, P. (2011). Climate change impacts and adaptation in cities: a review of the literature. *Climatic Change, 104*(1), 13–49.

Indian Network for Climate Change Assessment (INCCA). (2010). *Climate Change and India: A 4x4 Assessment, INCCA Report No. 2.* Indian Network for Climate

Change Assessment (INCCA), Ministry of Environment and Forests, Government of India, New Delhi, November 2010. Retrieved April 2, 2020 from: www.indi-aenvironmentportal.org.in/files/fin-rpt-incca.pdf.

Ingold, T. (2011). *Being Alive: Essays on Movement, Knowledge and Description.* Abingdon: Routledge.

Intergovernmental Panel on Climate Change (IPCC). (2014 *Climate Change 2014: Synthesis Report. Contribution of Working Groups I, II and III to the Fifth Assessment Report of the Intergovernmental Panel on Climate Change* [Core Writing Team, R. K. Pachauri and L. A. Meyer (eds.)]. Geneva, Switzerland: IPCC. Retrieved April 6, 2020 from: https://www.ipcc.ch/site/assets/uploads/2018/05/SYR_AR5_FINAL_full_wcover.pdf.

Jayaraj, J. & Muthukrishnan N. (22 May 2013). Management of white stem borer in coffee. *The Hindu.*

Levina, E. & Tirpak, D. (2006). *Adaptation to Climate Change. Key Terms. Draft Paper, Agenda Document 1.* OECD/IEA Project for the Annex I Expert Group on the UNFCCC, 5. Retrieved November 4, 2016 from: https://www.oecd.org/env/cc/36278739.pdf.

Macnaghten, P. (2006). Nature Theory. *Culture & Society 23*(2–3), 347–349.

Ministry of Agriculture. (2010). *National Mission for Sustainable Agriculture*, August 2010, II.

Ministry of Agriculture and Farmers' Welfare, Government of India. (2010). *National Mission for Sustainable Agriculture*. Retrieved April 5, 2020 from: www.agritech.tnau.ac.in/pdf/NMSA_GDL.pdf.

Ministry of Finance, Government of India. (2017–2018). Climate, climate change and agriculture. In *Economic Survey*, Vol. I, Chapter 6: 82–10. Retrieved April 6, 2020 from: http://mofapp.nic.in:8080/economicsurvey/pdf/082-01_Chapter_06_ENGLISH_Vol_01_2017-18.pdf.

Mishra, P. K. (1995). Is rainfall insurance a new idea? Pioneering work revisited. *Economic and Political Weekly, 30*(25), 84–88.

Moser, S. C., & Ekstrom, J. A. (2010). A framework to diagnose barriers to climate change adaptation. *Proceedings of the National Academy of Sciences, 107*(51), 22026–22031.

Nightingale, A. J. (2017). Power and politics in climate change adaptation efforts: struggles over authority and recognition in the context of political instability. *Geoforum, 84*, 11–20.

Prime Minister's Council on Climate Change, Government of India. (2008). *National Action Plan on Climate Change*. Retrieved April 5, 2020 from: https://www.indiawaterportal.org/sites/indiawaterportal.org/files/National%20Action%20Plan%20on%20Climate%20Change%20(NAPCC)_Prime%20Ministers%20Council%20on%20Climate%20Change_Government%20of%20India_%202008.pdf.

Pelling, M. (2010). *Adaptation to Climate Change: From Resilience to Transformation.* Abingdon and New York: Routledge.

Rao, K. N. (2010). Index based crop insurance. *Agriculture and Agricultural Science Procedia, 1*, 193–203.

Revi, A. (2008). Climate change risk: an adaptation and mitigation agenda for Indian cities. *Environment and Urbanization, 20*(1), 207–229.

Ribot, J. (2011). Vulnerability before adaptation: toward transformative climate action. *Global Environmental Change, 21*(4), 1160–1162.

Shah, T. (2009). Climate change and groundwater: India's opportunities for mitigation and adaptation. *Environmental Research Letters, 4*(3). Retrieved April 6, 2020 from: https://iopscience.iop.org/article/10.1088/1748-9326/4/3/035005/pdf.

Sharma, D., & Tomar, S. (2010). Mainstreaming climate change adaptation in Indian cities. *Environment and Urbanization, 22*(2), 451–465.

Smithers, J., & Smit, B. (1997). Human adaptation to climatic variability and change. *Global Environmental Change 7*(2), 129–146.

Sushma, N. (2018). Climate change could ravage Indian farmers. *Quartz India.* Retrieved September 19, 2019 from: https://qz.com/india/1191273/economic-survey-2018-climate-change-will-hammer-indias-agriculture-and-farmer-income.

Tol, R. S., Fankhauser, S., & Smith, J. B. (1998). The scope for adaptation to climate change: what can we learn from the impact literature? *Global Environmental Change, 8*(2), 109–123.

UNESCO. (2017). Retrieved April 5, 2020 from: https://whc.unesco.org/en/list/1342.

Vaast, P. (2011). *CAFNET Final Narrative Report. ENV/2006/114-382. January 2007 to September 2011: CAFNET: Connecting, Enhancing and Sustaining Environmental Services and Market Values of Coffee Agroforestry in Central America, East Africa and India.* Retrieved April 5, 2020 from https://agritrop.cirad.fr/562251/1/document_562251.pdf.

Walsh, Joseph M. (1894). *Coffee: Its History, Classification and Description.* Philadelphia, PA: The John C. Winston Co.

12 Disaster risk governance in northern Philippine communities

Issues and prospects in climate change talks

Leah Abayao

Introduction

Today's disaster risk governance is largely guided by the Sendai Framework for Disaster Risk Reduction (SFDRR) 2015–2030 (UNODRR, 2015). As a global framework of action for reducing disaster risks, official policies, programs and plans are expected to contribute to its targets and priorities. Priority 2 (Strengthening disaster risk governance to manage disaster risks) Secs E and F of this framework clearly articulate the necessity to adopt participatory and inclusive measures for local communities in disaster risk reduction (DRR). The provisions directs national and local governments to: 1) develop and strengthen, as appropriate, mechanisms to follow up, periodically assess and publicly report on progress of national and local plans; and promote public scrutiny and encourage institutional debates, including by parliamentarians and other relevant officials, on progress reports of local and national plans for disaster risk reduction (Sec. E); 2) assign, as appropriate, clear roles and tasks to community representatives within disaster risk management institutions and processes and decision-making through relevant legal frameworks, and undertake comprehensive public and community consultations during the development of such laws and regulations to support their implementation (Sec F). With this framework, policies, programs and plans are subject to public assessment. It is also the responsibility of local governments to establish a platform for community representatives to be assigned roles. This view is also being supported by studies that have looked into approaches and strategies for the integration of local and scientific knowledge in DRR. In terms of approach at the program level, the following are currently used in Asia: 1) community-based disaster risk reduction (CBDRR) which predicts high behavioral change (Izumi et al., 2019, 41), and, 2) participatory resilience and vulnerability (R&V) assessment tools for integrated understanding and engagement with communities (Ibid, 57).

In terms of strategic disaster risk reduction management, talks about local governance framework in Asia and the Pacific region are now looking at 'inclusiveness' and the 'effective participation' of communities. Shin (2017) suggests that "trust building and equitable participation" is a foundational element in disaster risk reduction. The value will be long term for institutional capacity building even if it requires a long process that is also contingent on local conditions and contexts. A recent study on 'Participatory three-dimensional maps for disaster risk reduction in Mindanao, Philippines' has been shown to be an effective platform for a deepened knowledge engagement of local communities such that contextualization and transferability, promoting trust and self help, has been achieved in the process (Kelman, Mercer and Gaillard, 2012). Community knowledge as a local strategy in tsunami disaster preparedness has also been acclaimed for its effectiveness in Indonesia (Yogaswara and Yulianto, 2005; Swafwina, 2014).

The 2019 UNODRR Global Report on Disaster Risk Reduction stated that little is known about the impact of the Sendai Framework on the implementation of local-level plans, noting that a global survey shows that only 27.4 percent of local governments have fully implemented DRR strategies (p. 329). The next section will discuss the case of the Philippines with respect to communities in the Cordilleras.

The Philippines has officially expressed commitment to climate change talks and contemporary resilience obligations. As a signatory to disaster management treaties such as the Hyogo Framework for Action (HFA) and the ASEAN Agreement on Disaster Management and Emergency Response (AADMER), the Philippines developed the Strategic National Action Plan for Disaster Risk Reduction 2009–2019 (SNAP). The SNAP serves as the 'road map' indicating the Philippines' vision and strategic objectives in compliance with the strategic goals of the HFA. Furthermore, the country has produced a legal framework on disaster management which includes the Philippine Disaster Risk Reduction and Management (DRRM) Act (RA10121)[1] together with its Implementing Rules and Regulation of 2010, and the National Disaster Risk Reduction and Management Plan (NDRRMP) 2011–2028.

Currently, the country has a strong policy framework for climate change and disaster resilience. The NDRRMP covered in Republic Act No. 10121, for example, is designed to build the adaptive capacity of communities, increasing the resilience of vulnerable sectors and optimizing disaster mitigation opportunities. Part of its strategy is to "strengthen the capacity of local government units (LGUs) to build the disaster resilience of communities and to institutionalize arrangements and measures for reducing disaster risks, including projected climate risks and enhancing disaster preparedness and response capabilities" (Republic Act No. 10121, 2010). Additionally, this provides that:

> the State will uphold the people's constitutional rights to life and property by addressing the root causes of vulnerabilities to disasters, strengthening the country's institutional capacity for disaster risk reduction and management and *building the resilience of local communities to disasters including climate change impacts.*
>
> (Emphasis in original)

Furthermore, Act 10121 mandates all local governments and communities to enforce DRR measures and to address their respective risks.[2] The Act has two fundamental provisions: 1) LGUs shall establish their own Local Disaster Risk Management Office (LDRRMO), and 2) LGUs are also bound to set aside not less than 5 percent of their estimated revenue as the Local Disaster Risk Reduction and Management Fund (LDRRMF). This act further provides that the LDRRMOs and the Barangay Development Councils are mandated to design, program and co-ordinate DRRM activities and to develop a Local Disaster Risk Management Plan (LDRRMP) to be consistent with the National DRR Framework. Thus, all local governments in the Philippines are required to address disaster risk reduction, especially those disasters that are climate change-induced.

Eight years has passed since the official implementation of DRRM in the Philippines. A recent assessment has observed that the Philippine disaster management system "tend[s] to rely on a response or reactive approach, in contrast to a more effective proactive approach, in which disasters are avoided".[3] Moreover, it argues that appropriate land-use planning and other disaster preventive measures are not given equal attention and that there is "a widespread emphasis on post-disaster relief and short-term preparedness (forecasting, evacuation, etc.) rather than mitigation or post-disaster support for economic recovery".[4] Apparently, the biased implementation towards post-disaster recovery is not a strategic response to sustainable development in Philippines' communities because it is only a reactive solution that allows communities to be dependent on social services. Such reliance becomes problematic in the long run. It fails to recognize the prospects of local community knowledge to form part of the solution. As a recent study remarked, "strengthening of local knowledge systems can work to counterbalance the drawbacks of centralized flood early warning systems" and therefore "the production and consumption of flood forecasting knowledge needs local and scientific communities to work together for reducing knowledge gaps at both ends" (Acharya and Prakash, 2018, 6). This view illustrates that risk governance requires both the experimentally sourced knowledge and experience sourced knowledge.

Looking at local conditions in the mountainous areas in the Northern Philippines, disaster concerns are very different from the lowland, metropolitan and coastal areas. Rain-induced hazards are multiple and the mitigation of these begins with a systematic inventory of hazards and assessments of risks using a process that is informed by indigenous and local knowledge. A recent work affirms the importance of local knowledge in DRRM. In studying the complex knowledge and practice of local communities to forecast floods and heavy rainfall in a river basin in Bihar, India, it was discerned that local predictive knowledge of floods is "profoundly rooted in livelihood struggles, memories of dislocation, gendered practices and an understanding that emerges out of living and surviving floods" (Acharya and Prakash, 2018, 6). This case also shows that local knowledge is very important in understanding and surviving climate change-related risks. Its contemporary relevance will have to be carefully studied to serve its purpose.

The Philippines is highly prone to natural disasters, such as earthquakes, volcanic eruptions, tropical cyclones and floods. The 2018 World Risk Report ranked the country third in the list of the top 15 countries with the highest exposure to risks. The Philippine Atmospheric, Geophysical and Astronomical Services Administration (PAGASA) reports that "rainfall is the most important climatic element in the Philippines and rainfall distribution throughout the country varies from one region to another". Climactic projections of PAGASA for rainfall in the provinces show that heavy rainfall will continue to become more frequent and that extreme rainfall is projected to increase in the Luzon and Visaya Islands.[5] PAGASA also established that the trends of tropical cyclone occurrence within the Philippine Area of Responsibility (PAR) show that an average of 20 tropical cyclones take place in the PAR every year.[6] Accordingly, the mean annual rainfall of the Philippines varies from 965 to 4,064mm annually with Baguio City (Benguet Province), Eastern Samar and Eastern Surigao receiving the largest amount of rainfall.[7] Widespread and heavy rainfall are experienced in Benguet Province from June to November, with the lowest rainfall in June being 475.8mm, and the highest rainfall in August being 905mm, giving an average of 3841.4mm annually.[8] Baguio City, Benguet has an annual rainfall of 3892mm (Nolasco-Javier and Kumar, 2018, 925). This falls in the upper range of the mean annual rainfall of the Philippines (965–4,065mm). Approximately 50 percent of this can be attributed to the passage of tropical cyclones (TCs) in Northern Luzon (Cinco et al., 2016). Aside from being one of the areas receiving the highest amount of rainfall annually in the country, Baguio City also records the most TC-related rainfall in the country (Cinco et al., 2016). This also explains the high vulnerability and susceptibility of the region to landslides and similar events. Proof of this are the high number of reported landslides in the Cordillera Administrative Region (CAR) during the passage of typhoons and tropical cyclones in the country. The majority of these landslides are deadly, leaving in their wake several deaths and huge damage to infrastructure and agriculture.

The majority of the reported landslides in Cordillera and Baguio City occur during the passage of tropical cyclones in the region (Nolasco-Javier and Kumar, 2018, 922; Nolasco-Javier, Kumar and Tengonciang, 2015). National and local reports often cite the high amount of rainfall in the City as the cause of rain-induced landslides in different parts of Benguet and CAR. A recent study on landslides illustrates that landslide occurrences in Benguet are triggered by "heavy and prolonged" rainfall due to its effect of increasing the water content and weight of the soil, therefore causing the slopes to weaken and, eventually, fail (Nolasco-Javier and Kumar, 2018, 922).

This chapter primarily used official reports and documents that are released to the public by the Commission of Audit (COA), the Department of Interior and Local Government (DILG)), PAGASA, and the National Disaster Risk Reduction Management Council (NDRRMC). I will first illustrate the state of local DRRM implementation then use the case of Benguet Province as a case in point. In terms of risks, this province is at high risk of rain-induced landslides as shown in the Benguet Provincial Rain-Induced Hazard map (see Figure 12.1).

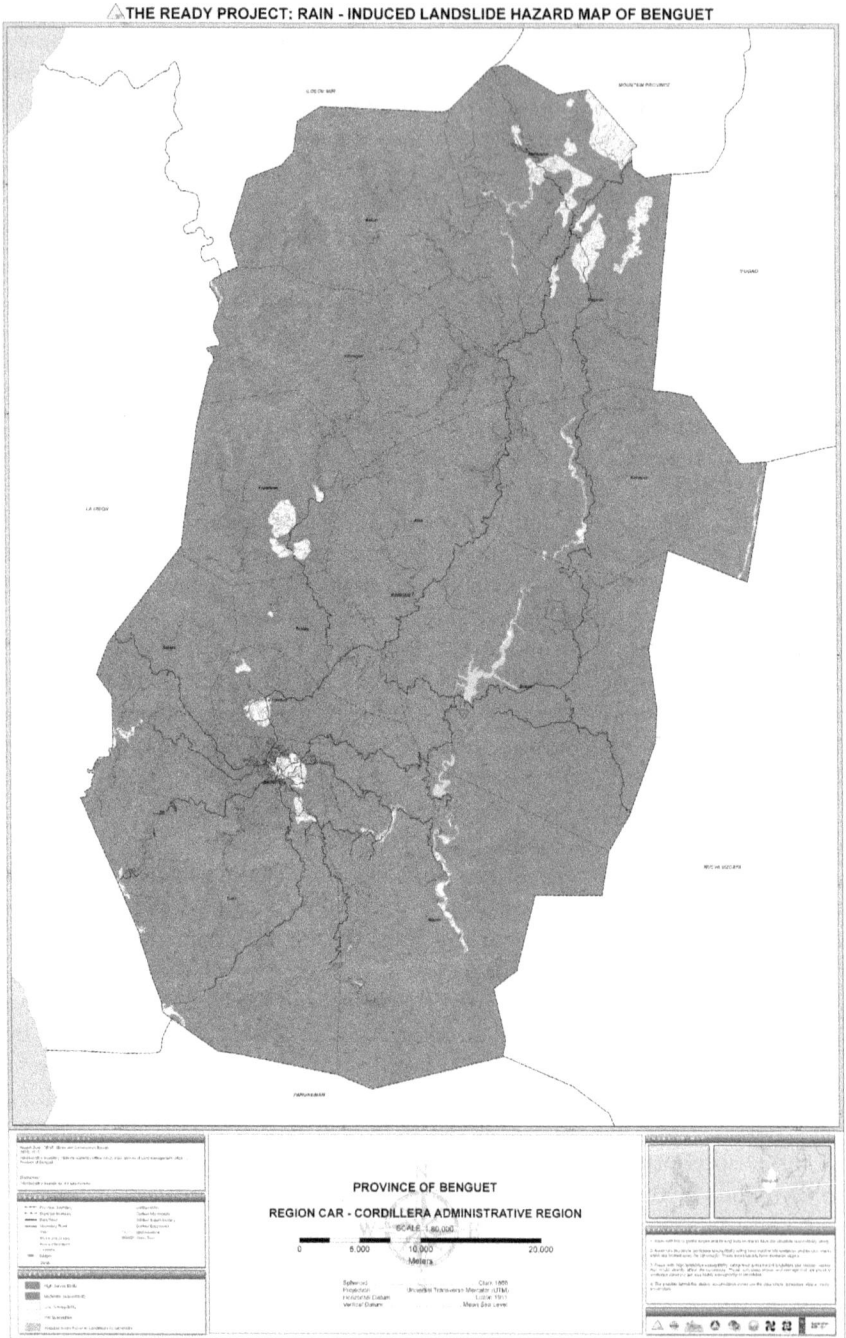

Figure 12.1 Rain-induced hazard map of the province of Benguet in the Philippine
        Cordilleras

Source: Office of Civil Defense. National Disaster Risk Reduction Management Council
(NDRRMC). http://ndrrmc.gov.ph/27-gmma-ready/hazard-maps/2852-benguet-provincial-rain-
induced-landslide-hazard.

DRRM is integral to the Philippines' plan of action for Climate Change Mitigation and Adaptation from 2011–2028. Mainstreaming DRR and CCA in the sectoral, regional and local development policies is an important target for the human security outcomes of this plan. The goal is for the Philippine communities to achieve climate change resilience.

State of local DRRM implementation

Legislative and organizational DRRM structures have been created at various governance levels in the Philippines. This strategy is consistent with the law called the 'Local Government Code of 1991' in which power is decentralized thus giving the local governments the authority and responsibility for almost any area of work. This section will present the governance structure and operation of the DRRM in the Philippines. It will present the case of Benguet Province as an illustrative example of the implementation of DRR at the local level.

The Provincial Disaster Coordinating Council was re-organized into the Provincial Disaster Risk Reduction Management Council (PDRRMC) by virtue of an Executive Order in 2011. Subsequently, a provincial DRRM plan was developed in 2012. The DRRM operations were formally separated from the Provincial Planning and Development Office (PPDO) through the issuance of an ordinance creating the DRRM Office and the appropriation of funds in 2013. It has established an organized communications protocol for response and relief operations and conducted information, education campaign (IEC) activities. An emergency operations center is also based in the provincial capital which houses the logistics needed in case of response and relief operations and serves as the incident command center during times of disasters. The provincial DRRM personnel have also been provided with incident command system (ICS) operations training, risk identification, emergency response, early warning, hazard mapping, basic life support, advance life support (ACLS) and other special skills. The provincial government was able to establish effective inter-department and inter-agency coordination during disasters and emergencies. This includes the active involvement of civil society organizations (CSOs) in DRRM activities.[9] The provincial government was also able to tap volunteers through its links to national government attached agencies like the Department of Social Welfare and Development (DSWD) for relief operations, Department of Public Works and Highways (DPWH) for road clearing operations and Department of Health (DOH) for doctors and nurses.

Each of the 16 municipalities of Benguet Province has created its respective Local Disaster Risk Reduction Management Council (LDRRMC) as mandated by Republic Act 10121. Legal bodies have issued ordinances and resolutions on DRRM. For instance, the Administrative Order No. 027-2016 was issued to reorganize the Municipal Disaster Risk Reduction Management Council (MDRRMC) for the Municipality of La Trinidad in recognition of the election results and of the need to replace members who no longer represent their organization. All the municipalities of Benguet Province have assigned an LDDRM officer.

While it is the government's policy to implement self-reliance among local officials and their constituents in responding to emergencies and disasters, apparently leadership and responsibility in DRRM is sometimes weak. Many elected officials, whose office term changes every three years, have yet to: 1) understand CCA and DRR, 2) learn the parameters for updating the DRRM plan, and 3) employ strategic governance to successfully address the needs of their respective communities.

Formulation of the LDRRM plans

All the barangays within the jurisdiction of the municipalities have been assisted to prepare their respective DRRM plans which they submit to the local DRRM council. This in turn was the basis for creating the Comprehensive Municipal DRRM Plan. These plans have been endorsed by the provincial DRRM council for information and inclusion in the provincial plans. The key components of the Comprehensive Municipal DRRM Plans are: 1) the Disaster and Hazard Profile (Risk Analysis and Vulnerability Assessment, State of DRRM, Vision, Mission and Goals in DRRM); 2) DRRM Plans and Strategic Actions; 3) Work and Financial Plan; Time Frame of Implementation), and 4) Project Monitoring and Evaluation.

All LGUs in Benguet have formulated DRRM plans that are in their early stages. Adding to this expression of national commitment, many Philippine communities, especially those with high vulnerability to disasters, have over the past ten years been receiving assistance from various national agencies,[10] non-government agencies, the United Nations (e.g. UN World Food Program) and other institutions. The assistance mostly comes in the form of capacity-building training, DRRM plan preparation, building or structural support.

Funding and governance for DRRM

By law, all LGUs are mandated to allocate funds for DRRM. The Joint Memorandum Circular No. 2013-1 of the National Disaster Risk Reduction and Management Council, the Department of Budget and Management and the Department of Interior and Local Government (DILG) stipulated official guidelines for the allocation and utilization of the LDRRM Fund. According to this circular, the 5 percent mandated LDRRM fund shall be allocated and used as follows:[11] 30 percent shall cover a lump-sum allocation for the quick response fund (QRF) to be used as a standby fund for relief and recovery programs, and 70 percent shall be provided for disaster prevention and mitigation, preparedness, response, rehabilitation and recovery programs such as: conduct of research, risk assessment, vulnerability analysis and other science-based technology and methodologies; capacity building activities to equip, organize, sustain and mainstream DRRM and climate change adaptation (CCA) at the local level; infrastructure and equipment support (flood control, procurement of early warning systems and other DRRM equipment); development of IEC and conduct of simulation exercises; development and implementation of standard operating procedures (SOP) and the institutionalization of early warning systems, evacuation and coordination; provision for relief operations (food and medicine stocks, tents and other temporary shelter facilities, etc.). The government has provided not only policy support but also a financial mechanism for the implementation of DRRM. However, a mechanism to implement the required conduct of relevant researches such as risk assessment and vulnerability analyses is absent.

The frequency of heavy rain, landslides and mudslides in Benguet obstruct efforts to reduce the incidence of poverty and increase economic resilience in many communities in the province. This situation is aggravated by environmental degradation resulting from direct human activities and/or climate change. A recent study on the institutional and sectoral implementation of NDRRMP in the Philippines highlighted several issues in the "institutional translation and resource allocation" of DRRM in the country (Domingo, 2016). The study contends that mainstreaming of the DRRM activities and agencies using a single fund source "signals the manifestation [of] effective institutionalization of DRRM within government" Domingo (2016, 16). Moreso, the study found that "much of the calamity funds were utilized for disaster response, recovery and rehabilitation", which are just two of the four pillars of DRRM. The study also raised the point that while the allocation of institutional funding for DRRM has increased dramatically—from approximately 1 billion pesos in 2009 to 4-4.5 billion in 2016, allocation has to be rationalized so that priority support will be given to the poorest LGUs in the Philippines (Ibid, 17). This view is useful but it could also be problematic because the poorest LGUs may not necessarily be the most vulnerable to disaster risks. Thus while local governments are mandated to address disaster risks, they are on the lookout to cut costs. This scenario is problematic for high-risk and high-vulnerability municipalities with very low internal budgets. LGUs will have to look to external financing schemes to help them address disaster preparedness.

The Commission on Audit (COA) accounts for several weaknesses and issues in the LGUs utilization of the LDRRM funds for the year 2017 (Commission On Audit, 2018, February). First, ten LGUs, three from the Cordillera Region, have not allocated funds for LDRRMF. As a consequence, they have failed to plan and implement DRRM programs and activities. Second, the majority (272) of the LGUs in the country did not successfully utilize their appropriated LDRRMF. There were also ten LGUs that were not able to allocate funds for LDRRM, resulting in a lack of LDRRM projects and programs in the respective areas. Third, 100 LGUs, six in the Cordillera Region, failed to submit their Local Disaster Risk Reduction Management and Fund Investment Plan (LDRRMFIP) for activities on disaster mitigation, prevention and preparedness. This plan could have delineated funds (30 percent allocation for the Quick Response Fund plus unexpended funds of previous year) for activities and expected outputs. Fourth, 11 LGUs have charged activities against the Quick Response Fund despite no evidence of a declaration of a 'state of calamity'. Fifth, expenses for LDRRM activities in 11 LGUs were charged from the wrong funding source, the Quick Response Fund (QRF). A total of 107 LGUs utilized large amounts from the LDRRMF appropriations for activities unrelated to disaster risk management. Moreover, the COA, in its annual audit for the year 2018, released similar observations as follows: 1) low rate of implementation of some 16 LGUs, 2 from the Cordillera Region, of their LDRRM plans, programs, projects and activities, 2) 39 LGUs, 1 from the Cordillera Region, did not utilize the allotment under the LDRRMF for different programs, projects and activities that are already budgeted, 3) 386 LGUs, 5 from the Cordillera Region, failed to submit the required reports on LDRRMF, d) funds were utilized for activities not related to DRRM (COA, 2019, August).

Utilization of mandated DRRM fund

Reports from the DILG full disclosure policy portal show that municipalities in Benguet have generally *not* utilized the 30 percent QRF from 2015–2017. The remaining 70 percent has been utilized for both infrastructure support (e.g. repair and rehabilitation of evacuation centers, road clearing, flood control and slope stabilization) and non-infrastructure related programs (e.g. capacity building training), purchase of equipment and IEC activities. Equipment purchases for DRRM are covered by DILG Memorandum Circular Number 2012-73. The Memorandum Circular provided a basic list of equipment that can be procured by the local government units for early warning systems, preparedness and response.[12] In Benguet's case, DRRM funds were in place, yet the LGU was not able to utilize it fully. Such a case is summarized in Table 12.1.

Table 12.1 shows LGUs allotted funds for DRRM but the utilization is low. The low utilization of this fund by most of the LGUs of Benguet says a lot about their capacity to develop programs on CCA and DRR that they can implement. In the case of Bokod, it had a limited number of PPAs identified for the four priority areas of DRR. Kabayan's reason could have been remedied by proper utilization of funds as the funds were accessible.

LGUs have yet to proactively request the assistance of technical experts in the academe and other institutions in the conduct of research, risk assessment, vulnerability analysis and other science-based technology and methodologies. This should have been the starting point of all activities so that LGUs had a good understanding of their respective situations which would have enabled them to design better programs and activities. The results of scientific research and assessments, which are unfortunately implemented in just two out of ten municipalities in Benguet, are important bases for program development in the DRRM.

PAGASA reports that heavy rainfall in Benguet, the lowest being 475.8mm in June, the highest being 905mm in August, with an average of 3841.4mm annually affect agricultural activities in the region[13] PAGASA's report on the agrometeorological situation in Benguet for August 1–10, 2018 shown in Table 12.2 shows the crops that are affected.[14]

There were incidences in the past when communities were heavily affected by typhoons during the months of June and July, the harvest months for rice which

Table 12.1 Utilization of the 70 percent mitigation fund from the LDRRMF in Benguet for the year 2017

Municipalities	Percentage use of the 70% mitigation fund from the LDRRMF	Reasons for low utilization of 70% mitigation fund from the LDRRMF	Implications as noted in the COA Annual Audit Report (AAR)
1. Kabayan	₱636,095.14 or 20.99%	Delayed project implementation	Delayed completion could result to further damage to the infrastructures
2. Bokod	₱ 4.635 million or 28.48%	Limited number of planned PPAs for disaster relief, recovery, risk reduction and management	The utilization of the LDRRMF was not maximized thus barring the attainment of the benefits
3. Atok	₱618,169.83 or 20%	Failure to submit monthly report on sources and utilization of DRRMF	Missed opportunities to address its DRRM objectives
4. Bakun	₱ 3,337,417.76 or 93.6%	NR	NR
5. Buguias	₱ PhP 3,865,374.14 or 89%	NR	NR
6. Kapangan	₱1,216,541.44 or 42.4%	NR	Work activities not related to disaster response were funded (₱ 80,000.00)
7. Tublay	₱1,978,295.96 or 46.1%	The identified PPAs in the municipality's MDRRM plan were not parallel with accomplished DRRM PPAs	LGU has not fulfilled the purposes of the DRRM fund for disaster preparedness, response and rehabilitation and recovery.
8. Kibungan	₱1,875,002.85 or 88.2%	NR	NR
9. La Trinidad	52% (117 out 225 projects)	NR	Desired socio-economic development and environmental benefits have not been achieved.
10. Sablan	₱2,397,574.00 or 28%	The LDRRMP was not properly implemented	NR

*NR- "Nothing reported in the official sources"

Source: Commission on Audit (COA). (2018). *Annual Audit Reports for the Year 2017* covering the Municipalities of Atok, Buguias, Bokod, Bakun, Kabayan, Kapangan, Kibungan, Tublay, La Trinidad, and Sablan.

Table 12.2 Agrometeorological situation in Benguet

Forecast rainfall (mm)	Rainy days (0.1mm or more)	Actual soil moisture condition	Crop phenology, situation and farm activities
50–150	2–6	Wet	Land preparation, planting and transplanting of upland rice and corn are on-going. Planting of lettuce, mustard, broccoli, carrots, potatoes, cauliflower, chayote, celery, green peas and ampalaya is underway.

Source: Philippine Atmospheric, Geophysical and Astronomical Services Administration (PAGASA). *Climatological Normals.* Retrieved March 5, 2019 from https://www1.pagasa. dost.gov.ph/index.php/climate-nl; and Philippine Atmospheric, Geophysical and Astronomical Services Administration (PAGASA). *Ten-Day Regional Agri Weather Information.* Retrieved March 5, 2019 from https://www1.pagasa.dost.gov.ph/index.php/agriculture/climate-information-for-agriculture.

is the staple food for people in this region. Such a situation has led communities to depend on rice produced from other areas such as the Cagayan Region and Pangasinan. Typhoons and prolonged rain in recent years have caused landslides in the regions. Their annual occurrence has led to deaths and the destruction of homes, crops, properties and public infrastructure. For many communities, the traumatic experience is worse than losing properties. In July 2018, the Regional Disaster Risk Reduction and Management Council (RDRRMC) said 14 families were preemptively evacuated in the provinces of Benguet and Mountain Province due to six major landslides, four in Benguet on July 28–29. In addition, damage to agriculture was pegged at P2.4 million.[15] Damages also extend to national roads that connect municipalities to cities and other provinces.

In Benguet, the legislative framework and institutional arrangements for DRRM have gained some footing. The provincial government has issued an administrative order reorganizing the Provincial Disaster Coordinating Council (PDCC) into the Provincial DRRM Council, thus providing the necessary support for the expanded functions of the Council. Although the province is finding ways to fully implement the provisions of the DRRM Act, work is developing on an effective interdepartmental and inter-agency coordination during disasters and emergencies.

Key initiatives

The Department of Interior and Local Government (DILG), the agency that has authority over all the LGUs in the country, is vice-chairperson for disaster preparedness. In 2015, this agency initiated a flagship program called *Operation L!sto* and launched its campaign for disaster preparedness by issuing a manual called *Local Government Units Disaster Preparedness Manual Checklist of*

Minimum Critical Preparations for Mayors (Department of Interior and Local Government, 2015). This manual served as a guide for LGUs to prepare their disaster preparedness plans. As the local chief executive, the mayor performs the following actions as presented in the manual: "create structures and systems that fit the locale, institutionalize policies and specific plans, build competencies of the personnel, and equip the LGU with technological facility such as hardware and supplies for DRRM" (Domingo and Manejar, 2018). *Operation L!sto*'s program goal is to build community self-reliance and resilience in the long term (Barrameda, 2018). Under this program, the LGUs of Benguet received training that aimed to improve their disaster preparedness, including the mainstreaming of climate change adaptation (CCA) and DRR in the comprehensive development planning and strengthening of DRR CCA operations. The province of Benguet and majority of its municipalities received significant support from international institutions, e.g., the United Nations World Food Program (UNWFP).

From May 2011 to December 2017, the United Nations World Food Programme (UNWFP) implemented a capacity building program on Disaster Preparedness and Response/Climate Change Adaptation with the goal of building the resilience of vulnerable communities (Mountfield, Balgos & Carnalan, 2017, December.) The UNWFP tapped universities and NGOs in the province of Benguet as partner institutions in delivering this program. The program activities were implemented to raise awareness of climate change and DRR, support community level mitigation projects, building emergency response infrastructures and conduct technical capacity building training. The UNWFP collaborated with the University of the Philippines Baguio (UP Baguio) in carrying out research, information dissemination and capacity building activities from 2013–2017. UP Baguio conducted research with communities, prepared nine information, education and communication (IEC) materials, nine training modules, three training designs on Climate Change and DRR, established the Knowledge Training and Resource Center (KTRC) and implemented CC and DPR trainings for LGUs in Benguet (Knowledge Training and Resource Center – UP Baguio, Project Report, 2016). The training modules contained topics that linked climate change to DRR.[16] The results eventually helped the LGUs in developing their municipal DRRM and updating of their contingency plans. Local communities were also provided with training on basic life support, mountain and water search and rescue and evacuation by UNWFP's partner from the government, e.g. Office of Civil Defense (OCD).

The UNWFP partnered with the People's Initiative for Learning and Community Development (PILCD), a non-profit organization working with marginalized communities and vulnerable sectors in Benguet, to support activities that aimed to strengthen public awareness and education in DPR. This organization prepared and handed out 14,000 poster-brochures on disaster preparedness and response to local government units in the province of Benguet (Philippine Information Agency (PIA) Cordillera Administrative Region (CAR) (2016, December 19)). Many of these posters have been visible in the LGU offices, schools and public venues since 2016.

Studies on climate change and agriculture were conducted by the Benguet State University (BSU). Agro-forestry projects were also implemented. In the municipalities of Tublay and Atok, tree planting activities, nurseries for planting materials were established and vetiver grass, a fast-growing plant with and a deeply penetrating root system known to prevent soil erosion, was planted in landslide prone areas (Knowledge Training and Resource Center – UP Baguio, Project Report, 2013). Physical structures that serve multipurpose functions have also been funded, built and used by LGUs as evacuation centers, for DRRM operations and as food and seed storage facilities. Apparently, the construction of physical structures are important for LGUs so they would have these in place to effectively coordinate and prepare for relief efforts during crises. Kapangan, Tublay, Bokod and Atok Municipalities were given financial support to construct their DRRM multi-purpose buildings. Logistic support and key emergency equipment such as communication/radio equipment, emergency and medical kits were also given to the project partners of these LGUs.

Finally, automated weather stations were also established by scientists from the University of the Philippines Baguio as part of the UNWFP disaster pre-paredness program. This helped the LGUs to use the rainfall threshold in the landslide early warning system. Automated weather stations were installed in the municipality of Tublay, Benguet and training sessions on how to interpret and use the data provided by weather stations were provided. Early warning devices, communication equipment and emergency rescue equipment were procured, and the LGU and community partners were given training in how to use and maintain them. Emergency operation guides and communications protocols have also been prepared.

The entire program on Disaster Preparedness and Response/Climate Change Adaptation implemented by the UNWFP appears to have advanced the knowl-edge and CCA and DRR practice of its LGUs. An external evaluation of this project provides an affirmation saying that "the DPR/CCA program achieved its objectives and has strengthened capacity within targeted LGUs" (Mountfield, Balgos & Carnalan, 2017). The report further states that "the program is seen as highly relevant by its stakeholders" and that it was "relevant and appropriate and well *aligned* (emphasis added) with Government policy" (Mountfield et al., 2017). While the objective of the program was revered during the years it was implemented, the implementation was largely influenced by national government policies and program practices. When the UNWFP aligned its program imple-mentation to government policies, it also meant that work was conducted with respect to official guidelines that were still being developed by the implementers of the NDMRRMC during the years of UNWFP implementation. Risk gover-nance was implemented within official platforms and may have inadvertently overlooked the efficacy of local community driven strategies.

Some LGUs in Benguet Province have been developing their landslide pro-tocols based on the available hazard maps produced from sample estimates and not on site specific and research-based risk assessments. Barangay officials, members of the Barangay DRRM Council (BDRRMC) and affected residents were tasked to conduct regular monitoring of areas where landslides frequently occur. Monitoring, especially for residential areas and highways, is conducted before the onset of the rainy season and before the occurrence of typhoons. Members of the BDRRMC conduct inspections on identified landslide-prone areas along major highways, road networks and residential areas within their jurisdiction. Locals were trained to measure cracks, gaps and dips (for sinking areas) of identified landslide-prone areas. They were also taught how to prepare reports or recordings of their observations. In the case of Tublay Municipality, the landslide-prone areas were included in the community maps (Barangay 3D maps with 1:6,000 scale). These were marked or identified as hazard areas and as an example, the monitoring of a landslide prone area along Halsema Highway was done by Barangay Ambassador.

The Municipality of La Trinidad, Benguet set up a municipal emergency response team (MERT) which is largely composed of volunteers. This LGU also passed an ordinance that sets guidelines for the accreditation, mobilization and protection of community disaster volunteers. This policy encourages various organizations and establishments to be part of the MERT. After accreditation, a Memorandum of Agreement (MOA) between the organization and the LGU was forged to ascertain the roles and responsibilities to be carried out during times of disaster. Members of the MERT include drivers and operators of transport groups and owners of business establishments (gasoline stations, grocery stores, etc.). This initiative was also replicated at the Barangay level hence the establishment of Barangay emergency response team. All accredited volunteers from the MERT receive logistical support when on duty and regular training is provided with annual insurance paid by the LGU from its DRRM fund. The LGU encourages anyone to be a volunteer and each one is given specific roles and duties to perform from cooking to driving to giving first aid.[17]

Issues and prospects for climate and disaster resilient communities

DRRM projects and initiatives have been implemented in Benguet and elsewhere in the Philippines. Activities are planned strategically and meaningful cooperation between various parties is encouraged. While externally initiated projects come to the LGUs with specific targets that need to be completed in a short timespan, LGUs need to strengthen their interventions and direct the projects towards specific goals of their own DRRM plans. The desired outcomes for each project should be clear and build on existing efforts or complement programs that are already in place. Additionally, while externally driven projects are implemented following their respective agenda, local governments can negotiate to include their local needs and actively engage with the project implementers to ensure that they share the benefits of the projects. This process begins with the conceptualization of the project when local governments and community leaders participate in compiling the project's content and implementation plan in response to identified priorities in the local DRRM plan. This is possible when local governments employ a platform that requires all externally funded DRR projects or programs to be reviewed in detail by the local disaster management board and ensure that the prospective projects merit implementation in the locality. Proposed projects should be amended to respond to a particular community need. This way, the projects can be designed to achieve a particular goal in the DRRM plan and ensure that expected outcomes can be measured.

DRRM plans can be improved to achieve a well-articulated vision, goals, strategies and outcomes. As LGUs are directed to build disaster resilient communities in the Philippines, the DRRM plans should embody a vision that will contribute to this. Specific goals should be determined following the stated vision and in consideration of local conditions and the prospect of cooperation or collaboration with external institutions. The goals should employ proactive and inclusive strategies designed to achieve the desired outcomes of specific communities and the country in general. LGUs may therefore express in their strategies a list of activities that they desire other institutions to perform, e.g. the academia, NGOs or others from which they could proactively request specific assistance. The carrying out of scientific studies, the formulation of maps and charts, the organization of workshops, the preparation of proposals and other needs should be specified in the plan. LGUs may need to proactively seek these forms of assistance a year in advance, especially as these require funding.

DRRM plans in Philippine communities will be most useful when they are informed by hazard vulnerability assessments. LGUs are mandated to use their DRRM funds to conduct such assessments, and it's crucial that they do this by tapping the expertise of professionals. Scientific studies conducted by universities in the Philippines and abroad that can be used by LGUs and communities are gratefully received. Local knowledge on the DRRM of communities has not been studied systematically for its potential contribution to DRRM. The recording and use of communities' own mechanisms in response to managing different disaster situations has not been done. Once conducted, the research may be a useful addition to the LGUs' contingency plans.

LGUs desire for continued capacity enrichment and an additional salaried workforce to deliver its official mandate in DRRM. While much autonomy is given to them, they can best carry out their full functions when their capacity is continuously enriched. LGUs would benefit from DRRM training programs that are designed with an effective training needs analysis (TNA) and not based on desired national targets only. Training sessions for DRRM could also address the importance of developing and implementing a municipal local DRRM investment plan which would facilitate the implementation of outcomes-driven activities for DRRM.

Previous studies have demonstrated the value of incorporating community-based knowledge in DRR. In the case of flood hazards in the Philippines, a flood hazard map model has been developed based on both scientific and the local knowledge of communities in Bohol, and has become a prototype for flood hazards in the Philippines that can serve to correctly forecast flood risks.[18] In Dagupan City plans have already been implemented to mainstream the community-based disaster risk management project (CBDRM) into good governance of the city. The Center for Disaster Preparedness (CDP) facilitated innovative efforts that resulted in the development of DRR plans or eight vulnerable communities and they have integrated these into the City Disaster Risk Reduction Plan.[19] In Indonesia, a community in the eastern part of Nusa Tenggara (south-eastern Indonesian islands) has developed its own food early warning system to prevent food shortage through a community-based disaster risk management (CBDRM) program implemented by a local NGO and the Community Association for Disaster Management.[20]

Studies have also confirmed that there is a "lack of technical assistance and personnel in conceptualizing and implementing local plans" (Domingo and Manejar, 2018, 45). Strategic initiatives will build disaster resilient communities. One that may be referred to as "people-centered" where locals know enough information about the likely impact of a hazard to effectively contribute to building community resilience (Anderson-Berry, 2018, 11). Scientific studies and technical knowledge combined with the local knowledge of communities foster effective implementation of an LDRRM. This is where strong partnerships can be forged between LGUs, academia and finance agencies. Recent studies reveal that disaster risk governance in Asia and the Pacific needs partnerships and efficient allocation of resources (Paolo et al., 2019).

Conclusions and suggestions

As LGUs are mandated to update their DRRM plans regularly, such amendments may include effective platforms for conversations, dialogues between local communities and government planners and co-production of knowledge about disaster risk reduction. Platforms can be diverse to suit the conditions and contexts of communities. The design has to carry the principles of collective equitability to effectively target the goal of risk resilience. This way, the local knowledge of communities is recognized and targeted for use from the beginning. The communities are also considered active players in a plan and are not recognized purely as disseminators of information but also as a source of knowledge for planning (Acharya and Prakash, 2018) and as an equally important support in implementing schemes. The indigenous and local knowledge and practices of local people are an important feature of DRRM plans alongside scientific knowledge.

The desired targets in disaster risk reduction require a strong workforce to assist in the development of programs and their implementation. Apparently, most LGUs lack this workforce. The development and implementation of DRRM plans appear to be limited. The policy articulation of goals and targets in DRR at the national level is strong but the capacity and workforce to implement them is weak. This needs to be improved and enhanced whenever a new need arises. A strategic plan for workforce development, e.g. adding to the workforce or developing the capacity of existing workforce, needs to be prioritized. Activities identified in the DRRM plans could be developed into programs that advance the desired outcomes of DRR. Strategies could be explored with stakeholders to address this gap. This study suggests the following in the interim:

a. A mechanism should be established by LGUs to require those bringing externally funded and initiated projects to communities to explain how their study can directly contribute to the improvement of the local climate change action plan or to update the local DRRM plan and propose long-term solutions that address disaster risk at the community level. This way, externally driven initiatives will be directed towards DRR and build the resilience of the communities involved.

b. LGUs may develop DRR programs as an important part of the DRRM plan, and hire or assign a designated person to each DRR program who will work directly with the DRRM officer. This will facilitate the implementation of DRR activities in the desired period of implementation. Program implementation is a full-time job that requires expertise to perform the various functions including collaboration with national and international agencies and private groups. The preparation of project proposals alone requires the attention and expertise of a focal person.

c. DRR programs may cover the following: 1) 'Vulnerability and risk assessments' to be developed and implemented in partnership with academic or research institutions that have the relevant expertise; 2) 'Resilient and sustainable livelihoods' that provide better alternatives for locals to engage in non-extractive activities and minimize the ill use of natural resources in local communities; 3) 'Ecology revitalization' aimed at strengthening cultural practices that support DRR; 4) ' Innovative technology' that promotes the use of scientific knowledge and technologies that cause no harm nor interfere in local cultural practices. For example, a rainwater harvesting and storage facility that, when installed, benefits the community by providing water resource storage.[21]

d. Contingency plans that are specific to hazards should be developed or improved in the long run. This is also consistent with the Philippines National Disaster Risk Reduction and Management Plan for 2011–2028. Outcome 7 of Thematic Area 2 (Disaster Preparedness) states the target is "increased awareness and enhanced capacity of the community to the threats and impacts of hazards". The contingency plans should be based on risk assessments. LGUs may request assistance from the academe especially studies and conduct of DRR workshops.

e. LGUs in the Philippines are to update their DRRM plans regularly, at least annually, on the premise that they should be amended in line with assessments and identified needs, done systematically. If the update of the DRRM plans is implemented well, the needs landscape should change every year. Desired outcomes may be assessed and new priorities may be identified. The assessment may address organizational and management difficulties and allow the implementers of DRRM to learn cumulatively as they go along. Scientific studies should be integral in the process of updating the DRRM plan.

The integration of local knowledge in DRR has to be developed on established platforms and on principles founded on local practices. The United Nations International Strategy for Disaster Reduction (UNISDR) suggests that policy support for the incorporation of such knowledge in paradigms for DRR should start with the identification of "specific entry points" that could be used as engines for the agenda setting of decision makers.[22] For many communities in the Philippines, indigenous knowledge (IK) on DRR will have to be systematically studied, understood well for its current value and confirmed as the most appropriate applicability in the varying community contexts before it can be officially adopted or mainstreamed in specific DRRM plans. It will be beneficial when LGUs and academic institutions collaborate to address such a need by conducting a multidisciplinary study to establish sound evidence for developing good DRR local governance in their respective regions and communities in the northern Philippine communities.

As DRRM is integral to the Philippines' plan of action for Climate Change Mitigation and Adaptation from 2011–2028, climate change governance practice may need to invest in a strategic program that effectively strengthens the capacity of LGUs to build resilient communities. Good climate change governance in Northern Philippines may be realized when the LGUs, as DRR frontliners, are equipped with an efficient manpower and the necessary technical knowledge and skills.

Notes

1 Republic Act No. 10121. (May 27, 2010) *An Act Strengthening the Philippine Disaster Risk Reduction and Management System, Providing for the National Disaster Risk Reduction and Management Framework and Institutionalizing the National Disaster Risk Reduction and Management Plan, Appropriating Funds therefor and for other Purposes.* Accessed May 20, 2018. www.officialgazette.gov. ph/2010/05/27/republic-act-no-10121

2 The *Joint Memorandum Circular No. 2013-1 of the National Disaster Risk Reduction and Management Council (NDRRMC), the Department of Budget and Management (DBM) and the Department of Interior and Local Government (DILG).* Accessed May 15, 2018. www.ndrrmc.gov.ph/attachments/ article/1323/JMC_No_2014-1_re_Implementing_Guidelines_for_the_Establishment_of_LDRRMOs_or_BDRRMCs_in_LGUs.pdf

3 The World Bank East Asia and Pacific Region Rural Development and National Development Council of the Philippines. *Natural Disaster Risk Management in the Philippines: Enhancing Poverty Alleviation through Disaster Reduction.* Accessed May 15, 2018. http://documents.worldbank.org/curated/en/975311468776739344/pdf /338220REPLACEM1aster0Report1combine.pdf

4 The World Bank East Asia and Pacific Region Rural Development and National Development Council of the Philippines. *Natural Disaster Risk Management in the Philippines: Enhancing Poverty Alleviation through Disaster Reduction.* Accessed May 15, 2018. http://documents.worldbank.org/curated/en/975311468776739344/pdf /338220REPLACEM1aster0Report1combine.pdf

5 Philippine Atmospheric, Geophysical and Astronomical Services Administration (PAGASA). *Climate Projections for Province.* Accessed February 22, 2019. http:// bagong.pagasa.dost.gov.ph/information/climate-change-in-the-philippines

6 Philippine Atmospheric, Geophysical and Astronomical Services Administration (PAGASA). *Current Climate Trends in the Philippines.* Accessed February 22, 2019. http://bagong.pagasa.dost.gov.ph/information/climate-change-in-the-philippines

7 Philippine Atmospheric, Geophysical and Astronomical Services Administration (PAGASA). *Climate of the Philippines.* Accessed May 15, 2018. https://www1.pagasa.dost.gov.ph/index.php/climate-of-the-philippines

8 Philippine Atmospheric, Geophysical and Astronomical Services Administration (PAGASA). *Climatological Normals.* Accessed July 20, 2018. https://www1.pagasa.dost.gov.ph/index.php/climate-nl

9 Earthquakes and Megacities Initiative (EMI) and World Food Programme (2011). *Final Report: Capacity Needs Assessment for Disaster Preparedness and Response.* Earthquakes and Megacities Initiative (EMI) and World Food Programme (WFP).

10 Carlito Dar. *Cordillera DRRM Council Synchronizes Disaster Preparedness Efforts.* Philippines Information Agency. http://pia.gov.ph/news/articles/1008189. May 22, 2018.

11 The *Joint Memorandum Circular No. 2013-1 of the National Disaster Risk Reduction and Management Council, the Department of Budget and Management and the Department of Interior and Local Government.* Accessed August 10, 2018. www.ndrrmc.gov.ph/attachments/article/1320/JMC_No_2013-1_re_Allocation_and_Utilization_of_LDRRMF.pdf

12 See http://old.dilgcar.com/images/DisasterPrepCorner/Issuances/DILG%20MC%20No%202012-73%20dtd%20Apr%2017%202012%20%20-Utilization%20of%20LDRRMF.pdf

13 Philippine Atmospheric, Geophysical and Astronomical Services Administration (PAGASA). *Climatological Normals.* Accessed July 20, 2018. https://www1.pagasa.dost.gov.ph/index.php/climate-nl

14 Philippine Atmospheric, Geophysical and Astronomical Services Administration (PAGASA). *Ten-Day Regional Agri Weather Inforamtion.* Accessed August 10, 2018. https://www1.pagasa.dost.gov.ph/index.php/agriculture/climate-information-for-agriculture

15 Frank Cimatu. *Typhoon Gorio: Closed roads, landslides in Cordillera and Ilocos* (July 30, 2017) Accessed August 10, 2018. https://www.rappler.com/nation/177163-typhoon-gorio-closed-roads-landslides-cordillera-ilocos

16 The nine training modules covered the following topics and concerns: Understanding Climate Change; The Legal Framework of the Disaster Risk Reduction and Management Policy; Enhancing Disaster Preparedness; Psychosocial Response to Disaster; Post-Disaster Rehabilitation and Recovery; Risk Communication in Disaster Risk Reduction and Management; Media Management in Disaster Risk Reduction and Management; Alliance Building in Disaster Risk Reduction and Management; Business Continuity Planning in Disaster Risk Reduction and Management; DRRM Risk Communication Planning.

17 Interview with MDRRMO Yoshio Labi, April 2017.

18 Catherin Abon, Carlos David, & Guillermo Tabios III, (2012) Community-based monitoring for flood early warning system: An example in central Bicol River Basin, Philippines. *Disaster Prevention and Management: An International Journal*, 21(1), 85–96. https://doi.org/10.1108/09653561211202728

19 Center for Disaster Preparedness (CDP). Mainstreaming community-based mitigation in city governance" in *Building Disaster Resilient Communities: Good Practices and Lessons Learned: A Publication of the "Global Network of NGOs" for Disaster Risk Reduction.* UNISDR, Geneva: United Nations International Strategy for Disaster Reduction: 2007.

20 PMPB (Community Association for Disaster Management). Combining Science and Indigenous Knowledge to Build a Community Early Warning System" in *Building Disaster Resilient Communities: Good Practices and Lessons Learned: A Publication of the "Global Network of NGOs" for Disaster Risk Reduction.* UNISDR, Geneva: United Nations International Strategy for Disaster Reduction: 2007.
21 The need for facilities such as rain water harvesting has been identified as a need in the communities. This facility will capture water during periods of heavy rains and become a good source of clean water for households, offices and the agricultural industry, especially during the months of January to May when water is scarce.
22 Rajib Shaw et.al. *Indigenous Knowledge Disaster Risk Reduction: Policy Note.* United Nations ISDR. Accessed March 15, 2019. https://www.unisdr.org/files/8853_IKPolicyNote.pdf

References

Abon, C., David, C., & Tabios, G. (2012). Community-based monitoring for flood early warning system: An example in central Bicol River Basin, Philippines. *Disaster Prevention and Management: An International Journal, 21*(1), 85–96. Retrieved April 19, 2019 from https://doi.org/10.1108/09653561211202728

Acharya, Amitangshu, & Anjal Prakash. (2018, December). When the river talks to its people: Local knowledge-based flood forecasting in Gandak River Basin, India. *Environmental Development, 31,* 55–67. Retrieved April 19, 2019, from https://www.sciencedirect.com/science/article/pii/S2211464518301738?via%3Dihub.

Anderson-Berry, Linda, Achilles, Tamsin, Panchuk, Shannon, Mackie, Brenda, Canterford, Shelby, Leck, Amanda, Bird, Deanne. (2018, September). Sending a message: How significant events have influenced the warnings landscape in Australia. *International Journal of Disaster Risk Reduction, 30,* Part A, 5–17. Retrieved from https://doi.org/10.1016/j.ijdrr.2018.03.005

Barrameda, Silvertre. (2018). Operation L!sto, Building Our Story of Resilience. Retrieved April 7, 2020 from https://www.preventionweb.net/files/56219_1th02philippinesamcdrroperationlist.pdf

Center for Disaster Preparedness. (2007). *Mainstreaming Community-Based Mitigation in City Governance. Building Disaster Resilient Communities: Good Practices and Lessons Learned: A Publication of the "Global Network of NGOs" for Disaster Risk Reduction.* Geneva: United Nations International Strategy for Disaster Reduction.

Cimatu, Frank. (2017, July 30). *Typhoon Gorio: Closed Roads, Landslides in Cordillera and Ilocos.* Retrieved from https://www.rappler.com/nation/177163-typhoon-gorio-closed-roads-landslides-cordillera-ilocos

Cinco, T. A., Guzman, R. G., Ortiz, A. M., Delfino, R. J., Lasco, R. D., Hilario, F. D., Juanillo, E. L., Barba, R., & Ares, E. D. (2016). Observed trends and impacts of tropical cyclones in the Philippines. *International Journal of Climatology., 36*: 4638–4650. doi:10.1002/joc.4659

Commission on Audit (COA). (2018). *Annual Audit Reports for the Year 2017Covering the Municipalities of Atok, Buguias, Bokod, Bakun, Kabayan, Kapangan, Kibungan, Tublay, La Trinidad, and Sablan.* Retrieved July 31, 2018 from https://www.coa.gov.ph/index.php/reports/annual-audit-report

Commission on Audit. (2018, February). *Consolidated Report on the Audit of the Disaster Risk Reduction Management Fund for the Year Ended December 31, 2016.* Retrieved November 15, 2018 from https://www.coa.gov.ph/phocadownloadpap/userupload/DRRM/Consolidated_Audit_Report-DRRM_Fund_CY2016.pdf

Commission on Audit. (2019, August). *Consolidated Report on the Audit of the Disaster Risk Reduction Management Fund for the Year Ended December 31, 2017.* Retrieved April 13, 2020 from https://www.coa.gov.ph/index.php/reports/disaster-risk-reduction-and-management-reports.

Dar, Carlito. Cordillera DRRM Council Synchronizes Disaster Preparedness Efforts. Philippines Information Agency. Retrieved May 22, 2018 from http://pia.gov.ph/news/articles/1008189.

Department of Interior and Local Government (DILG). (2012, April 17). *Memorandum Circular 2012–73 on the Utilization of the Local Disaster Risk Management Fund (LDRRMF)* Retrieved May 15, 2019 from http://old.dilgcar.com/images/DisasterPrepCorner/Issuances/DILG%20MC%20No%202012-73%20dtd%20Apr%2017%202012%20-Utilization%20of%20LDRRMF.pdf

Department of Interior and Local Government (DILG). (2014). *1st Semester FY 2014 Narrative Report.* Retrieved from https://www.dilgcar.com/images/LGRRC_CAR/12_1_DILG%20CAR%201st%20sem%202014%20Narrative%20Report_wd_LGRRC.pdf

Department of Interior and Local Government (DILG). (2015) *LGUs Disaster Preparedness Manual Checklist of Minimum Critical Preparations for Mayors.* Retrieved April 7, 2020 from https://lga.gov.ph/media/uploads/2/Publications%20PDF/Book/LGA-OL-Critical-Preparations-for-Mayors.pdf

Domingo, Sonny N. (2016, December). *An Assessment of the Sectoral and Institutional Implementation of the NDRRMP.* Discussion Paper Series No. 2016-49. Philippine Institute for Development Studies. Quezon City: PIDS.

Domingo, Sonny N., & Manejar, Arvie Joy A. (2018, December). *Disaster Preparedness and Local Governance in the Philippines.* Retrieved September 1, 2019, from http://hdl.handle.net/11540/9530

Earthquakes and Megacities Initiative (EMI) and World Food Programme (2011). *Final Report: Capacity Needs Assessment for Disaster Preparedness and Response.* Earthquakes and Megacities Initiative (EMI) and World Food Programme (WFP). Retrieved May 15, 2018 from https://emi-megacities.org/wp-content/uploads/2015/04/CNA-Final-Report-1-May-2011.pdf

Final World Food Program Disaster Preparedness Response Programme. (2012). *National Report,* 2012.

Izumi, T., Shaw, R., Ishiwatari, M., Djalante, R., & Komino, T. (2019, March 30). *Innovations for Disaster Risk Reduction.* IRIDeS, Keio University, the University of Tokyo, UNU-IAS, CWS Japan. Retrieved October 1, 2019 from https://apru.org/wp-content/uploads/2019/03/30-Innovations-for-Disaster-Risk-Reduction_final.pdf.

Knowledge Training and Resource Center (KTRC) – UP Baguio (2017), Project Report for 2016.

Knowledge Training and Resource Center (KTRC) – UP Baguio (2017), Project Report for 2013.

Mountfield, Ben, Balgos, Benigno, & Carnalan, Darlyn. (2017). *Decentralized Evaluation. Final Evaluation of Disaster Preparedness and Response/Climate Change Adaptation Activities under the Office of Foreign Disaster Assistance Fund in the Philippines,*

May 2011 to September 2017. Retrieved April 7, 2020 from https://docs.wfp.org/api/documents/WFP-0000063555/download/?_ga=2.133316171.459162673. 1586397486-75185682.1586397486

National Disaster Risk Reduction and Management Council (NDRRMC). (2013, March 25). *Joint Memorandum Circular No. 2013-1 of the National Disaster Risk Reduction and Management Council, the Department of Budget and Management (DBM) and the Department of Interior and Local Government (DILG) Regarding Allocation and Utilization of Local Disaster Risk Reduction Management Fund (LDRRMF)* Retrieved August 10, 2018 from www.ndrrmc.gov.ph/attachments/article/1320/JMC_No_2013-1_re_Allocation_and_Utilization_of_LDRRMF.pdf

National Disaster Risk Reduction and Management Council (NDRRMC). (2014, April 4). *Joint Memorandum Circular No. 2014-1 of the National Disaster Risk Reduction and Management Council, the Department of Interior and Local Government (DILG), the Department of Budget and Management (DBM) and the Civil Service Commission* (CSC) *Regarding Implementing Guidelines for the Establishment of Local DRRM Offices (LDRRMOs) or Barangay DRRM Committees (BDRRMCs) in LGUs.* Retrieved May 15, 2018 from www.ndrrmc.gov.ph/attachments/article/1323/JMC_No_2014-1_re_Implementing_Guidelines_for_the_Establishment_of_LDRRMOs_or_BDRRMCs_in_LGUs.pdf

Nolasco-Javier, D. and Kumar, L. (2018). Deriving the rainfall threshold for shallow landslide early warning during tropical cyclones: A case study in northern Philippines. *Natural Hazards, 90*(2), 921. Retrieved September 1, 2019 from https://doi.org/10.1007/s11069-017-3081-2

Nolasco-Javier, D., Kumar, L., & Tengonciang, A. M. P. (2015). Rapid appraisal of rainfall threshold and selected landslides in Baguio, Philippines. *Natural Hazards, 78,* 1591–1594. Retrieved September 1, 2019 from https://doi.org/10.1007/s11069-015-1790-y

Kelman, I., Mercer, J., & Gaillard, J. (2012). Indigenous knowledge and disaster risk reduction. *Geography, 97*(1), 12–21. Retrieved from www.jstor.org/stable/24412175

Philippine Atmospheric, Geophysical and Astronomical Services Administration (PAGASA). *Climate Projections for Province.* Retrieved March 5, 2019 from http://bagong.pagasa.dost.gov.ph/information/climate-change-in-the-philippines

Philippine Atmospheric, Geophysical and Astronomical Services Administration (PAGASA). *Climatological Normals.* Retrieved March 5, 2019 from https://www1.pagasa.dost.gov.ph/index.php/climate-nl

Philippine Atmospheric, Geophysical and Astronomical Services Administration (PAGASA). *Ten-Day Regional Agri Weather Information.* Retrieved March 5, 2019 from https://www1.pagasa.dost.gov.ph/index.php/agriculture/climate-information-for-agriculture

Philippine Atmospheric, Geophysical and Astronomical Services Administration (PAGASA). *Current Climate Trends in the Philippines.* Retrieved February 22, 2019 from http://bagong.pagasa.dost.gov.ph/information/climate-change-in-the-philippines

Philippine Atmospheric, Geophysical and Astronomical Services Administration (PAGASA). *Climate of the Philippines.* Retrieved May 15, 2018 from https://www1.pagasa.dost.gov.ph/index.php/climate-of-the-philippines

Philippine Information Agency (PIA) Cordillera Administrative Region (CAR). (2016, December 19). *PILCID Distributes IEC Materials on Disaster Preparedness* to *LHUs and Agencies*. Retrieved April 12, 2020 from https://reliefweb.int/report/philippines/pilcd-distributes-iec-materials-disaster-preparedness-lgus-agencies

PMPB (Community Association for Disaster Management in partnership with Yayasan Pikul). (2007). Combining science and indigenous knowledge to build a community early warning system. *Building Disaster Resilient Communities: Good Practices and Lessons Learned: A Publication of the "Global Network of NGOs" for Disaster Risk Reduction*. Geneva: United Nations International Strategy for Disaster Reduction (UNISDR).

Republic Act No. 10121. (2010, May 27). An Act Strengthening the Philippine Disaster Risk Reduction and Management System, Providing for the National Disaster Risk Reduction and Management Framework and Institutionalizing the National Disaster Risk Reduction and Management Plan, Appropriating Funds Therefor and for Other Purposes. Retrieved March 1, 2019 from www.officialgazette.gov.ph/2010/05/27/republic-act-no-10121

Shaw, Rajib, Takeuchi, Yukiko, Uy, Noralene, & Sharma, Anshu. (2009) *Indigenous Knowledge Disaster Risk Reduction: Policy Note*. Retrieved March 15, 2019, from https://www.unisdr.org/files/8853_IKPolicyNote.pdf

Shin, Yunjung. (2017). *Integrating Local Knowledge into Disaster Risk Reduction: Current Challenges and Recommendations for Future Frameworks in the Asia-Pacific*. Lund, Sweden: Division of Risk Management and Societal Safety, Lund University.

Syafwina. (2014). Recognizing indigenous knowledge for disaster management: Smong, early warning system from Simeulue Island, Aceh. *Procedia Environmental Sciences, 20*: 573–582. Retrieved June 5, 2019 from https://doi.org/10.1016/j.proenv.2014.03.070

Trias, Angelo Paolo, Lassa, Jonathan, & Surjan, Akhilesh. (2019, February). Connecting the actors, discovering the ties: Exploring disaster risk governance network in Asia and the Pacific. International Journal for Disaster Risk Reduction, *33*, 217–228. Retrieved August 15, 2019 from https://doi.org/10.1016/j.ijdrr.2018.10.007

United Nations International Strategy for Disaster Reduction. (2007). *Building Disaster Resilient Communities: Good Practices and Lessons Learned: A Publication of the "Global Network of NGOs" for Disaster Risk Reduction*. Geneva: United Nations International Strategy for Disaster Reduction. Retrieved February 22, 2019, from https://www.unisdr.org/files/596_10307.pdf

United Nations Office for Disaster Risk Reduction. (2019). *Global Assessment Report on Disaster Risk Reduction*. Geneva, Switzerland: UNDRR. Retrieved February 20, 2020 from https://gar.unisdr.org/sites/default/files/reports/2019-05/full_gar_report.pdf

United Nations Office for Disaster Risk Reduction. (2015). *Sendai Framework for Disaster Risk Reduction 2015–2030*. Retrieved February 10, 2019 from https://www.undrr.org/publication/sendai-framework-disaster-risk-reduction-2015-2030

Weather Division, PAGASA-DOST. (2018, September 20). *Typhoon Ompong (Mangkhut/1822) Summary Report*. Retrieved March 5, 2019 from https://pubfiles.pagasa.dost.gov.ph/pagasaweb/files/tamss/weather/tcsummary/TY_Ompong_Mangkhut.pdf

World Bank East Asia and Pacific Region Rural Development and National Development Council of the Philippines. Natural Disaster Risk Management in the Philippines:

Enhancing Poverty Alleviation through Disaster Reduction. Retrieved May 15, 2018 from http://documents.worldbank.org/curated/en/975311468776739344/pdf/3 38220REPLACEM1aster0Report1combine.pdf

Yogaswara H., & Yulianto, E. (2005). *Smong: Local Knowledge and Strategies on Tsunami Preparedness in Simeulue Island, Nangroe Aceh Darusallam.* Jakarta: UNESCO and LIPI.

13 Assessing climate governance of Tainan City through stakeholder networks and text mining

Roger S. Chen and Ho-Ching Lee

Introduction

Urban governance is increasingly recognized as a vital component of the human response to climate change. However, a diverse urban ecosystem and the associated climate hazards impose great challenges on an urban climate response and its assessment. Tainan, a city in southern Taiwan, experiences these challenges and is the focus of the present chapter. We propose to apply social network analysis (SNA) and text mining techniques to comprehend the complexity of the social ecosystem and assess its stakeholder interactions with the goal of offering suggestions for Tainan's climate governance.

SNA at the level of subgroup interactions offers an analytic framework for interpreting how a network structure might influence resource management and environmental governance (Bodin and Crona, 2009; Vance-Borland and Holley, 2011; Keskitalo et al., 2014; Fischer et al., 2016). Among various indicators at the subgroup level of analysis, bridging and bonding effects have been extremely useful as measurements of stakeholder relationships and multilevel interactions. This chapter aims to enrich the applicability of SNA to environmental studies by utilizing text mining to remedy the interpretative vagueness of bridging and bonding ties, especially between stakeholders in urban settings.

Text mining is a system that processes, organizes, and analyzes large quantities of natural-language text to detect lexical or linguistic usage patterns, with the wider aim of extracting useful (but usually latent) information (Sullivan, 2001; Witten, 2005). Text mining is by nature an interdisciplinary field that brings together techniques from data mining, natural language processing, machine learning, and information retrieval, among many others (Zanasi, 2007; Miner et al., 2012). This chapter uses the technique's potential to: (1) facilitate the theoretical interpretation of SNA data in environmental investigations, and (2) improve the data reliability of stakeholder analyses conducted in urban settings. Our case study for text mining's application in both of these areas comprises Tainan's ongoing efforts to mitigate the potential losses from water-related disasters between 2011 and 2015.

Tainan's social ecosystem and climate hazards

The urban social ecosystem

Urban governance is increasingly recognized as a vital part of the human response to climate change (Anguelovski and Carmin, 2011; Bulkeley and Betsill, 2005; Romero-Lamkao, 2014). Cities are hubs of economic activities and technological development. Furthermore, on a symbolic and cultural level, "they are the places where humankind realizes ambitions, aspirations and dreams, fulfills yearning needs, and turns ideas into realities" (UN Habitat, 2013, p. x). While cities now hold the majority of the world's population, many cities, particularly coastal cities, are vulnerable to climate change-induced disasters, including land subsidence, water shortages, floods, and storms (Aerts et al. 2009, 2012). Accordingly, Tainan City, Taiwan, suffers from the impact of extreme weather resulting from climate change.

Located in southwestern Taiwan, Tainan has a large population (nearly 1.9 million) and a wide range of ecological conditions and economic activities. In 2010, Tainan was accorded the status of a special municipality due to continuous urbanization, with the former Tainan County and Tainan City being combined into a single jurisdiction of more than 2,100 square kilometers. In this vast area, the landforms include coastal wetlands, plains, and mountains. These diverse landscapes support numerous economic activities and an uneven distribution of population. Tourism, commerce, and service industries flourish in the densely populated and heritage-rich southwestern core of the city, where Dutch colonial settlement was located nearly 400 years ago. The National Southern Taiwan Science Park and three other city-run industrial parks have been built within or adjacent to the core urban area. Meanwhile, aquaculture and fishing have expanded in the western wetlands and bay area, and agriculture is the main economic activity in the Chianan Plain that lies between the western shore and eastern mountains. The plain contains 94,476 hectares of farmland producing agricultural goods such as rice, fruit, vegetables, and flowers. Due to the plentiful water coming from the eastern mountain area, Tainan has the most complex hydrographic system in Taiwan, including five major rivers, a canal, six reservoirs, and three large impoundments. With the hydrographic system and a 56-kilometer-long coastline, Tainan has a large fishery breeding ground of 17,591 hectares. This diversity of landscapes and economic activities poses a great challenge to Tainan's climate governance.

Water-related disasters in Taiwan

According to a series of interviews with Tainan residents conducted by the authors in 2014 as a fieldwork of Taiwan Climate Change Adaptation Program sponsored by Taiwan's Ministry of Science and Technology, the most perceptually salient threats caused by extreme weather were floods (56 percent) and droughts/heat (38 percent). This chapter thus focuses on the water-related disasters induced by climate change. In Taiwan, rainfall normally intensifies during the monsoon season of May to June and the typhoon months from July to September, as seen in the top left frame of Figure 13.1. The country-level rainfall patterns approximately reflect local and regional rainfall trends within Taiwan, as shown in the smaller frames of Figure 13.1. In southern Taiwan, the tropical region where Tainan is located, nearly 90 percent of precipitation occurs in the rainy seasons. This extreme precipitation is a result of dramatic climate change over the last 50 years (Liu et al., 2009). Typhoon Morakot, which wreaked havoc in Taiwan in August 2009, exemplified these trends. It produced extreme torrential rain, with the three-day accumulated rainfall exceeding 2,000 mm at many gauges in southern Taiwan (Chu, Chen, and Lin, 2014).

Due to the extreme weather, Tainan has been devastated periodically by water-related disasters, including floods, mudflows, landslides, and droughts. In fact, Tainan covers one-third of Taiwan's potential inundation areas and is recognized as the most flood-prone city in the country. While former Mayor William Lai targeted Tainan's Flood-prevention Reinforcement Project to a simulated once-in-10-year scenario as shown in Figure 13.2a, it was certainly far from sufficient in light of the physical flood-line left by Typhoon Morakot in 2009 (Figure 13.2b), which exceeded once-in-100-year estimates (Figure 13.2c). This startling disparity highlights the severity of climate hazards that Tainan has suffered.

Furthermore, heavy precipitation often occurs adjacent to periods of heat and drought. Tainan currently lacks appropriate technology and infrastructure to cope with water shortages, which can impose severe restrictions on residents' day-to-day lives. Figure 13.3 maps the frequency of media exposure regarding Tainan's flood and drought events. The research suggests that flood events in Tainan are consistent with the rainy seasons, but the annual double-peak frequency of flooding has become normal. Meanwhile, drought events occurred once every two years from 2011 to 2015, and their relief relied on monsoon rainfall that in turn brings flooding (Figure 13.3). The interactive effects of floods and droughts intensify the impact of climate hazards on Tainan, suggesting an urgent need for effective adaptation and resilience measures.

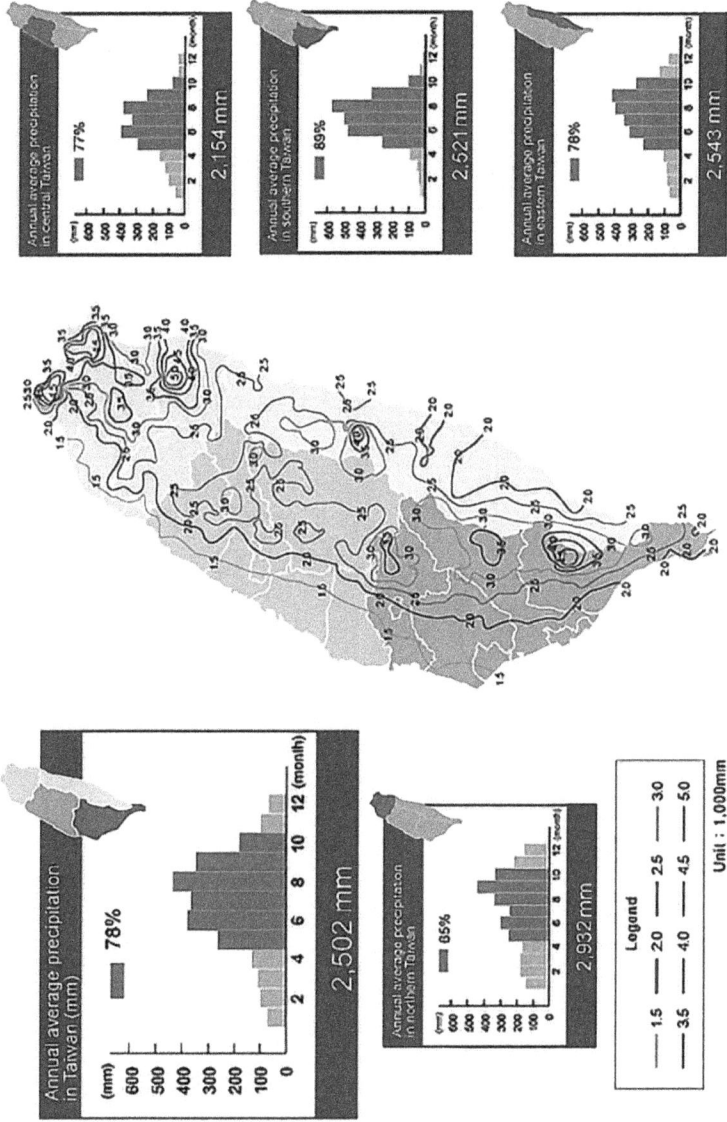

Figure 13.1 Annual average precipitation in Taiwan

Source: https://eng.wra.gov.tw/7618/7664/7718/7719/7720/12622

2a 1:10 years scenario of flood line

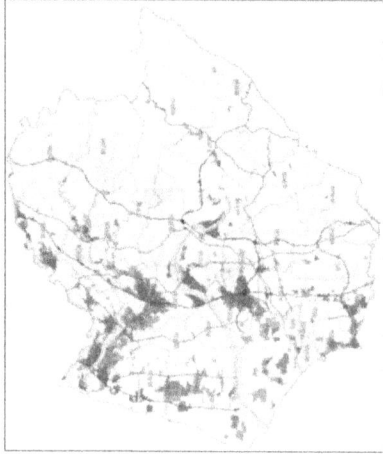

2b Flood zone of Morakot

2c 1:100 years scenario of flood line

Figure 13.2 Tainan's maximum depth of flooding in different cases

Source: Tainan City, 2013

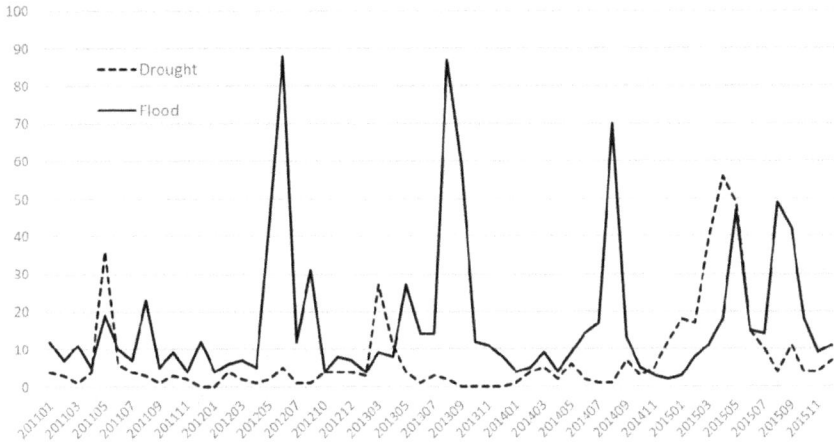

Figure 13.3 Media exposure of Tainan's flood and drought events between 2011 and 2015

Source: *United Daily News*, 2016 and *Liberty Times*, 2016.

Stakeholder overview and perceptions

Challenges to climate governance and its assessment

The diverse urban ecosystem and associated climate hazards impose great challenges on climate response and its assessment. Through the lens of urban vulnerability, defined as the degree to which a system is susceptible to adverse effects from one or more hazards or stresses, it is clear that adaptation and resilience measures in Tainan should consider not only the governance system's exposure to climate hazards but also its self-assessment and institutional ability to cope (Romero-Lamkao, 2014). To assess Tainan's institutional ability, it is important to identify all the relevant stakeholders and their perception of Tainan's climate governance. In 2014, the authors supported by the state-sponsored Taiwan Climate Change Adaptation Technology (TaiCCAT) Program conducted a series of in-depth, semi-structured interviews to identify stakeholders and collect their views. The TaiCCAT initiative aimed to bring in, combine, and utilize the input of people with many different perspectives, an approach inspired by the principle of public service co-design (Bradwell and Marr, 2008).

The interviewees were selected from the participants of the Tainan Citizen Roundtable Forum, held on June 27, 2014, who were identified by relevant city bureaus as primary stakeholders in Tainan's adaptation planning. Following the conference, snowball sampling was used to recruit more interviewees. The process resulted in a total of 48 interviewees, their affiliations including 2 Taiwan central government officials, 8 Tainan municipal government officials, 2 district officials, 8 private sector and industry representatives, 6 academic experts, 9 NGO representatives, and 13 individual citizens. Although individual background may not be the most precise method of stakeholder identification in the long run, it is the simplest and clearest way to recognize stakeholder membership in the context of local flood/drought prevention projects. Consequently, these affiliation attributes are used as criteria for stakeholder identification in the rest of the current chapter.

Overview on Tainan's climate governance

The interview results offer an insight into Tainan's climate governance by suggesting analytical directions for our stakeholder assessment and elucidating the stakeholder perceptions of the configuration of local climate response. The views regarding Tainan's overall climate governance are outlined as follows. Although climate governance is somewhat new to Tainan's policy-makers, flood protection and drought mitigation have been a concern since the Japanese colonial period (1895–1945) and have routinely been a focus of local authorities. In 2010, when William Lai took office as the first mayor of the Tainan municipality, he set up the Disaster Prevention and Response Office and placed flood prevention and water resource management as top policy priorities. In 2013, in a national appraisal of disaster prevention, Tainan took the lead in good governance and performance. Mayor Lai allocated a total of NT$ 25.34 billion in flood prevention funds from the central government, a disproportionately large amount, before leaving the mayor's office to become the premier minister of the Democratic Progressive Party government in 2017. It has therefore been intriguing to see how various stakeholders perceive and react to municipal flood responses, since dissonance is commonly seen in agenda setting and resource prioritization.

Despite these early forays into climate governance, different stakeholders in Tainan commonly experienced dissonance with regard to agenda setting and resource prioritization. The flood- and drought-control measures adopted by Mayor Lai had largely been based on conventional, passive engineering solutions, such as the construction of dikes, levees, and flood walls. The creation of a mobile app for flood early warning was an exceptionally innovative but unrepresentative initiative. There existed neither an official conduit for the 'bottom-up' communication of residents' views and concerns, nor any international partnerships that could offer policy learning opportunities. Mayor Lai and his administration heavily relied on their own communication and negotiation skills to address problems and defend their actions. Considering the intractable threats of floods and droughts, his leadership strategy demonstrates the need to bring in multiple actors to support the municipal measures, and, sometimes, the need to appeal to relevant individual citizens. According to Grimble and Wellard (1997, p. 175), it is methodologically warranted to classify the "unorganized" individuals as an important type of stakeholder when they emerge from local adaptation events. Stakeholder analysis can shed light on the possibilities for better collaborative governance.

Local stakeholder perceptions

Understanding stakeholders' perception of climate threats can further elucidate the configuration of Tainan's climate governance. In the 2014 fieldwork, interviewees were asked to rate statements using a coding spectrum from 0 to 10 (0 being lowest, 10 being highest), the results of which are found in Figures 13.4 and 13.5.

By aggregating and normalizing the scores in terms of actor affiliations, Figure 13.4 shows that most actors feel susceptible to the impact of climate hazards, as indicated by scores ranging from 6 to 10 for the "self-perception of being affected," as measured on the horizontal axis. In particular, local district officials considered themselves to be the most affected, suggesting that they experience increasing stress at the forefront of hazard prevention and abatement. Although the number of interviewees varies, the scale of "self-perception of influence" along the vertical axis in Figure 13.4 demonstrates stakeholders' assessment of their roles in adaptation and resilience construction. The results from district and municipal officials imply a municipality-driven adaptation strategy in Tainan.

Municipality-driven adaptation is further evident when observing Figure 13.5. In Figure 13.5, the vertical values on the scale "self-perception of influence" are fixed to the values in Figure 13.4. However, the horizontal

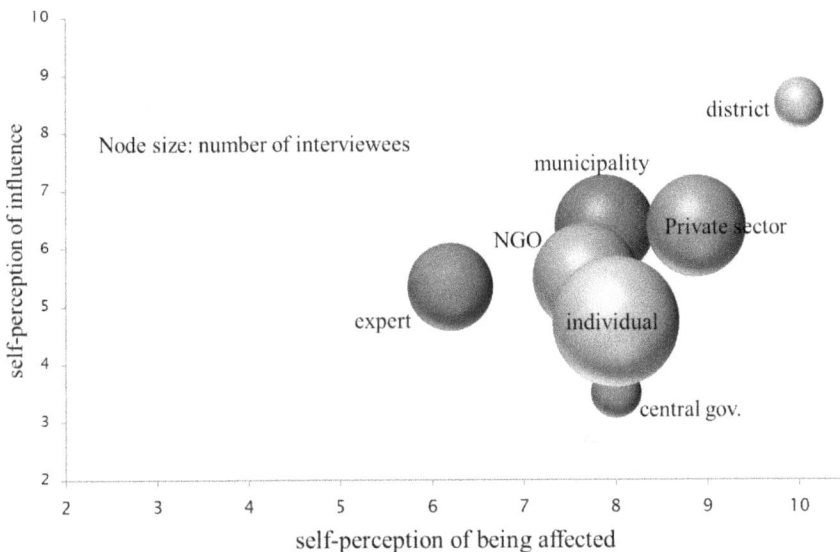

Figure 13.4 Stakeholder self-perception of being affected by Tainan's water hazards

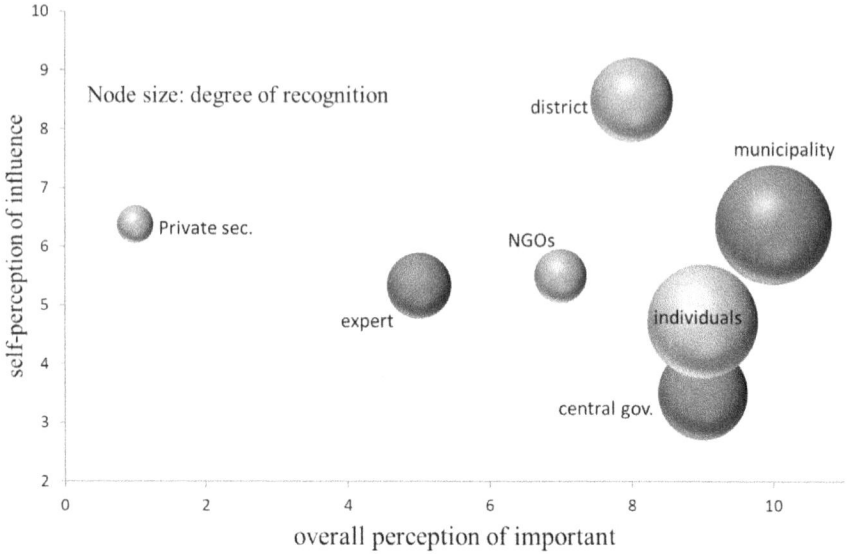

Figure 13.5 Overall perception of important stakeholders of Tainan climate governance

axis aggregates the perception from all sectors with respect to how important a role each sector has played. The high perception of importance accorded to the municipality and district offices in Figure 13.5 substantiate their dominant role in climate policy implementation. Additionally, there are some disparities between these two figures. The most notable disparities arise between the self-perceived and collectively-assessed roles played by the private sector and the central government. The private sector representatives expressed strong anxiety about the impact of the climate and an obligation to play a role in governance, but their self-assessment did not seem to gain other stakeholders' recognition. On the other hand, the central government rated itself as having substantially less influence and importance than other stakeholders suggested.

According to the data collected thus far, the above disparities may be associated with the institutional shortcomings of climate governance. For example, Figure 13.5 has a nodal size representing the degree to which the stakeholders gain recognition from each other. The data suggests that the private sector self-perceives a moderate to high level of influence, but may either be neglected by other sectors or offer less of a contribution to local adaptation and resilience projects. In either case, stakeholder management so far has failed to diminish the disparity by providing ways of involving the private sector with other actors. It seems that the private sector can certainly play a much active role. Further evidence of institutional shortcomings can be seen in the role of individual citizens, who have a highly valued position in Tainan's adaptation program but lack a formal mechanism to participate in climate governance. In sum, rather than a central control system or community-based governance, the characteristics of Tainan's climate governance more or less reflect district and municipal-led governance, an asymmetric power structure that has long been recognized in environmental management (Johnson et al., 2004; McAllister et al., 2008; Crona and Bodin, 2010).

Text mining and stakeholder network

Research questions

Based on the analysis in the preceding sections, there are two fundamental issues with Tainan climate governance. Tainan needs to develop an improved adaptation and resilience program and also to mitigate areas of institutional shortcoming. The current municipal-led governance structure tends to neglect other actors and impede an effective examination of climate hazards. Thus, the chapter proposes an SNA approach to assess stakeholders' subgroup interactions (Bodin and Crona, 2009; Vance-Borland and Holley, 2011; Keskitalo et al., 2014; Fischer et al., 2016), with the aim of outlining arenas for improved cooperation and communication in climate governance.

Data on hazard event are useful to grasp the dynamics of stakeholder interactions and to validate or challenge the dynamics that came to light in the interviews. Interviews are usually required to cover all related stakeholders, but could therefore distort the real-world asymmetry of policy stakeholders. Since self-reported information suffers from a subjective selection by the stakeholders (Varvasovszky and Brugha, 2000; Reed et al., 2009; Frooman, 2010), and can be affected by certain underlying principles of stakeholder analysis, such as treating all stakeholders equally or putting the last first (Hart and Sharma, 2004), we apply a text mining technique to digital event data to avoid distortion from research bias. Moreover, text mining is designed to capture the multiple dimensions of stakeholder interactions, and as such, to allow a clearer inspection and interpretation of sub-network analysis.

Social network analysis

Social network analysis (SNA) is increasingly applied to various types of environment-related research. It has been widely acknowledged that the cross-scale and cross-level interactions in most social ecosystems tend to be characterized by multiple stakeholders and heterogeneous actors, and this has underpinned the popularity of SNA in environmental studies (Bulkeley, 2005; Janssen et al., 2006; Cumming et al., 2010; Vance-Borland and Holley, 2011; Guerrero et al., 2013; Matous, 2015).

Generally, SNA data comprises a given number of nodes, their interrelationships, and information regarding their size and other attributes (Scott, 1991;

Wasserman and Faust, 1994). Considered to be a relational approach, SNA has an inherent capacity to entangle common-pool problems both theoretically and practically, resulting in its enduring popularity in environmental research and governance (Rydin and Falleth, 2006; Bodin et al., 2011). More importantly, a growing awareness of the diversity and complexity of ecosystems (and of the limits of corresponding human knowledge) has given rise to an intellectual movement that embraces multi-party involvement and participatory governance in both research and practice (Folke et al., 2005; Olsson et al., 2007; Armitage et al., 2009; Crona et al., 2011). Amid this paradigm shift, SNA and its relational approach are frequently used to identify and facilitate social interaction and collaborative action.

Bridging and bonding effects

Among the various indicators in SNA, bridging and bonding ties have been extremely useful as measurements of stakeholder relationships and multilevel interactions (Bodin and Crona, 2009; Vance-Borland and Holley, 2011; Keskitalo et al., 2014; Fischer et al., 2016). Bonding ties operate within rather than between subgroups (Jones et al., 2009). In the context of environmental research, they refer to a form of social capital that promotes communication, collective action, knowledge transfer, and mutual understanding, and thus tend to create cohesive structures based on trust and common norms (Fischer et al., 2016). On the other hand, having more bridging ties – which operate among different subgroups rather than within them – implies diversity in governance and management, which facilitates exchange of information and resources and enhances the overall capacity for mobilization, collaboration, and adaptation in the face of changes and challenges (Woolcock, 2001; Hahn et al., 2006).

However, in a given network composed of various subgroups, the mere identification or enumeration of bridging and bonding ties is not automatically useful. Indeed, as Newman and Dale (2005) observed, social capital could also have a negative effect on governance that may be ignored if researchers focus on quantity rather than quality of ties. A number of studies have endeavored to clarify the dual aspects of network characteristics in ecology-related settings (Bodin, Crona, and Ernstson, 2006; Bodin and Crona, 2009, Prell, Hubacek, and Reed, 2009). There is broad consensus that bridging ties indicate a network's social diversity and the fragile nature of dispersive links among subgroups, while bonding ties are likely to represent both a network's cohesiveness and excess homogeneity (Reagans and Zuckerman, 2001, Borgatti and Cross, 2003, Reagans and McEvily, 2003). The adverse effect of excess bridging or bonding as a result of the dual aspects are referred to here as "latent effects."

Searching for networks with specific meanings

We suggest that this dual characteristic of bridging/bonding ties results from the fundamentally multivalent nature of meaning in human interactions.

Considering the multiplicity of meanings embedded in even a single human interaction, which are notoriously difficult (not to say expensive) to capture via self-reported information in interviews and surveys, we propose to use text mining to construct networks with specific meanings for a subgroup analysis in order to reduce the possible interpretative vagueness. In other words, assigning a network's ties a definite meaning can constrain as well as clarify the measurement and interpretation of bridging/bonding effects.

Table 13.1 uses a simple bridging/bonding ratio, similar to Krackhardt's external-internal (E-I) link index (1988), to illustrate the argument set forth above. A ratio > 0 or < 0 could be referred to as the predominance of bridging or bonding ties respectively, while at a ratio of 0, no particular bridging/bonding tendency is discernible. Let each tie be further differentiated as having either a positive or a negative meaning, yielding the total range of four manifest interpretations shown in Table 13.1. Given that positive or negative meanings are already assigned to ties, it becomes fairly straightforward to distinguish the negative aspects of bridging/bonding ties (C and D) from the positive ones (A and B). Additionally, because bridging-bonding tendencies are inversely associated based on the values of bridging-bonding ratios, which now can be statistically verified with advanced SNA tools, such as Exponential Random Graph Model (ERGM) (Robins et al., 2007), only a bridging (A or C) or a bonding effect (B or D) manifests at any given time (Table 13.1).

Furthermore, the dual characteristic of bridging or bonding ties gives a concurrent validity to both manifest effect A and latent effect E, and the same goes for effects B/F, C/G, and D/H (Table 13.1). Now, in the meaning-fixed networks, the manifest effects prevail and their chances of transforming into latent effects are so rare that they can be ignored in day-to-day implications. For example, in a normal positive network once the ratio of bridging and bonding justifies the prevalence of bridging effects (A and E), which excludes

Table 13.1 Typology of bridging and bonding effects in a meaning specified network

Meaning of network	E-I index	Manifest effects	Latent effects
Positive ties	E-I>0	(A) Positive bridging effect (e.g., diversity)	(E) Harmful bridging effect (e.g., dispersion)
	E-I<0	(B) Positive bonding effect (e.g., cohesiveness)	(F) Harmful bonding effect (e.g., homogeneity)
Negative ties	E-I>0	(C) Negative bridging effect (e.g.,dispersion)	(G) Beneficial bridging effect (e.g., diversity)
	E-I<0	(D) Negative bonding effect (e.g., homogeneity)	(H) Beneficial bonding effect (e.g.,cohensiveness)

Note: E-I > 0 and < 0 denote the predominance of bridging and bonding ties respectively.

bonding effects, liking ties largely decrease the chances that excess bridging ties of liking can result in harmful effects (E). If an abnormal and adverse effect occurs in a liking network, it is reasonable to treat it as a special case. As such, networks with specific meanings shift a quantitative perspective of research foci and data validity to a qualitative one, and largely diminish interpretative vagueness often intertwined with the manifest/latent effects. Even if a network meaning is more complex than a positive or negative one, such as a conflict network, the same principle applies. That is, the implication of bridging/bonding ties should be consistent with the meaning of the network under investigation. Crucially, the above clarification offers an interpretative lucidity for easily exploring the dual aspects of bridging/bonding ties in real-world situations without neglecting any effect or resorting to an arbitrary interpretation.

In this chapter, document classification – one of the major functions of text mining – is used to identify stakeholder networks that feature interactions with specific meanings, thus distinguishing manifest actions from latent ones and improving the interpretation of bridging/bonding indications. When text mining is applied to solid and traceable information extracted from digital event data such as news reports, it is not only able to identify stakeholders and their interactions, but also to remedy problems of data reliability that often occur in investigations on an urban scale.

Method and data

Text mining and event data

A text mining approach to data processing requires collecting and analyzing digitized textual data in place of information from surveys or interviews. The text documents that are used normally remain in an unstructured (i.e., natural language) format. The digitization of text documents has benefited from a massive growth in data storage and computing tools, and is generally agreed to be easily accessible, cost-effective, and chronologically traceable (Miner et al., 2012). While many sources of digital data can be explored, the present chapter regards press reports as a public communication platform that is especially useful for examining multiple interactions and the views of diverse actors before, during, and after disasters.

For decades, environmental and policy studies have stressed the importance of communication in the public sphere wherein various actors construct their world views and the meanings of their actions (Kingdon, 1984; Fischer and Forester, 1993; Hannigan, 1995; Cox, 2006). Without communication, individual perceptions are meaningless; if actor reciprocity is not recognizable, it loses its social meaning and has no impact whatsoever (Luhmann, 1989; Knoke and Yang, 2008). The news articles studied depicted not only the scenes of floods and droughts, but also numerous actors and actions: revealing various resilience activities including planning and implementation, reactions to emergencies, calls for information, knowledge sharing, negotiations, conflicts, and resolutions. Consequently, news texts reflect a broad and inclusive range of salient activities by stakeholders.

In the present case study, the application of text mining techniques to digital press data had several advantages that helped overcome the difficulties of data collection in an urban setting. The data retrieved from reports of events, following an instrumental approach to stakeholder identification (Reed et al., 2009), assured us that the identified stakeholders were relevant to the focus of our investigation and provided a means of defining a network boundary for emerging stakeholders. Chronological and traceable information further secured the reliability of the data that captured stakeholder interaction. The co-occurrence frequency of stakeholders with hazard events provided the actors with different interactional weights, and thus reflected the asymmetry of power structures among stakeholders. Concurrently, our text mining procedure was able to aggregate unorganized and fringe stakeholders into discernible status units for further analysis. In this way, a balance was maintained between reflecting actual power disparities and avoiding underrepresentation.

More importantly, text mining's document classification function was able to sort stakeholder relations into different meanings in terms of the specific events the stakeholders were involved in. The sub-networks that we identified based on classified relational information represented different dimensions of stakeholder interactions, and provided greater theoretical clarity and interpretability vis-à-vis the measurement of bridging/bonding ties and their implications for subgroup analysis. In other words, text mining of event data not only helped achieve a more rounded understanding of urban complexity, but also provided a flexible approach to data processing in support of the SNA of urban environmental governance. In the following sections, Tainan's water-related disaster responses are assessed using SNA of stakeholder subgroup relations, with a view to better understanding the interactions between sectors in the execution of Tainan's plans for disaster prevention and recovery.

Text mining procedure

Since the reliability of multimedia mining for thematic applications remains less developed, only digital news sources were used in the present chapter. To limit bias resulting from institutional and editorial idiosyncrasies of single news sources, two major Taiwanese news sources – *United Daily News* and *Liberty Times* – were selected as the primary sources for analysis (*Liberty Times*, 2016, *United Daily News*, 2016). Together, they had the largest amount of coverage of Tainan's water-related disasters and complemented each other in that they advocated rival ideological positions on most issues.

Using the keywords 淹水 (flooding) and 缺水 (water shortage) to retrieve news reports from both media sources from 2011 through 2015, we collected 1,333 news documents. The retrieved articles covered a variety of incidents and activities pertaining to floods and droughts in Tainan, including not only catastrophic scenes, but also hazard planning, management, implementation, negotiation, contention, and education, reflecting the complexity and stakeholder diversity of Tainan's climate responses. The standard procedure of text mining was followed, beginning with the execution of word segmentation. Since Chinese is a language usually written without punctuation, words and phrases, rather than characters, provide more accurate indications of meaning and were therefore set as the targeted corpora for analysis.

The segmentation was conducted in an R software environment using the jiebaR package to extract and tag keywords with the help of stop words and a TF-IDF algorithm (Qin, 2016). Additionally, jiebaR allowed the use of a custom dictionary. Hence, the list of stakeholder institutions and organizations derived from our fieldwork was added to the lexicon to ensure that fringe actors were not omitted. To ensure effective classification operations, sparse terms were removed from the resultant document-term matrix at a threshold of 99 percent, leaving 2,367 keywords. These remaining keywords were seen as a set of attributes depicting the characteristics of each document. The $1,333 \times 2,367$ document-term matrix became the vector space for exploring the stakeholders' relationships in Tainan's process of building resilience to water-related disasters. Within these keywords, those containing stakeholder information about actors and organizations were the main foci for analysis. It was found that even in this primary matrix, the frequency of mentions of the municipal agencies and central government agencies was manifest, indicating that Tainan's building of local climate resilience was characterized by municipal-led governance.

Stakeholder identification

Based on the present chapter's research focus and on the set of actors that emerged from our event data, stakeholders were categorized into seven sectors: central government, municipal government, district, private sector, NGO, academic expert, and individual citizen (non-affiliation individuals). The governmental actors in the vector space were the easiest to identify, while NGOs were among the least conspicuous. The rarity of NGOs' involvement in local flooding and drought events meant that their full titles were mostly absent from the vector space following our procedure for removing sparse terms. However, their smaller morphemes that contained a generalized and yet explicit reference to a particular type of stakeholder could still be identified after word segmentation, and therefore were grouped into the unit of NGOs. For example, 基金會 (foundation), 聯盟 (union), and 團體 (group) were categorized as units of NGOs. The same logic was applied to other types of actors that were distinguishable enough to be sorted as private-sector businesses, experts, and citizens. To reflect the role of unorganized citizens in Tainan's disaster-resilience practices, certain keywords pertinent to the general public were assigned to this group, including 災民 (victims),居民 (residents), 網友 (netizens), and 市民 (citizens). Actors with a high frequency of occurrences in the data, such as governmental departments, could be tokenized with more than one title because of their different abbreviated forms or representative figures. In such cases, manual processing was applied to aggregate distinct synonyms into single units, e.g., CWB (Central Weather Bureau) included 中央氣象局 (CWB), 氣象局 (Weather Bureau), 鄭明典 (Cheng Ming-Dean, director of the CWB's Weather Forecast Center). In the end, the vector space acquired from local flood and drought disaster reports contained 59 salient stakeholders.

To aid interpretation of the interactions between sectors, we classified and extracted sub-networks involved in different dimensions of resilience activities based on the linguistic information in the original event data and the document-term matrix. Three stable meaning dimensions of Tainan's stakeholder interactions emerged: resource exchange, cooperation, and contention. The resource exchange network focuses solely on interactions based on money, manpower, and devices, while the cooperation network refers to relationships attributed to agreements, negotiations, and reciprocal support. The classification was done through text mining, i.e., document classification via supervised machine learning. This process began with the building of classification models from training data, a dataset that was randomly selected from the original news data and manually classified in accordance with the purpose of the investigation and activities emerging from event data. Next, the trained models (classifiers) were applied to the targeted data to arrive at a set of automatically labeled documents, the labels of which were in accordance with the training data's designated categories (Feldman and Sanger, 2007; Miner et al., 2012). The supervised machine learning procedure minimized subjective interventions and reduced document classification costs in terms of both time and labor.

Grounding the document classification process in textual meaning is a demanding task. An ensemble learning scheme of the specific type known as 'bagging' was used as the machine learning classifier for this purpose. Ensemble learning makes a decision more reliable by combining the output of several different models, while its bagging variant utilizes the prediction result based on the averaged weights received from such models (Witten, Frank, and Hall, 2011). The classification operation employed Weka (Hall et al., 2009), and the classifier evaluations applied tenfold cross-validation (Kohavi, 1995). The classifier evaluations were determined by the degree of precision and F-measures displayed in a confusion matrix (Kohavi and Provost, 1998).

As shown in Table 13.2, the degrees of precision of the three models ranged from 84.6 percent to 91.2 percent, indicating that the classifiers' ability to predict

Table 13.2 Results of classifier evaluations and classified documents

Sub-network	Precision	F-measure	Number of documents
Resource exchange	84.6%	84.6%	421
Cooperation	91.2%	90.8%	160
Contention	85.6%	85.0%	203

Note: Weka's bagging classifier module was applied.

unlabeled documents were strong. The accuracy outcomes of classifier evaluations were fair, considering the many ramifications of meaning that might be found in just one document and the challenging task of decomposing such meanings' multiplicity into one specific dimension across 1,333 documents. For each event, the classifiers extracted three subsets of documents that represented the resource-exchange, cooperation, and contention dimensions of Tainan's resilience practices, with the number of classified documents being 421, 160, and 203 respectively (Table 13.2). Those subsets of data were then converted into three networks each with a specific meaning, from which we identified stakeholders and their co-occurrence relationships. The resulting stakeholder networks were placed under SNA using the UCINET and the ERGM packages of R (Borgatti, Everett, and Freeman, 2002; Lusher, Koskinen, and Robins, 2013).

Assessment of stakeholder networks

The vector space derived from our event data comprised the basic structure of the stakeholder interaction network pertinent to Tainan's efforts to build resilience against water-related disasters. In these networks, links denoted each specific incident of stakeholder participation in various formal events (e.g., decisions in/ after official meetings) as well as informal ones (e.g., appealing for help at a disaster scene). While this gave substance to stakeholder interactions, it also constrained such interactions in the form of co-occurrence (undirected) links.

Figure 13.6 displays the overall network of interactions during the five-year period covered by our data and reveals the basic structure of subgroup interactions in Tainan's resilience governance for water-related disasters. The

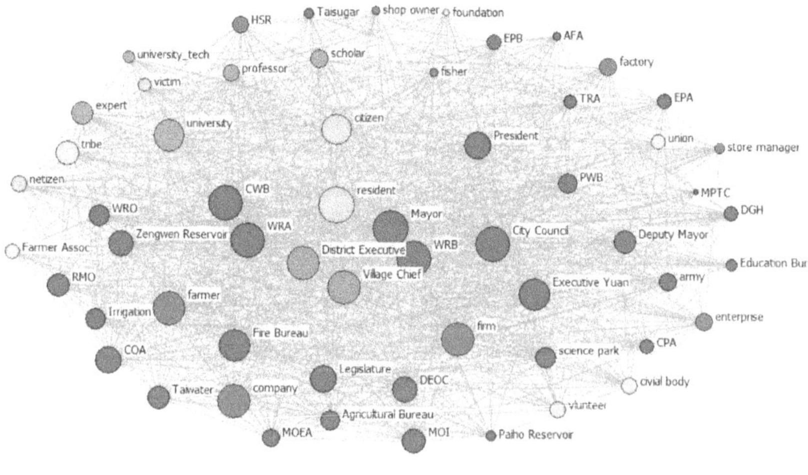

Figure 13.6 Stakeholder co-occurrence network in Tainan's climate governance, 2011–2015

Note: Node size = degree centrality; WRA: Water Resource Agency; CWB: Central Weather Bureau; COA: Council of Agriculture; WRB: Water Resource Bureau.

municipality-led style of governance that we had observed during the fieldwork phase is again evident here, insofar as the core actors in the resilience-building interaction network were the Mayor, City Council, and Water Resources Bureau (WRB). The other core stakeholders, as indicated by degree centrality in Figure 13.6, were: (1) the Executive Yuan, Water Resources Agency (WRA), and CWB, representing central authorities; (2) village chiefs and district executives, highlighting the role of the district; (3) firms and farmers from the private sector, reflecting the main economic activities of the city; and (4) residents, whose mobilization and influence had been amplified by democratization. The districts seemed to create a buffer and linkages between governmental and non-governmental sectors, but the sector has often been ignored in the theory of municipal-led governance. The stakeholder network implies less active roles for NGOs and experts than one might expect, but highlights the role of aboriginal tribal groups, which have endured continuous struggles in the face of climate-related threats.

The bridging and bonding effects that result from interaction between the different sectors were our primary research concerns. A better understanding of these effects in Tainan's stakeholder network paves the way for future adjustment to planning and governance around water-related disasters. Table 13.3 compares the number of external and internal ties in each sector to measure the extent to which the bridging/bonding effects among sectors affected the configuration of the three sub-networks with specific meanings. Specifically, this was achieved through the application of the valued Exponential Random Graph

Table 13.3 ERGM homophily test on stakeholder interactions in meaning specified networks, including resource exchange, cooperation, and contention

	Model I *Resource*	*Model II* *Cooperation*	*Model III* *Contention*
Central gov.	0.079**	0.094#	−0.090
Municipality	0.625***	0.502***	0.402***
District	2.402***	1.754***	2.391***
Private sector	−0.152	−0.195	0.005
NGOs	−0.827*	−0.790	−0.847#
Experts	0.307*	0.479**	−0.326
Individuals	0.180	0.334	0.262
Transitiveweights	0.487***	0.851***	0.876***
Sum (intercept)	0.739***	−0.248*	0.066
Nonzero	−3.357***	−2.598***	−3.200***
AIC	−3127	−1068	−1478
BIC	−3072	-1014	−1423

Note: ***$p < 0.001$, ** $p < 0.01$, * $p < 0.05$, # $p < 0.10$.

Models (ERGMs) – statistical models for network structure which permit inferences about how network ties are patterned and provide insight into the underlying process that creates and sustains the network-based social systems (Lusher, Koskinen, and Robins, 2013; Robins et al., 2007). Unlike an E-I Index, which relies on an absolute ratio of external and internal links between subgroups, an ERGM allows us to test the hypothesis that stakeholders prefer to interact with members of their own sector. An ERGM takes into account the statistical probability of actor interactions and assumes that each sector has a different "homophily" tendency to prefer actors similar to themselves (McPherson, Smith-Lovin, and Cook, 2001). Since the homophily effect mainly depends on sector affiliations, a "transitiveweights" term is added in an ERGM to detect a potential structure of "heterophily," the logical opposite of homophily, without resorting to affiliations. In addition, we assign a "nonzero" term in the valued ERGM to control the sparseness of the network (Krivitsky, 2012).

The results of an ERGM test are shown in Table 13.3 in which three models, reflecting resource exchange, cooperation, and contention interactions, all reach models' goodness of fit. Here, the "sum" term (intercept) is the probability of a dyadic tie value between any two members of different groups, indicating the overall bonding tendency in the network. Hence, the positive coefficient (> 0) of a given subgroup relative to the intercept denotes an homophily tendency, that is a bonding effect, for that subgroup, while a negative coefficient (< 0) signifies a bridging effect. In Table 13.3, Model I and Model II denote that in the events of Tainan's water disasters and their responses, resource exchange and cooperation networks have very similar patterns of stakeholder interaction, where the central government, municipality, district, and experts possess a clear homophily tendency, meaning the prevalence of within-sector interactions. Based on the fixed and positive meanings of networks that were described previously, the homophily tendency implies that the four sectors enjoy a positive and manifest bonding effect. That is, they acted cohesively when planning and implementing local resilience-building, which supports the sustainability of the networks. Apparently, the municipal-led governance is much more robust than expected, considering the embeddedness of central departments and experts as estimated by the ERGM (Table 13.3). The role played by the district sector is especially significant, since the odds of a successful formation of a resource exchange network for its within-sector interactions was 11.045 ($\exp^{2.402} = 11.045$) times that of interaction between district and other sectors, while the odds ratio in the cooperation network is 5.778 ($\exp^{1.754} = 5.778$) (Table 13.3). This data suggests that district executives, considered as an auxiliary arm of the city authorities, have emerged as internally robust pillars for controlling and managing hazard damage at a community level.

Due to the inverse relationship between bridging and bonding effects, the predominance of bonding effects in both resource exchange and cooperation networks implies that coordination among different sectors has yet to be nurtured in Tainan's climate governance. The exceptional performance of NGOs in resource exchange networks, with the coefficient -0.827, demonstrates their important role in reconciling different stakeholders' demands and facilitating negotiations among sectors (Model 1 of Table 13.3). Indeed, NGOs act like a boundary spanner in supporting resource exchange and management in Tainan's resilience-building, though no such actor exists in the cooperation network. The structural heterophily factor, "transitive-weights," in Table 13.3 signals a positive influence on both resource exchange and cooperation networks, denoting a rather open and collaborative partnership underlying both networks without any affiliation prerequisite. The open structure betokens a promising likelihood of creating a much more active interplay between sectors.

In the context of contentious events, as shown in Table 13.3, the municipality and district officials had the strongest bonding effects among all the sectors, denoted in Table 13.3 by the coefficients 0.402 and 2.391 respectively. The significant bonding ties in the contention network should not be interpreted as positively as the bonding ties observed in the resource exchange and cooperation networks. Instead, the strong preference for within-group bonding suggests that the close inward interactions of these two sectors have developed alongside continuous contention, which indicates the ineffectiveness of their problem-solving activities in the sphere of disaster mitigation. In other words, the municipality and district offices may have been caught in a network inertia that results in ineffectual responses to natural hazards over the long term. As such, the problem is not how the municipal-led governance is consolidated, but how these actors can widen public engagement and bring in innovative planning ideas. Moreover, because the model does not show evidence of conflict between the municipality and district offices and other sectors, there is clear potential for future partnership building – especially compared to NGOs, which have struggled to negotiate with various sectors in the course of various controversial events (with coefficient -0.847 in Table 13.3) – yet it remains fragile and disempowered. The NGOs can and should play a much more supportive role in the planning and execution of water disaster prevention and recovery in Tainan.

Tainan may need to continue implementing innovative resilience measures, such as dry dams, levee setbacks, storm-water runoff tunnels, and water-reuse facilities, if it is to successfully mitigate the impact of future disasters. Since these measures involve drastic changes to land use, the operation of different sectors as a collaborative network is essential. As can be seen from the overall network information in Table 13.3, the roles played by the private sector, NGOs, experts, and citizens have so far been relatively passive. All three models in Table 13.3 show the private sectors and individual citizens have no particular impact, certifying again the lack of useful platforms for participation. Therefore, if Tainan is to meet its goal of improved resilience, it will need to foster partnerships between the municipality, central agencies, and the local business sector; make use of third-party (including NGO) facilitators; seek assistance and ideas from scientific experts; and encourage citizens' participation.

Conclusion

In assessing the specific case of Tainan's resilience-building against water-related disasters in the period 2011–2015, this chapter incorporated text mining into SNA. We highlighted the severity of climate hazards that Tainan has suffered and pointed out that adopting more innovative and effective measurements are needed in Tainan's climate governance. That, however, requires a broader consensus and collaboration between heterogeneous actors, the failure of which is the weakest aspect of Tainan's municipal-led governance model. Our research found that with the embeddedness of central departments and expert systems, Tainan's municipal-led governance is far more robust than expected, but still lacks multi-sector coordination. It is suggested that district officials should function like pillars in controlling flood and drought damage at a community level, while NGOs could play a much more supportive role in the planning and execution of disaster prevention and recovery in Tainan. All three models (resource, cooperation, and contention) in the ERGM estimation show that the private sector and individual citizens have no particular impact, confirming the lack of an effective platform for their participation in Tainan's climate governance.

Climate governance in Tainan can be better understood by looking into its institutional frameworks. There has been an institutional shift from a low-carbon city to a climate-resilient municipality. In 2012, Tainan City Council passed a "self-governance statute for a low-carbon city" as a legal basis for detailed implementation. Mayor William Lai declared 2012 to be "carbon neutral year one." And, an inventory on greenhouse gas (GHG) emissions, led by the low-carbon project office was conducted between 2001 and 2013. The inventory data sets indicated the total carbon dioxide equivalent (CO2e) was 25,680,000 metric tons, with the industrial energy services sector being the biggest contributor at 67.17 percent. It was followed by transport services, at 12.26 percent; and residential and commercial services, at 11.80 percent. The remaining sectors are industrial manufacturing, waste management, and agriculture, accounting for 4.47 percent, 1.77 percent, and 0.54 percent of total emissions, respectively.

On a city level, Tainan's emission reduction target is set to cut Tainan's GHG emissions by 20 percent below 2005 emission levels by 2030, equivalent to cutting 3.9 million metric tons of CO2e emissions; and a long-term reduction of 50 percent equivalent to cutting 9.76 million metric tons of emissions. Meeting Tainan's mid-term and long-term reduction targets will not be an easy task and it is premature to determine whether or not it will be achieved. Arguably, Tainan's municipal-led policy frameworks are complex and often have conflicting interests. The low-carbon project office that serves as a coordinating platform is made up of the land administration, economic development, urban development, environmental protection, water resources, public works, health, transportation, education, tourism, and agriculture bureaus. Managing water resources turned out to be the most challenging as shown by the outcome of the 2014 Tai-CCAT round-table. Three issues emerged and still require immediate policy attention, agricultural vs. industrial use of water, climate-induced disasters, and the engineering-only approach.

Internationally, Tainan has begun to actively participate in a global city alliance. In 2015, Tainan joined the Compact of Mayors. Since 2015, when the Paris Agreement was signed, Tainan has continuously participated in climate change summits (known as the Conference of the Parties to the UN Framework Convention on Climate Change, COPs) as a non-party stakeholder. Along with the passing of 17 sustainable development goals, Tainan established a set of low-carbon adaptation and sustainable development indicators. In 2017 and 2018, Tainan received the certificate of registration of ISO 37120 platinum level, an indicator of city services and the quality of life, developed by the World Council on City Data. However, since then, for Tainan City Government, climate change mitigation has been re-framed from low-carbon and disaster mitigation to a much larger context of climate adaptation, climate resilience, energy, education, economy, finance, health, recreation, shelter, and other key elements of overall city performance. After all, climate change has a "multiplier effect" involving multiple state actors, non-state actors, and sub-state actors.

With regard to the method and theory applied in this chapter, we have addressed two interconnected limitations in the application of SNA to the study of urban environmental governance: (1) vagueness in the theoretical interpretation of bridging and bonding effects, and (2) the difficulties of data collection in urban settings, and proposed text mining as a technique with the potential to remedy both of them. Text mining of digitized news reports proved a useful alternative for both data collection and data processing. Specifically, text mining is able to cope with the complexity of an urban social ecosystem and to construct stakeholder networks with specific interactional themes. These meaning-fixed networks demonstrate that the latent aspects of bridging/bonding ties can be excluded from interpreting the effects of subgroup linkages, and that the inverse implications of bridging and bonding ties can also be more distinctly revealed. Our proposed clarification strategy therefore improves the interpretive power of SNA when assessing environmental governance and management, especially at the level of subgroup interactions.

Text mining was used in this chapter to identify stakeholders, extract their interactions, and capture the various meanings of stakeholders' actions. Although all these tasks can be carried out by conventional methods of data collection, text mining is more easily accessible, more cost-effective, and offers chronological traceability through event data. More importantly, unlike conventional methods of data collection in which the reliability of data and the meanings of stakeholders' actions are heavily dependent on self-reports, text mining is able to extract sub-networks with specific interactional meanings by taking into account the context of events that actors were involved in. That is, text mining can distinguish a specific interaction network from the multiple meanings of stakeholders' behaviors, and therefore supports better interpretations of bridging/bonding ties and improved subgroup analysis.

Finally, there are some issues that will need to be addressed if better results are to be obtained from text-mining enhanced environmental SNA in the future. First, to obtain clearer and more useful policy and managerial implications, as well as to cope with the increasing complexity of social ecosystems, it will be essential to explore the ability of text mining to detect multiple attributes in event data. Building such capabilities may require improvement to text mining's named-entity recognition, geographical notations, sentimental differentiations, and other capabilities. Second, it should be noted that mining technology is still at an experimental stage (Witten, Frank, and Hall, 2011). Along with advances in mining methods, it will eventually be possible to apply semantic classifiers with higher levels of precision and to assemble powerful data from multimedia sources, such as text, audio, and video, as well as from multiple platforms including the press, social media, and web pages. Given the dramatic changes ongoing in communication technology and the ever-increasing application of intelligent technology, SNA and environmental studies will inevitably face both challenges and opportunities posed by the modes of actor interaction and data collection methods. This chapter brings us a step nearer to responding to these challenges.

References

Aerts, J., D. C. Major, M. J. Bowman, P. Dircke, and M. A. Marfai. (2009). *Connecting Delta Cities. Coastal Cities, Flood Risk Management and Adaptation to Climate Change*. Amsterdam: VU University Press.

Aerts, J., W. Botzen. M. Bowman, P. J. Ward, and P. Dircket. (2012). *Climate Adaptation and Flood Risk in Coastal Cities*. Abingdon, UK: Earthscan.

Anguelovski, I, and J. Carmin. (2011). Something borrowed, everything new: Innovation and institutionalization in urban climate governance. *Current Opinion in Environmental Sustainability* 3(3): 169–75.

Armitage, D. R., R. Plummer, F. Berkes, R. I. Arthur, A. T. Charles, I. J. Davidson-Hunt, A. P. Diduck, N. C. Doubleday, D. S. Johnson, M. Marschke, P. McConney, E. W. Pinkerton, and E. K. Wollenberg. (2009). Adaptive co-management for social-ecological complexity. *Frontiers in Ecology and the Environment* 7(2): 95–102.

Bodin, Ö., B. Crona, and H. Ernstson. (2006). Social networks in natural resource management: What is there to learn from a structural perspective? *Ecology and Society* 11(2): r2. [Online] URL: https://www.ecologyandsociety.org/vol11/iss2/resp2.

Bodin, Ö., and B. I. Crona. (2009). The role of social networks in natural resource governance: What relational patterns make a difference? *Global Environmental Change-Human and Policy Dimensions* 19(3): 366–374.

Bodin, Ö., S. Ramirez-Sanchez, H. Ernstson, and C. Prell. (2011). A social relational approach to natural resource governance. *In:* Bodin, Ö., and C. Prell, editors. *Social Networks and Natural Resource Management: Uncovering the Social Fabric of Environmental Governance*. New York: Cambridge University Press.

Borgatti, S. P., and R. Cross. (2003). A relational view of information seeking and learning in social networks. *Management Science* 49(4): 432–445.

Borgatti, S. P., M. G. Everett, and L. C. Freeman. (2002). *Ucinet for Windows: Software for Social Network Analysis.* Harvard, MA: Analytic Technologies.

Bradwell, P., and S. Marr. (2008). *Making the Most of Collaboration: An International Survey of Public Service Co-Design.* London: Demos.

Bulkeley, H. (2005). Reconfiguring environmental governance: Towards a politics of scales and networks. *Political Geography 24*(8): 875–902.

Bulkeley, H, and M. M. Betsill. (2005). Rethinking sustainable cities: Multilevel governance and the 'urban' politics of climate change. *Environmental Politics 14*(1): 42–63.

Chu, P.-S., D.J. Chen, and P.-L. Lin. (2014). Trends in precipitation extremes during the typhoon season in Taiwan over the last 60 years. *Atmospheric Science Letters 15*(1): 37–43. https://doi.org/10.1002/asl2.464.

Cox, R. (2006). *Environmental Communication and the Public Sphere.* Thousand Oaks, CA: Sage Publications.

Crona, B., and O. Bodin. (2010). Power asymmetries in small-scale fisheries: A barrier to governance transformability? *Ecology and Society 15*(4): 32. [Online] URL: https://www.ecologyandsociety.org/vol15/iss4/art32.

Crona, B., C. Prell, M. Reed, and K. Huback. (2011). Combining social network approaches with social theories to improve understanding of natural resource governance. *In:* Bodin, Ö., and C. Prell, editors. *Social Networks and Natural Resource Management: Uncovering the Social Fabric of Environmental Governance.* New York: Cambridge University Press.

Cumming, G. S., T. Bodin, H. Ernstson, and T. Elmqvist. (2010). Network analysis in conservation biogeography: Challenges and opportunities. *Diversity and Distributions 16*(3): 414–425.

Feldman, R., and J. Sanger. (2007). *The Text Mining Handbook: Advanced Approaches in Analyzing Unstructured Data.* New York: Cambridge University Press.

Fischer, A. P., K. Vance-Borland, L. Jasny, K. E. Grimm, and S. Charnley. (2016). A network approach to assessing social capacity for landscape planning: The case of fire-prone forests in Oregon, USA. *Landscape and Urban Planning 147*: 18–27.

Fischer, F., and J. Forester. (1993). *The Argumentative Turn in Policy Analysis and Planning.* Durham, NC: Duke University Press.

Folke, C., T. Hahn, P. Olsson, and J. Norberg. (2005). Adaptive governance of social-ecological systems. *Annual Review of Environment and Resources 30*, 441–473.

Frooman, J. (2010). The issue network: Reshaping the stakeholder model. *Canadian Journal of Administrative Sciences 27*(2): 161–173.

Grimble, R., and K. Wellard. (1997). Stakeholder methodologies in natural resource management: A review of principles, contexts, experiences and opportunities. *Agricultural Systems 55*(2): 173–193.

Guerrero, A. M., R. R. J. McAllister, J. Corcoran, and K. A. Wilson. (2013). Scale mismatches, conservation planning, and the value of social-network analyses. *Conservation Biology 27*(1): 35–44.

Hahn, T., P. Olsson, C. Folke, and K. Johansson. (2006). Trust-building, knowledge generation and organizational innovations: The role of a bridging organization for adaptive comanagement of a wetland landscape around Kristianstad, Sweden. *Human Ecology 34*(4): 573–592.

Hall, M., E. Frank, G. Holmes, B. Pfahringer, P. Reutemann, and I. H. Witten. (2009). The WEKA data mining software: An update. *SIGKDD Explorations 11*(1): 10–18.

Hannigan, J. A. (1995). *Environmental Sociology: A Social Constructionist Perspective.* London, Routledge.

Hart, S. L., and S. Sharma. (2004). Engaging fringe stakeholders for competitive imagination. *Academy of Management Executive 18*(1): 7–18.

Janssen, M. A., Ö. Bodin, J. M. Anderies, T. Elmqvist, H. Ernstson, R. R. J. McAllister, P. Olsson, and P. Ryan. (2006). Toward a network perspective of the study of resilience in social-ecological systems. *Ecology and Society 11*(1): 15. [Online] URL: www.ecologyandsociety.org/vol11/iss1/art15.

Johnson, N., N. Lilja, J. A. Ashby, and J. A. Garcia. (2004). Practice of participatory research and gender analysis in natural resource management. *Natural Resources Forum 28*: 189–200.

Jones, N., C. M. Sophoulis, T. Iosifides, I. Botetzagias, and K. Evangelinos. (2009). The influence of social capital on environmental policy instruments. *Environmental Politics 18*(4): 595–611.

Keskitalo, E. C. H., J. Baird, E. L. Ambjornsson, and R. Plummer. (2014). Social network analysis of multi-level linkages: A Swedish case study on northern forest-based sectors. *Ambio 43*(6): 745–758.

Kingdon, J. (1984). *Agendas, Alternatives, and Public Policies.* Boston: Little Brown.

Knoke, D., and S. Yang. (2008). *Social Network Analysis.* Los Angeles: Sage.

Kohavi, R. (1995). A study of cross-validation and bootstrap for accuracy estimation and model selection. *Proceedings of the 14th International Joint Conference on Artificial Intelligence 2*: 1137–1143.

Kohavi, R., and F. Provost. (1998). Glossary of terms. *Machine Learning 30*(2–3): 271–274.

Krackhardt, D. (1988). Predicting with networks: Nonparametric regression analysis of dyadic data with networks. *Social Networks 10*: 359–381.

Krivitsky, P. N. (2012). Exponential-family random graph models for valued networks. *Electronic Journal of Statistics 6*: 1100–1128.

Liberty Times. (2016). *Liberty Times Net.* [Online] URL: http://news.ltn.com.tw/search.

Liu, S., C. C. Fu, C.-J. Shiu, J.-P. Chen, and F. Wu. (2009). Temperature dependence of global precipitation extremes. *Geophysical Research Letters 36*: L17702.). https://doi.org/10.1029/2009GL040218.

Luhmann, N. (1989). *Ecological Communication.* Cambridge: Polity Press.

Lusher, D., J. Koskinen, and G. Robins. (2013). *Exponential Random Graph Models for Social Networks: Theory, Methods, and Applications.* New York: Cambridge University Press.

Matous, P. (2015). Social networks and environmental management at multiple levels: Soil conservation in Sumatra. *Ecology and Society 20*(3): 37. [Online] URL: https://www.ecologyandsociety.org/vol20/iss3/art37.

McAllister, R. R. J., B. Cheers, T. Darbas, J. Davies, C. Richards, C. J. Robinson, M. Ashley, D. Fernando, and Y. T. Maru. (2008). Social networks in arid Australia: A review of concepts and evidence. *Rangeland Journal 30*(1): 167–176.

McPherson, M., L. Smith-Lovin, and J. M. Cook. (2001). Birds of a feather: Homophily in social setworks. *Annual Review of Sociology 27*: 415–444.

Miner, G., D. Deen, J. Elder, A. Fast, T. Hill, and R. A. Nisbet, editors. (2012). *Practical Text Mining and Statistical Analysis for Non-structured Text Data Applications.* Waltham, MA: Elsevier/Academic Press.

Newman, L., and A. Dale. (2005). Network structure, diversity, and proactive resilience building: A response to Tompkins and Adger. *Ecology and Society 10*(1): r2. [Online] URL: www.ecologyandsociety.org/vol10/iss1/resp2.

Olsson, P., C. Folke, V. Galaz, T. Hahn, and L. Schultz. (2007). Enhancing the fit through adaptive co-management: Creating and maintaining bridging functions for matching scales in the Kristianstads Vattenrike Biosphere Reserve, Sweden. *Ecology and Society* *12*(1): 28. [Online] URL: https://www.ecologyandsociety.org/vol12/iss1/art28.

Prell, C., K. Hubacek, and M. Reed. (2009). Stakeholder analysis and social network analysis in natural resource management. *Society & Natural Resources 22*(6): 501–518.

Qin, W. (2016). Package 'JiebaR'. [Online] URL: https://github.com/qinwf/jiebaR.

Reagans, R., and B. McEvily. (2003). Network structure and knowledge transfer: The effects of cohesion and range. *Administrative Science Quarterly 48*(2): 240–267, 355–356.

Reagans, R., and E. W. Zuckerman. (2001). Networks, diversity, and productivity: The social capital of corporate R&D Teams. *Organization Science 12*(4): 502–517.

Reed, M. S., A. Graves, N. Dandy, H. Posthumus, K. Hubacek, J. Morris, C. Prell, C. H. Quinn, and L. C. Stringer. (2009). Who's in and why? A typology of stakeholder analysis methods for natural resource management. *Journal of Environmental Management 90*(5): 1933–1949.

Robins, G., T. Snijders, P. Wang, M. Handcock, and P. Pattison. (2007). Recent developments in exponential random graph (p*) models for social networks. *Social Networks 29*(2): 192–215.

Romero-Lamkao, P. (2014). Governing carbon and climate in the cities: An overview of policy and planning challenges and options. *In*: Priemus, H. and S. Davoudi, editors. *Climate Change and Sustainable Cities*. London: Routledge.

Rydin, Y., and E. Falleth. (2006). *Networks and Institutions in Natural Resource Management*. Cheltenham: Edward Elgar.

Scott, J. (1991). *Social Network Analysis: A Handbook*. London: Sage.

Sullivan, D. (2001). *Document Warehousing and Text Mmining Techniques for Improving Business Operations Marketing and Sales*. New York: John Wiley & Sons.

Tainan City. (2013). *Ti Nan Shi Di Qu Zai Hai Fang Jiu Ji Hua (Tainan City Disaster Prevention and Protection Act)*. Tainan: Tainan City Government.

Tainan City Government. (2016). *Tainan Climate Action*. [Online] URL: http://tainan.carbon.net.tw/LowCarbon_Tainan_English/FileDownload/2016%20Tainan%20Climate%20action.pdf.

UN Habitat. (2013). *State of the World's Cities 2012/2013*. New York: Routledge.

United Daily News. (2016). Udndata.com. [Online] URL: http://udndata.com. autorpa.pccu.edu.tw/library.

Vance-Borland, K., and J. Holley. (2011). Conservation stakeholder network mapping, analysis, and weaving. *Conservation Letters 4*(4): 278–288.

Varvasovszky, Z., and R. Brugha. (2000). How to do (or not to do) …: A stakeholder analysis. *Health Policy and Planning 15*(3): 338–345.

Wasserman, S., and K. Faust. (1994). *Social Network Analysis: Methods and Application*. Cambridge: Cambridge University Press.

Witten, I. H. (2005). Text mining. *In*: Singh, M. P., editor. *Practical Handbook of Internet Computing*. Boca Raton, FL: Chapman & Hall/CRC Press.

Witten, I. H., E. Frank, and M. A. Hall. (2011). *Data Mining: Practical Machine Learning Tools and Techniques*. Burlington, MA: Morgan Kaufmann.

Woolcock , M. (2001) The place of social capital in underdstanding social and economic outcomes. *Canadian Journal of Policy Research 2*(1): 11–17.

Zanasi, A. (2007). *Text Mining and Its Applications to Intelligence, CRM and Knowledge Management*. Southampton: WIT Press.

Index

For Product Safety Concerns and Information please contact our EU
representative GPSR@taylorandfrancis.com
Taylor & Francis Verlag GmbH, Kaufingerstraße 24, 80331 München, Germany